W9-AOT-748

Animal Cell Biotechnology

METHODS IN BIOTECHNOLOGY™

John M. Walker, *Series Editor*

12. **Environmental Monitoring of Bacteria,** edited by *Clive Edwards, 1999*
11. **Aqueous Two-Phase Systems,** edited by *Rajni Hatti-Kaul, 1999*
10. **Carbohydrate Biotechnology Protocols,** edited by *Christopher Bucke, 1999*
9. **Downstream Processing Methods,** edited by *Mohamed Desai, 1999*
8. **Animal Cell Biotechnology: *Methods and Protocols*,** edited by *Nigel Jenkins, 1999*
7. **Affinity Biosensors: *Techniques and Protocols,*** edited by *Kim R. Rogers and Ashok Mulchandani, 1998*
6. **Enzyme and Microbial Biosensors: *Techniques and Protocols,*** edited by *Ashok Mulchandani and Kim R. Rogers, 1998*
5. **Biopesticides: *Use and Delivery,*** edited by *Franklin R. Hall and Julius J. Menn, 1998*
4. **Natural Products Isolation,** edited by *Richard J. P. Cannell, 1998*
3. **Recombinant Proteins from Plants: *Production and Isolation of Clinically Useful Compounds,*** edited by *Charles Cunningham and Andrew J. R. Porter, 1998*
2. **Bioremediation Protocols,** edited by *David Sheehan, 1997*
1. **Immobilization of Enzymes and Cells,** edited by *Gordon F. Bickerstaff, 1997*

METHODS IN BIOTECHNOLOGY™

Animal Cell Biotechnology

Methods and Protocols

Edited by

Nigel Jenkins

Bioprocess Research and Development
Eli Lilly & Co., Indianapolis, IN

Totowa, New Jersey

999 Riverview Drive, Suite 208
Totowa, New Jersey 07512

This publication is printed on acid-free paper. ∞
ANSI Z39.48-1984 (American Standards Institute) Permanence of Paper for Printed Library Materials.

Cover design by Patricia F. Cleary.

For additional copies, pricing for bulk purchases, and/or information about other Humana titles, contact Humana at the above address or at any of the following numbers: Tel: 973-256-1699; Fax: 973-256-8341; E-mail: humana@humanapr.com, or visit our Website at www.humanapress.com

Printed in the United States of America. 10 9 8 7 6 5 4 3 2 1

Library of Congress Cataloging in Publication Data

Main entry under title:

Methods in biotechnology™.

Animal cell biotechnology / edited by Nigel Jenkins.
p. cm.—(Methods in biotechnology: 8)
Includes index.
ISBN 0-89603-547-6 (alk. paper)
1. Animal cell biotechnology. 2. Cell culture. I. Jenkins, Nigel, 1954– . II. Series.
TP248.27.A53A542 1999 98-30567
660.6—dc21 CIP

Preface

In recent years mammalian cell biotechnology has expanded greatly from a relatively limited base, mainly concerned with vaccine production, to a major source of biopharmaceuticals, which now accounts for over 40% of the total recombinant biopharmaceutical market and 2% of total pharmaceutical sales. Examples of products that are used clinically are monoclonal antibodies; vaccines; drugs for cardiovascular, respiratory, and immune diseases; and several anticancer agents. Although the current stock of recombinant proteins produced in animal cells that have made it to the clinic is still relatively limited (between 25 and 30), many more are in the pipeline and will undoubtedly play an increasing role in the armory of therapies in the future. In addition to producing recombinant proteins, animal cells are increasingly being used as a treatment in themselves, for example, in *ex vivo* therapy, hematopoiesis, and other forms of cell replacement. With the advent of gene therapy trials, animal cells are also used as vectors for the introduction of therapeutic genes.

The technology behind these new products can be very specialized, and differs significantly from the techniques used in small-scale cell culture by scientists studying fundamental cell biology. There is a thus need for a volume on *Animal Cell Biotechnology: Methods and Protocols*, which brings together the diverse techniques used in both industry and research laboratories that are focused on the expression of recombinant proteins and other commercial uses of animal cells.

Animal Cell Biotechnology: Methods and Protocols is divided into parts that reflect the processes required for different stages of production. In Part I the basic culture facilities and methods for the successful initiation and propagation of cell lines are described. In Chapter 1, emphasis is placed on providing modular culture facilities with the flexibility for multiple purposes and cell types. Issues of Good Laboratory Practice (GLP) compliance and operator safety are also addressed.

Methodologies for cell adaptation [principally Chinese hamster ovary (CHO) cells] to both suspension and serum-free culture are described in Chapter 2. In addition, methods for the crucial adaptation to high-density cultures are summarized (including development of cell tolerance to such inhibitory substances as lactate and ammonia). Although many of the formulations currently used in serum and protein-free media are proprietary, useful hints are given on the critical culture media components required, and extensive cross-referencing to related studies is given.

Probably the most difficult source of cell infection to be monitored and controlled is viral infection, and mammalian cell lines differ in their susceptibility to each virus. These differences are highlighted in Chapter 3, and assays to detect viruses, based on PCR, immunofluorescence, or cross-contamination of other cells, are described. The regulatory environment controlling viral safety issues is also addressed by the author.

In Part II the molecular methods required for gene transfection, cell immortalization, and cell fusion are described, together with the molecular techniques used to characterize the cell's chromosomes and lineage, as well as the preparation of cell banks. When creating new recombinant cell lines, it is often necessary to increase the level of gene amplification to ensure maximum protein production. This issue is addressed in Chapter 4, which outlines an alternative method to the common amplification by marker selection and drug treatment, i.e., amplification in vitro by constructing multigene concatenamers. Creating highly gene-amplified cell lines raises issues of genetic stability, in both the presence or absence of selection agents. Cytogenetic techniques that can be used to trace the stability and integration of recombinant genes in cell lines, such as Fluorescence *in situ* Hybridization (FISH), are described in Chapter 5.

Besides recombinant protein production, there is a need to create new mammalian cell lines, for both drug and toxicology testing and as gene vectors. For these studies it is important to create immortalized cell lines with unlimited growth potential, while preserving other phenotypic characteristics of the cells. Chapter 6 describes the use of an oncogene (SV40 large T antigen) to immortalize cells and also compares different methods of plasmid transfection.

Many of the proteins used for research, diagnostics, and therapy are monoclonal antibodies, and the relevant protocols are described in Chapter 7. Crucial stages in antigen preparation, immunization, cell fusion, and selection are highlighted, along with enzyme-linked immunosorbant assays (ELISA) for the evaluation of cell productivity. The selection of individual clones by limiting dilution, plating in soft agar, and using feeder cells is also covered, as is the detection and elimination of mycoplasma contamination and methods for its elimination.

Precise, well-documented cell banking and cell characterization procedures are not only required by regulatory authorities controlling the licensing of biopharmaceuticals, but are also crucial for the successful maintenance of cell stocks. Chapter 8 describes the key methodologies for cell cryopreservation, recovery, and sterility testing, and outlines methods for testing genetic stability that complement those described in Chapters 5 and 9. Specific protocols for DNA fingerprinting of animal cells are covered in Chapter 8, including a comparison of the use of multilocus and single-locus DNA probes. These methods are used to confirm the identity of cell lines (since cultures can become cross-contaminated in a multipurpose facility), and also to track cell line stability during cell selection, adaptation, and long-term culture.

In Part III, detailed protocols are provided for the study of cell growth, viability, metabolism, and productivity. The first step is to determine the cell number and viability, and both direct cell counting methods and indirect parameters (such as glucose uptake) are covered in Chapter 10. Different cell viability parameters, such as dye exclusion, release of lactate dehydrogenase, intracellular energy charge, and protein synthesis rates, are also compared in this chapter. Flow cytometry has become a very useful method for assessing cell growth and protein productivity, and Chapter 11 outlines the relevant protocols for cell cycle analysis, performance prediction, cell viability, and stability measurements, as well as for assessing the IgG content of recombinant cells. During culture, cells can die by either necrosis or apoptosis, the latter being a quick and controlled process that can curtail cell productivity. Chapter 12 describes histochemical, flow cytometric, and DNA labeling protocols for evaluating the rates of cell death by both apoptosis and necrosis. NMR spectroscopy has the unique ability simultaneously to generate extensive information on a wide range of important metabolites, such as nucleotide–phosphates, NAD, and creatine phosphate. Protocols for measuring these cell energetics parameters in perfusion bioreactors are described in Chapter 13.

In Part IV, the more specialist techniques (either in methods of cell culture or in cultivating specialist cell lines) are described. Chapter 14 outlines the construction and maintenance of a perfusion bioreactor, using a fluidized bed containing cells anchored to microspheres. There is a small group of proteins that are anchored to the cell's plasma membrane via glycosylphosphatidyl inositol (GPI). This enables the protein to be harvested easily by cleavage of the anchor using phospholipase C from cells grown in suspension or on microspheres. Besides naturally occurring anchored proteins, the GPI pre-anchor sequence can be engineered into recombinant products by creating fusion proteins, thus facilitating the harvesting of proteins that are not normally secreted (Chapters 15 and 16).

Increasingly, human hematopoietic cells are being grown in vitro for cell transplantation and as gene therapy vectors. Modifications to the basic cell culture, characterization, and selection and expansion techniques are required, since the initial cultures contain a heterogeneous population of different cell lineages. These specialized techniques, which include protocols for retroviral-mediated transgene infection of blood cells, are described in Chapters 17 and 18. A similar problem of culturing diverse cell types occurs in cell-based cytotoxicity testing. The relevant assays used to monitor toxic responses, which include dye release, dye uptake, radioisotope release, and metabolite assays, are the subject of Chapter 19.

In Part V, the effects of cell culture conditions on the posttranslational modification of proteins are described. Many proteins require extensive posttranslational modification to achieve full biological activity. Indeed, authentic posttranslational machinery is often cited as the main reason why many proteins used therapeutically are manufactured using mammalian, as opposed to microbial, cells. Methods used to detect the extent of protein folding and the kinetics of protein secretion are described in Chapter 20. These methods involve radioisotopes used in a pulse-chase regime, followed by resolution on acrylamide gels and autoradiography. Proteolysis of the expressed protein during biosynthesis is a common phenomenon, either in the culture media itself or during downstream processing. In some cases the action of such specific proteases as prohormone convertase is a prerequisite to achieving full biological activity. More commonly, however, proteolysis is an unwanted and degradative step that can be minimized by accurate monitoring and the use of protease inhibitors. Protocols for manipulating these protease-related events are described in Chapter 21.

Glycosylation is the most extensive and variable posttranslational modification made to proteins expressed in mammalian cells. It occurs in most secreted and membrane-bound mammalian proteins. Detailed protocols for protein glycoform analysis using capillary electrophoresis and mass spectrometry are described in Chapter 23. In addition to biosynthetic variations, glycan heterogeneity can arise from glycosidases released from cells during culture. The measurement of glycosidase activity using fluorescent substrates, its effects on glycan heterogeneity, and measures used to minimize glycosidase-dependent degradation are described in Chapter 23.

In summary, this volume constitutes a comprehensive manual of techniques that are essential for the setting up of a cell culture laboratory, the routine maintenance of cell lines, and the optimization of critical parameters for cell culture. Inevitably, some omissions will occur in the text, but the authors have sought to avoid duplication by extensive cross-referencing to chapters in other volumes of this series and elsewhere. We hope the volume provides a useful compendium of techniques for scientists working in industrial and research laboratories that use mammalian cells for biotechnology purposes. The editor is grateful for the support of all the contributors and the publishers who have made this volume possible.

Nigel Jenkins

Contents

Contributors

S. R. ADAMSON • *Genetics Institute, Andover, UK*
MOHAMED AL-RUBEI • *Department of Biochemical Engineering, Birmingham University, Birmingham, UK*
MARIA L. ANTHONY • *Biochemistry Department, University of Cambridge, UK*
CATHERINE M. BENTLEY • *Cell Culture Develpment, Glaxo-Wellcome, Kent, UK*
KEVIN M. BRINDLE • *Biochemistry Department, University of Cambridge, Cambridge, UK*
LORRAINE D. BUCKBERRY • *Biomolecular Science, De Montfort University, Leicester, UK*
NEIL J. BULLEID • *School of Biological Sciences, University of Manchester, Manchester, UK*
MICHAEL BUTLER • *Department of Microbiology, University of Manitoba, Manitoba, Canada*
MARIA G. CASTRO • *Molecular Medicine Unit, University of Manchester, Manchester, UK*
THOMAS G. COTTER • *Department of Biochemistry, University College Cork, Ireland*
DENIS DRAPEAU • *Genetics Institute, Andover, UK*
STEPHEN J. FROUD • *Lonza Biologics plc, Berkshire, UK*
MARGARET GOODALL • *Department of Immunology, University of Birmingham, Birmingham, UK*
MICHAEL J. GRAMER • *Cellex Biosciences Inc., South San Francisco, CA*
DIANE L. HEVEHAN • *Northwestern University, Evanston, IL*
ANDREW D. HOOKER • *Research School of Biosciences, University of Kent, UK*
DAVID C. JAMES • *Research School of Biosciences, University of Kent, UK*
WILFRED A. JEFFERIES • *Biotechnology Laboratory, University of British Columbia, Vancouver, BC, Canada*
NIGEL JENKINS• *Bioprocess Research and Development, Eli Lilly & Co., Indianapolis, IN*
MALCOLM L. KENNARD • *Biotechnology Laboratory, University of British Columbia, Vancouver, BC, Canada*
GREGORY A. LIZEE • *Biotechnology Laboratory, University of British Columbia, Vancouver, BC, Canada*

Carolynne T. Marshall • *Cell Culture Development, Glaxo-Wellcome, Kent, UK*
Todd A. McAdams • *Chemical Engineering Department, Northwestern University, Evanston, IL*
John Stephen McLean • *University of Paisley, Paisley, Scotland*
Lucia Monaco • *DIBIT, Istituto Scientifico San Raffaele, Milano, Italy*
William M. Miller • *Chemical Engineering Department, Northwestern University, Evanston, IL*
Sigma S. Mostafa • *Chemical Engineering Department, Northwestern University, Evanston, IL*
E. Terry Papoutsakis • *Chemical Engineering Department, Northwestern University, Evanston, IL*
Marcelo J. Perone • *Molecular Medicine Unit, University of Manchester, UK*
James M. Piret • *University of British Columbia, Vancouver, BC, Canada*
Afshin Samali • *Department of Biochemistry, University College Cork, Ireland*
Dietmar Schiffmann • *Institut de Genie Chimique, Ecole Polytechnique Federale de Lausanne, Lausanne Switzerland*
Michelle F. Scott • *Cell Culture Development, Glaxo Wellcome, Kent, UK*
Alasdair J. Shepherd • *Q-One Biotech Ltd., Glasgow, Scotland, UK*
Martin S. Sinacore • *Genetics Institute, Andover, UK*
Kenneth T. Smith • *Q-One Biotech Ltd., Glasgow, Scotland, UK*
Glyn N. Stacey • *National Institute of Biological Standards and Control, South Mimms, UK*
Adrian R. Walmsley • *School of Biological Sciences, University of Manchester, UK*
Annette Waugh • *British Biotech Pharmaceuticals Ltd., Oxford, UK*
Shane N. O. Williams • *Department of Biochemistry, Cambridge, UK*
Simon Windeatt • *Molecular Medicine Unit, University of Manchester, UK*
Florian M. Wurm • *Institut de Genie Chimique, Ecole Polytechnique Federale de Lausanne, Lausanne Switzerland*

I

Enabling Technologies

1

Setting Up a New Cell Culture Laboratory

Michelle F. Scott, Catherine M. Bentley, and Carolynne T. Marshall

1. Introduction

Animal cells are cultured in vitro for a variety of uses. Advances in recombinant DNA technologies have resulted in an array of heterologous proteins being produced from cultured cells for use as research tools, potential biotherapeutics *(1)*, or analytical reagents. Cells are also used as products in their own right in receptor research, to study xenobiotic metabolism and toxicity in vitro *(2)*, in high throughput screening, and for biopharmaceutical biological activity assays.

The design of a new cell culture laboratory should focus on the proposed function of the facility. With rapid changes in all areas of biotechnology, particularly the emergence of genomics and the subsequent explosion in the need for cell culture facilities, it is important to design a flexible facility which will support quality science and be amenable to modifications as functionality changes. The aim therefore is for a modular facility with the following attributes:

1. Can be used for small scale culture of any cell type.
2. Has all the necessary equipment to support cell culture.
3. Minimizes the risk of cross-contamination with other cell lines or adventitious agents.
4. Can be used safely.

It is suggested that several small self-contained laboratories be incorporated within the facility. Each laboratory may be assigned to a single cell line and should contain dedicated equipment. Additional centralized support facilities are provided for non-dedicated activities and equipment.

Here, we describe the design features of a laboratory and its ancillary facilities for the routine small-scale culture of nonpathogenic animal cells.

From: *Methods in Biotechnology, Vol. 8: Animal Cell Biotechnology*
Edited by: N. Jenkins © Humana Press Inc., Totowa, NJ

2. The Modular Facility

In addition to the cell culture laboratories, the modular cell culture facility may include areas for sample analysis, storage of disposables and glassware, cryopreservation, and decontamination/wash-up. These support facilities should be located in close proximity to the cell culture laboratories to minimize the movement of potentially hazardous biological material (**Fig. 1**).

3. Cell Culture Equipment

3.1. Inside the Laboratory

The essential items of dedicated equipment for each laboratory are those which enable safe aseptic manipulation and examination of the culture, controlled incubation and cell separation (**Table 1**). Each laboratory may also contain a mobile trolley for storing disposable items such as pipets and sample tubes.

3.2. Outside the Laboratory

The centralized support facilities should contain equipment for analyzing dead-end samples (i.e., those that will not be reintroduced into the culture), cold storage, cryopreservation, wash-up (if glassware is used), and decontamination/sterilization.

Advances in analytical technologies mean that there are many pieces of equipment used routinely to analyze cells or their products. Such equipment include those used to analyze metabolites (amino acid analyzer, glucose/lactate analyzer), physicochemical properties (blood gas analyzer, pH meter, osmometer), cells (automatic cell counter, cell sorter, flow cytometer), and product (ELISA, HPLC, and nephelometer). Equipment for monitoring cell and product quality and activity may also be necessary.

Central cold storage (+4°C) and freezing (–20°C and –70°C) facilities should be available for the storage of cell culture media components, reagents and samples (*see* Chapter 8 for further details).

The cryosuite should contain equipment for freezing and storing cells. Cell freezing is generally carried out in a controlled rate or constant temperature (–70°C to –150°C) freezer. Cell storage should be below –130°C *(3)* and may be in ultracold electrical freezers or liquid nitrogen (liquid or vapor phase). Consideration should be given to potential cross-contamination for cells stored in the liquid nitrogen *(4)*. Suitable safety equipment (oxygen monitor, thermal gauntlets, face shield) must be available in the cryosuite.

Autoclaves may be required for decontamination of waste and sterilization of some equipment. Sinks will be required for disposal of decontaminated waste and for wash-up.

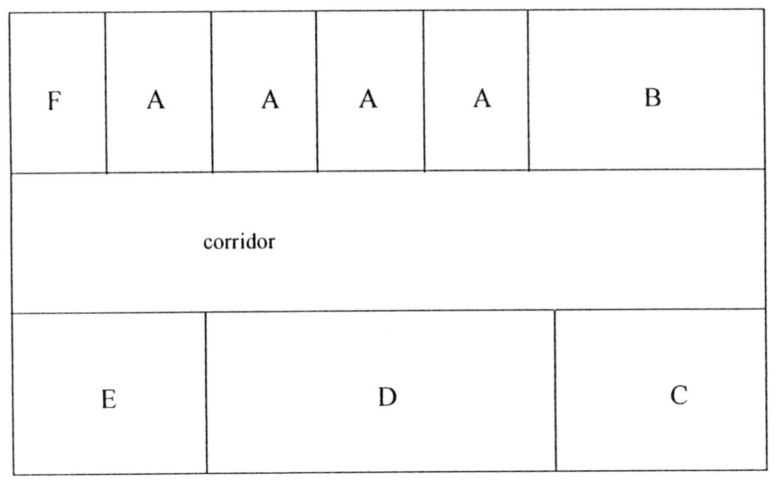

A - Cell Culture laboratory
B - Cryopreservation laboratory
C - Decontamination/Autoclave room
D - Analytical laboratory
E - Storage Area
F - Refridgerator and -20°C freezer

Fig. 1. The cell culture laboratories and supporting facilities.

4. Laboratory Services

The services required for a cell culture facility include gas supplies, water, and power. A vacuum source may be necessary in the laboratories for filtration work, liquid nitrogen must be supplied to the cryosuite, and steam will be required for any autoclave areas.

The following gases should be provided to the laboratories:

1. 100% Carbon dioxide for use with CO_2 incubators.
2. 5% Carbon dioxide in air to adjust the pH of cell cultures.
3. Air to aerate high density cultures.

These should be supplied to all laboratories from a central location outside the laboratories. To ensure constant supply, cylinders should be duplicated and fitted with automatic changeover units.

Water and steam are required in the centralized facility for wash-up and decontamination, respectively. Water must also be provided to each cell culture laboratory for hand-washing ***(5)***.

Power is required for the operation of equipment such as incubators, microbiological safety cabinets, microscopes, and centrifuges. In addition, there may be a need to use small pieces of electrical equipment such as pipetting devices in the laboratory. Manufacturers will specify the precise requirements for individual pieces of equipment and this may include the necessity for a three-phase power supply.

Table 1
Essential Cell Culture Equipment

Equipment	Features
Microbiological safety cabinet	Provides operator protection Visible airflow meter and audible alarm to monitor performance Formaldehyde addition box and venting hose for decontamination Easy to clean
Binocular microscope	Magnification of x400 for use with a hemocytometer
Inverted Microscope	Magnification such that a single well on a 96-well plate can be seen in one field of view
Bench-top centrifuge	Temperature controlled and refrigerated Volumes from 5 mL to 1 L (4 × 250 mL)
Bench-top shaking incubator	Good temperature control Can use 'sticky stuff' to maximize the number of flasks
Floor-standing shaking incubator	Wide range of temperatures Foot pedal to raise the lid (floor-standing)
Static CO_2 incubator	Water jacketed to maintain temperature consistently Copper walls to minimize fungal contamination Magnetic stirrer platform at the bottom for stirred culture vessels Humidified for use with cloning plates and small volumes

5. The Cell Culture Laboratory

5.1. Design

The layout of the equipment in the cell culture laboratory is important to ensure the correct and safe use of the facility. A suggested floor plan is presented in **Fig. 2**. It is essential that there is adequate space for all equipment and staff, and doorways must be large enough for every piece of equipment to pass through. Floor and work surface materials must be compatible with cleaning regimes and safety regulations.

The largest piece of equipment is usually the microbiological safety cabinets (generally Class II Laminar flow). Placement of these are critical, as disturbance of the local environment may cause excessive movement of air, which could have an adverse effect on the operation of the cabinet. Manufacturers' specifications provide guidance on this. In general, the cabinet should be located away from the door and any through-ways in the laboratory.

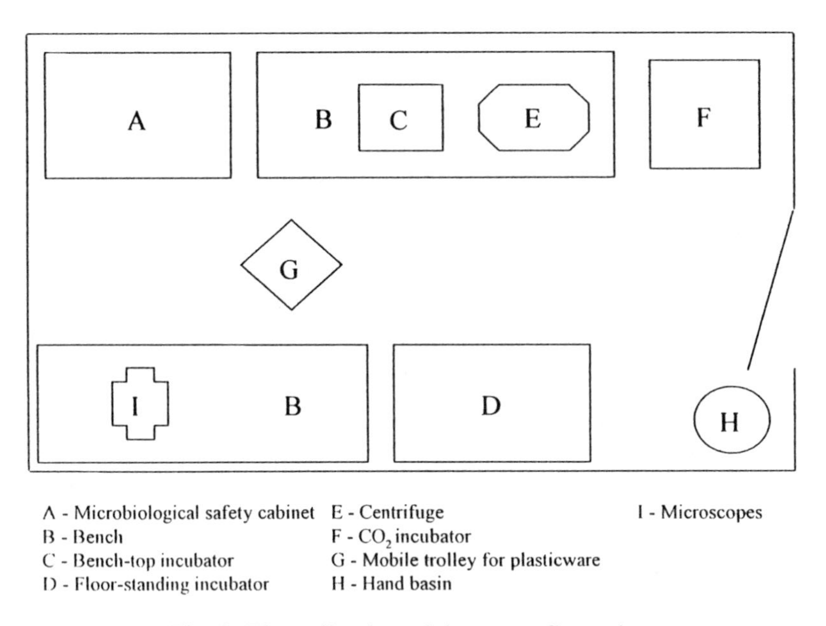

Fig. 2. The cell culture laboratory floor plan.

Centrifuges should be positioned on a separate bench from the microscopes to avoid centrifuge vibrations disturbing the microscopy examinations. Centrifuges also cause air turbulence and, thus, should not be located in front of the microbiological safety cabinet.

5.2. Use

Ideally, each laboratory should be dedicated to a single cell line at any one time. Access to each laboratory should be limited to authorized personnel, and the number of staff working in each laboratory should be minimized. Laboratory coats should be dedicated to a specific laboratory to reduce the possibility of cross-contamination. Equipment should not be moved between laboratories without disinfection.

Cleaning of the laboratories is important to maintain a suitable working environment. Disinfection of work tops and microbiological safety cabinets should be carried out at least weekly, using a variety of broad-spectrum disinfectants on a rotation basis.

6. Safety and Environment

6.1. Environment

The cell culture laboratories should be safe, appropriate for purpose, and pleasant to inhabit. Environmental factors influencing comfort include the thermal, visual, and acoustic conditions. Guidance should be sought from government health and safety authorities for local regulations concerning workplace environmental conditions.

Provision should be made for a sufficient quantity (as per government regulations) of fresh or purified air. Ventilation should be mechanical, as the opening of windows will increase the risk of contamination and will disturb the environment around the microbiological safety cabinet.

Thermal comfort is subjective and dependent on individuals, amount of clothing, level of physical activity, air temperature, air movement, and humidity. From experience, the temperature should be controlled between 17°C and 24°C and the humidity between 35% and 70%. This provides comfort for the occupants and is suitable for the operation of the majority of laboratory equipment.

Noise levels are regulated by the government and must be monitored to show compliance. In some cases, ducting microbiological safety cabinets can substantially reduce noise levels.

6.2. Air Handling

Genetically modified organisms should be handled in such a way as to prevent accidental release into the environment ***(5,6)***. This will include maintaining laboratory air pressures which are negative with respect to atmospheric pressure and filtering the air extracted from the laboratory. A pressure differential of 15 Pa is sufficient for this purpose. Ideally, a positive pressure lobby should be positioned at each end of the facility, the corridor pressure should be negative to atmosphere, and the pressure in each laboratory should be negative to the corridor. To reduce the ingress of contaminants from the corridor, the laboratory air extracts should be positioned such that all air entering from the corridor is immediately extracted from the room.

6.3. Spillages and Waste

A procedure for the management of spillages in the cell culture laboratories should be established. Features such as benches surrounded by a 2 cm lip should be considered for spillage containment, and spillage kits should be provided in, or near, each laboratory.

A decontamination policy for waste produced from the cell culture laboratories should be devised. This may include decontamination by autoclaving or chemical disinfection. The effectiveness of these procedures should be confirmed.

References

1. Lubinecki, A. S. and Vargo, S. A. (1994) Introduction to regulatory practice in novel biotechnology, in *Regulatory Practice for Biopharmaceutical Production* (Lubinecki, A. S. and Vargo, S. A., eds.), Wiley-Liss, New York, p. 1.

2. MacDonald, C., Reid, F., Anderson, K., et al. (1997) Metabolism and toxicological studies in immortalised rat hepatocyte cell lines, in *Animal Cell Technology* (Carrondo, M. J. T., Griffiths, B., and Moreira, J. L. P., eds.), Kluwer Academic Publishers Dortrecht, Netherlands, p. 73.
3. Doyle, A., Griffiths, J. B., Morris, C. B., and Newell, D. G (1990) Culture and maintenance, in *Animal Cells* (Doyle, A., Hay, R., and Kirsop, B. E., eds.), Cambridge University Press, Cambridge, UK, p. 63.
4. Tedder, R. S., Zuckerman, M. A., Goldstone, A. H., et al. (1995) Hepatitis B transmission from contaminated cryopreservation. *Lancet* **346,** 137.
5. Advisory Committee on Dangerous Pathogens (1995) *Categorisation Of Biological Agents According To Hazard and Categories Containment.* 4th ed. HSE Books.
6. Advisory Committee on Genetic Modification (1997) *Compendium Of Guidance From The Health And Safety Commission's Advisory Committee On Genetic Modification.* Health and Safety Executive.

2

Adaptation of Mammalian Cells to Growth in Serum-Free Media

Martin S. Sinacore, Denis Drapeau, and S. R. Adamson

1. Introduction

The intent of this chapter is to discuss mammalian cell adaptation procedures useful in the development of cell phenotypes compatible with extended cultivation in suspension culture, without serum, and at high cell densities. Such phenotypes are useful for the production of recombinant proteins and monoclonal antibodies for therapeutic or diagnostic applications. Emphasis will be placed on Chinese hamster ovary (CHO) cells due to their widespread use throughout the biotechnology industry. However, the general principles of cell adaptation can be applied to other animal cells, including baby hamster kidney (BHK) and hybridoma cells.

The regulatory and economic rationale for elimination of serum from mammalian cell growth medium has been well established *(1)*. The use of animal sera in cell-culture processes brings along with it the potential for introduction of adventitious agents such as viruses and other transmissible agents (e.g., bovine spongiform encephalopathy). Additionally, the use of animal sera as a raw material impacts negatively on the economics of large-scale cell-culture processes. Finally, because protein biotherapeutics produced by mammalian cells are secreted into the conditioned media, the use of serum-free medium greatly simplifies development and robust execution of downstream protein purification processes.

Without the use of adaptation procedures such as those described here, CHO and BHK cells tend to be both anchorage dependent and serum dependent. Hybridoma cells, although not anchorage dependent, are also serum dependent. In addition, once a serum-free suspension-adapted lineage is established,

From: *Methods in Biotechnology, Vol. 8: Animal Cell Biotechnology*
Edited by: N. Jenkins © Humana Press Inc., Totowa, NJ

the volumetric productivity of the cell-culture process can be optimized by achieving relatively high final cell densities ($\geq 4.0 \times 10^6$ viable cells/mL). Obviously, in order to support these high cell densities, the concentration of nutrients in the culture medium must be proportionately increased. Furthermore, cells can often adapt to high cell densities by developing tolerance to growth-inhibiting substances released by the cells like lactic acid and ammonium ***(2–6)***.

The adaptation to serum-free suspension or high-cell-density conditions is generally performed using cell lineages that have already been engineered for expression of the desired protein biotherapeutic through recombinant technologies or technologies based on cell fusion ***(7–13)***. However, reports have been made of CHO ***(14–16)***, murine myeloma ***(17,18)*** and Namalwa ***(19)*** cell lineages that have been adapted to growth in serum-free suspension culture prior to serving as host cells. This approach greatly reduces or eliminates the need for further adaptation once expression of the desired protein is achieved in these preadapted mammalian cell host lineages.

On a cautionary note, the transition to a high-cell-density serum-free suspension culture can lead to changes in the growth performance of cells and the structural characteristics of secreted proteins. Some decline in growth performance (i.e., decreased growth rate or cell viability) following serum withdrawal and/or removal of cell attachment substrata has been reported ***(15,20)***. This response to changes in culture conditions is apparently due to a perturbation of events associated with cell cycle progression and entry into the S-phase ***(21)***. In the most severe case, activation of apoptotic pathways can occur ***(22,23)***. Furthermore, significant differences in the carbohydrate structures of secreted proteins have been documented ***(24–29)***. Finally, cell populations undergoing a "growth crisis" (i.e., a significant decline in growth rate and cell viability) during the adaptation process may be vulnerable to outgrowth of nonrepresentative or undesirable subpopulations of cells ***(15,30)*** that may not have optimum performance characteristics (i.e., growth rate, cellular productivity, cell density ceiling, genetic stability, etc.). Accordingly, the general growth performance characteristics of the resultant cell lineages, as well as the structural and functional integrity of expressed proteins, should be closely monitored throughout the cell line adaptation process.

2. Materials

2.1. Serum-Free Media Formulations

It is not within the scope of this chapter to undertake an extensive review of serum-free media formulations useful in the large-scale production of protein biotherapeutics. Several serum-free media formulations have been reported for CHO, BHK, and hybridoma cells ***(1,31–35)*** in addition to several commercially available formulations. Identification of a suitable serum-free formulation for a given animal cell line and culture condition should be determined empirically or by information gained from the literature.

A serum-free medium into which CHO cells can be generally adapted is composed of DME : F12 (50 : 50) (JRH Biosciences, Lenexa, KS or equivalent) supplemented with additional L-glutamine (2–4 m*M*), recombinant human insulin (10 µg/mL) (Eli Lilly and Company), hydrocortisone (0.072 µg/mL), putrescine (2.0 µg/mL), sodium selenite (0.010 µg/mL), ferrous sulfate (0.91 µg/mL), and polyvinyl alcohol surfactant (2.4 mg/mL). Adaptation of recombinant CHO cells to high-cell-density conditions requires a more nutritionally rich medium formulation like the one shown in **Tables 1** and **2**. If applicable to the recombinant cell line in question, the use of selective agents (e.g., methotrexate, G418, etc.) in the medium should be continued to ensure that the genotype and phenotype of the resultant lineages are not compromised during the adaptation process.

Cryopreservation medium for serum-free adapted lineages, in many instances, can be formulated by supplementing the serum-free growth medium to which cells have adapted with dimethyl sulfoxide (10%, v/v). Additional proteinaceous (e.g., bovine serum albumen) or nonproteinaceous supplements (e.g., pluronics, methyl cellulose) may be necessary to improve recovery of viable cells from cryopreserved stocks ***(36)***.

2.2. Cell-Culture Equipment

Commercially available cell-culture plasticware by Corning, Falcon, and Nunc (or equivalent) is recommended for maintenance of monolayer cell cultures. Spinner flask glassware (50 and 100 mL working volume) and four or six-position magnetic stirrers manufactured by Bellco, Vineland, NJ (or the equivalent from another supplier) will be required for maintaining suspension cultures. Cultures are kept in a humidified, temperature-controlled CO_2 incubator under conditions appropriate for the particular cell line to be adapted (usually 37°C, 7% CO_2 in air).

Adaptation to high-cell-density conditions is performed in bioreactors that control dissolved oxygen and pH at nonlimiting levels (Applikon, Foster City, CA, or equivalent). They must supply oxygen (e.g., by sparging small oxygen bubbles into the culture) in response to a drop in the signal from a dissolved oxygen electrode. Likewise, they must supply a base (e.g., by injecting a sodium bicarbonate/sodium carbonate solution into the culture) in response to a drop in the signal from a pH electrode. In addition, they must maintain temperature at an appropriate level (generally 37°C) and they must provide agitation at a rate sufficient to prevent cell settling but yet low enough to avoid cell lysis. Using growth media supplemented with 0.24% polyvinyl alcohol (**Tables**

Table 1
Inorganic and Amino Acid Components (μg/mL) in a High-Nutrient, Low-Serum Medium Used in the Adaptation of a CHO Cell Line to High-Cell-Density Conditions and in a Similar Serum-Free Medium

Component	Low-serum medium	Serum-free medium
Sodium chloride	4600	4400
Potassium chloride	624	310
Calcium chloride, anhydrous	232	58
Sodium phosphate, dibasic, anhydrous	142	—
Sodium phosphate, monobasic, hydrate	125	130
Magnesium chloride, anhydrous	57	—
Magnesium sulfate, anhydrous	98	84
Cupric sulfate, anhydrous	0.0016	0.0018
Ferrous sulfate, anhydrous	0.68	0.91
Ferric nitrate, nonahydrate	0.10	—
Zinc sulfate, septahydrate	0.86	0.92
Sodium selenite	0.010	0.010
Sodium bicarbonate	2440	2400
L-Alanine	36	71
L-Arginine	600	760
L-Asparagine hydrate	180	540
L-Aspartic acid	133	270
L-Cysteine hydrochloride hydrate	282	700
L-Cystine dihydrochloride	125	—
L-Glutamic acid	59	120
L-Glutamine	1168	1200
Glycine	60	60
L-Histidine hydrochloride hydrate	126	290
L-Isoleucine	210	470
L-Leucine	260	680
L-Lysine hydrochloride	291	730
L-Methionine	104	240
L-Phenylalanine	165	330
L-Proline	138	280
L-Serine	315	630
L-Threonine	190	380
L-Tryptophan	33	130
L-Tyrosine disodium dihydrate	262	420
L-Valine	187	370

1 and **2**), we have found that a tip speed of 2000–6000 cm/min is optimal. The tip speed is calculated using the equation

$$\text{Tip speed} = \pi ND,$$

Table 2
Vitamin and Other Organic Components (μg/mL) in a High-Nutrient, Low-Serum Medium Used in the Adaptation of a CHO Cell Line to High-Cell-Density Conditions and in a Similar Serum-Free Medium

Component	Low-serum medium	Serum-free medium
Biotin	0.41	1.6
D-Calcium pantothenate	4.5	18
Choline chloride	18	72
Folic acid	5.3	21
i-Inositol	25	100
Nicotinamide	4.0	16
Pyridoxine hydrochloride	0.062	16
Pyridoxal hydrochloride	4.0	—
Riboflavin	0.44	1.8
Thiamine hydrochloride	4.3	18
Vitamin B12	1.45	5.6
D-Glucose	6000	6200
Sodium pyruvate	110	—
Linoleic acid	0.084	0.17
Thioctic acid	0.21	0.42
Putrescine dihydrochloride	2.2	2.0
Polyvinyl alcohol	2400	2400
Insulin or Nucellin	10	10
Hydrocortisone	0.072	0.072
Methotrexate[a]	1.3	1.3
Soybean phospholipid[b]	10	—
Fetal bovine serum	0.5% (v/v)	—

[a]The use of particular concentrations of methotrexate (or other drugs) as a selection agent is cell line dependent.

[b]The benefit associated with the use of soybean phospholipid supplement is cell line dependent and should be determined empirically.

where *N* is the revolutions per minute (rpm) of the bioreactor impeller and *D* is the impeller diameter. For example a 1.0-L Applikon bioreactor fitted with a 4.5-cm Rushton impeller should be operated at 150–450 rpm to provide the appropriate level of agitation.

3. Methods

In order to minimize any negative impact of the adaptation process on the growth performance (i.e., growth rate, cell viability, and cellular productivity) of anchorage- and serum-dependent mammalian cell cultures, we have found that a three-phase adaptation process works most consistently. The first phase adapts the cells to suspension culture, the second adapts them to serum-free conditions, and the third adapts them to high-cell-density conditions. At the start of each adaptation phase, it is important to ensure that cells are growing exponentially at the time the adaptation process is initiated. To ensure against catastrophic loss of adapted cell lineages, it is critical that backup cultures be maintained using growth conditions in place prior to the adaption step in progress. Furthermore lineages should be cryopreserved at several points during the adaptation process to provide additional protection against catastrophic loss of lineages.

3.1. Adaptation to Suspension Culture

1. The first phase of the adaptation process begins with monolayer cultures in standard cell culture plasticware (T75 or T150 flasks or culture dishes) using fetal bovine serum (FBS)-supplemented growth medium. Cell cultures are passaged at a frequency appropriate for the mammalian cell in question (usually every 3–4 days) by trypsinization. At each passage cycle, cell monolayers are washed twice with sterile phosphate-buffered saline (free of divalent cations) and are then overlayed with a minimal volume of 0.25% trypsin (Sigma Chemical Co., St. Louis, MO or recombinant equivalent from Eli Lilly & Co.). Cultures are incubated at 37°C for 5–10 min and the trypsin solution is quenched with 5–10 mL of FBS-supplemented growth medium. The cell suspension is then used to inoculate another monolayer culture ($1–3 \times 10^4$ cells/cm^2) or to initiate a suspension culture.
2. Suspension cultures are initiated by pelleting the trypsinized cells by centrifugation at 200*g* for 5 min. The cell pellet is gently resuspended in FBS-supplemented growth medium and the viable cell density is determined by counting cells in a hemocytometer using the Trypan Blue dye exclusion cell viability assay. If cell viability is ≥90% the cells are used to establish a suspension culture in a 50- or 100-mL working volume spinner flask at a density of $(1.0–3.0) \times 10^5$ viable cells/mL FBS-supplemented medium. The spinner flask may be pretreated with a siliconizing solution if attachment of cells to the glass surface of the spinner flask is a concern. We have evaluated the use of subpopulations of nonadherent CHO cells, which often arise in monolayer cultures to initiate suspension cultures. In our experience initiation of adaptation to growth in suspension culture with nonadherent cells does not consistently improve the outcome.
3. Spinner flask suspension cultures are magnetically agitated and maintained in a CO_2 incubator under the same conditions of temperature, humidity, and atmosphere optimal for monolayer cultures. Every 2–3 days, the viable cell density of the culture is determined and adjusted to $(1–3) \times 10^5$ cell/mL medium by dilution with fresh medium. If growth is insufficient to allow dilution by a factor of at least 4:1, the cells should be centrifuged and gently resuspended in fresh medium.
4. The growth rate and cell viability of the culture should be monitored closely during the first few weeks of culture. In our experience with CHO cells, the

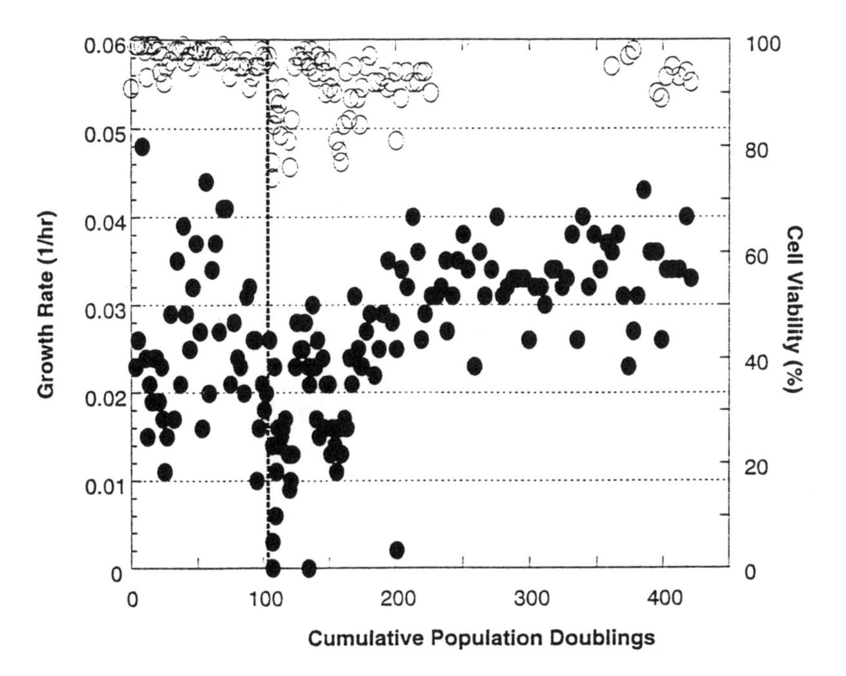

Fig. 1. Adaptation of CHO DUKX B11 cells to growth in serum-free suspension culture. CHO DUKX B 11 cells were released from monolayer culture by trypsinization and resuspended in growth medium containing 10% FBS (cumulative population doublings [CPD] = 0). Cells were continuously maintained in suspension culture and monitored for growth rate and cell viability at each passage cycle. The FBS concentration was reduced to 1% after 10 CPD in suspension culture using 10% FBS-supplemented medium. The dashed line represents the point at which the lineage was transferred into serum-free medium (Reprinted from **ref.** ***14***).

establishment of suspension cultures using FBS-supplemented medium does not greatly impact the growth performance of the culture. However, a "crisis" manifested by a drop in cell viability and/or growth rate may develop during the initial passages of the suspension culture (*see* **Fig. 1**). Generally, the culture will spontaneously recover from such a crisis and good growth will resume within a few weeks.

3.2. Adaptation to Serum-Free Medium

1. Once the growth performance of the suspension culture has stabilized, the culture will be ready to proceed to the serum-reduction/elimination phase of the adaptation process. This consists of a series of steps in which the serum concentration is successively reduced. The medium at each step will be the target serum-free or low-serum medium (i.e., with all growth-promoting additives), to which serum has been added at the appropriate concentration.

2. When good growth has been established following a given serum-reduction step, the next step is taken. When the serum concentration is still greater than 1%, each step is typically a 50% reduction, but when serum concentration has been lowered to 1% or less, the next step is generally directly into serum-free medium. In some cases, however, we have found that an initial step consisting of an 80–100% reduction in serum concentration can be successful for CHO cell cultures. If a step leads to a severe growth crisis, the resulting culture should be discarded and a parallel culture that has not been subjected to this step should be subjected to a smaller step. Once a serum-free lineage has been established, the growth performance phenotype of lineages may benefit from continued adaptation in serum-free medium over a period of months.

3.3. Adaptation to High-Cell-Density Conditions

Cells can often adapt to high-cell-density conditions by developing tolerance of growth-inhibiting substances released by the cells. Two such substances are lactic acid and ammonia. Tolerance of these substances may be developed by adding them to the fresh medium in progressively higher concentrations during the adaptation program ***(2–5)***. However, tolerance of these substances, as well as other undefined cell-generated growth-inhibiting substances, may be developed by ensuring that no other factors besides inhibitory substances limit growth during the adaptation program ***(6)***. This can be achieved as follows:

1. The cells should be cultured for at least 5 passages and preferably 10 or more passages, each of which lasts between 1 and 5 days. Throughout each passage, dissolved oxygen, pH, and nutrients should be maintained at nonlimiting levels. Each passage should be initiated by diluting the culture with an amount of fresh medium that results in an initial cell density that is high enough that the cells will encounter growth inhibition during the passage due to secreted substances.
2. The appropriate dilution ratio can be calculated on the basis of the duration of the passage *(t)* and the maximum specific growth rate of the cell line when growth is not inhibited by any extracellular factors (μ_{max}). The formula is:

$$\text{Dilution ratio} = e^{\mu t}$$

where μ is a value between 60% of μ_{max} and 90% of μ_{max}. For CHO cells, μ_{max} is typically approximately 0.030/h. Therefore, appropriate dilution ratios for initiating 3-day and 4-day passages would be between 3.7 and 7.0, and 5.6 and 13.4, respectively.

An example involving a recombinant CHO cell line expressing recombinant human macrophage colony-stimulating factor (rhM-CSF) which had already been adapted to low-serum (0.5%) suspension culture in spinner flasks is represented in **Table 3**. The use of low-serum growth medium in this example was specific to this particular CHO cell lineage. Mammalian cells adapted to growth in serum-free medium can be further adapted to high-cell-density growth in a

Table 3
High-Cell-Density Adaptation of a rhM-CSF-Expressing CHO Cell Line

Passage no.	Passage length (days)	Dilution ratio	Initial cell density (10^6/mL)	Final cell density (10^6/mL)	Final product titer (μg/mL)
1	4	—	0.12	1.24	11.6
2	3	5.4	0.23	1.96	14.3
3	3	6.3	0.31	3.00	16.5
4	3	5.1	0.59	2.44	14.9
5	4	12.2	0.20	1.79	—
6	4	6.0	0.30	3.50	—
7	3	5.0	0.70	2.25	12.2
8	3	5.2	0.43	2.70	15.6
9	4	12.3	0.22	4.30	20.2
10	4	14.3	0.30	5.90	29.2
11	3	5.9	1.00	5.70	33.5
12	3	11.4	0.50	4.90	32.6
13	4	16.3	0.30	5.30	34.2

similar fashion. The rhM-CSF-expressing recombinant CHO cells were transferred to a 2-L bioreactor where they were diluted with fresh medium to an initial cell density of 1.2×10^5 cells/mL. They were then cultured for eleven 3-day or 4-day passages in the 2-L bioreactor (passages 1 through 11 in **Table 3**) in the nutrient-rich, low-serum medium (**Tables 1** and **2**). During each passage, pH was maintained at between 7.0 and 7.2 by the addition of sodium carbonate and sodium bicarbonate, and dissolved oxygen was maintained at between 20% and 60% of air saturation by the introduction of oxygen bubbles. Each 3-day passage was started by diluting the culture from the preceding passage by a factor between 5.1 and 6.3, whereas each 4-day passage was started by diluting the culture from the preceding passage by a factor between 6.0 and 14.3.

The beneficial effect on the cell line was evident during two subsequent passages (passages 12 and 13 in **Table 3**). For example, in passage, 12, which was started at a density of 5.0×10^5 cells/mL, cell density reached 4.90×10^6 cells/mL and rhM-CSF titer reached 32.6 μg/mL. In contrast, in passage 4, which had been started at a higher density (5.9×10^5 cells/mL), the cell density had reached only 2.44×10^6 cells/mL and the rhM-CSF titer had reached only 14.9 μg/mL.

The success of the high-cell-density adaptation approach is cell line dependent and should be evaluated on a case-by-case basis. In those instances where improvements in the final cell density and final product titers were observed using this adaptation protocol, the changes in the high-cell-density phenotype have been stable. Furthermore, in experiments in which direct comparisons between cell stocks cryopreserved before and after the high cell adaptation was carried out, no qualitative change in the recombinant human proteins expressed by CHO cells was detected subsequent to the adaptation.

4. Summary

Methods for the adaptation of mammalian cells to growth in a serum-free suspension culture have been disclosed. A three-step adaptation protocol was described in which mammalian cell cultures were first established in suspension culture using serum-supplemented medium. In the second step, the serum concentrations in the medium were incrementally reduced and then completely replaced with defined growth-promoting factors (recombinant insulin, hydrocortisone, putrescine, selenium, and ferrous sulfate) and polyvinyl alcohol surfactant. And finally, in the third step cells are adapted to high cell density conditions.

This adaptation protocol has been successfully used at Genetics Institute, Inc. in the development of large-scale cell-culture processes for the manufacture of recombinant protein biotherapeutics. The absence of animal-derived medium components in the resultant cell-culture process eliminates the potential for introduction of adventitious process contaminants and greatly simplifies downstream protein purification processes.

References

1. Lambert, K. J. and Birch, J. R. (1985) Cell growth media, in *Animal Cell Biotechnology*. Volume I. (Spier, R. E. and Griffiths, J. B., eds.), Academic Press, pp. 85–112.
2. Miller, W. M., Wilke, C. R., and Blanch, H. W. (1988) Transient responses of hybridoma cells to lactate and ammonia pulse and step changes in continuous culture. *Bioprocess Eng.* **3,** 113–122.
3. Matsumura, M., Shimoda, M., Arii, T., and Kataoka, H. (1991) Adaptation of hybridoma cells to higher ammonia concentration. *Cytotechnology* **7,** 103–112.
4. Schumpp, B. and Schlaeger, E.-J. (1992) Growth study of lactate and ammonia double-resistant clones of HL-60 cells. *Cytotechnology* **8,** 39–44.
5. Inlow, D., Maiorella, B., and Shauger, A. E. (1992) Methods for adapting cells for increased product production through exposure to ammonia, US Patent 5,156,964.
6. Adamson, S. R., Drapeau, D., Luan, Y.-T., and Miller, D. A. (1996) Adaptation of mammalian cell lines to high cell densities, US Statutory Invention Registration H1532.
7. Hayter, P. M., Curling, M. A., Baines, A. T., Jenkins, N., Salmon, I., Strange, P. G., and Bull, A. T. (1991) Chinese hamster ovary cell growth and interferon production kinetics in stirred batch culture. *Appl. Microbiol. Biotechnol.* **34,** 559–564.
8. Murata, M., Eto, Y., and Shibai, H. (1988) Large-scale production of erythroid differentiation factor (EDF) by gene-engineered Chinese hamster ovary (CHO) cells in suspension culture. *J. Ferment. Technol.* **66(5),** 501–507.

9. Berg, D. T., McClure, D. B., and Grinnell, B. W. (1993) High-level expression of secreted proteins from cells adapted to serum-free suspension culture. *BioTechniques* **14(6),** 972–978.
10. Broad, D., Boraston, R., and Rhodes, M. (1991) Production of recombinant proteins in serum-free media. *Cytotechnology* **5,** 57–55.
11. Mather, J. P. (1990) Optimizing cell and culture environment for production of recombinant proteins. *Methods Enzymol.* **185,** 567–577.
12. Keen, M. J. and Nicholas, T. R. (1995) Development of a serum-free culture medium for the large scale production of recombinant protein from a Chinese hamster ovary cell line. *Cytotechnology* **17,** 153–163.
13. Perrin, P., Madbusudana, S., Gontier-Jallet, C., Petres, S., Tordo, N., and Merten, O.-W. (1995) An experimental rabies vaccine produced with a new BHK-21 suspension cell culture process: use of serum-free medium and perfusion-reactor system. *Vaccine* **13,** 1244–1250.
14. Sinacore, M. S., Charlebois, T. C., Harrison, S., Brennan, S., Richards, T., Hamilton, M., Scott, S., Brodeur, S., Oakes, P., Leonard, M., Switzer, M., Anagnostopoulos, A., Foster, B., Harris, A., Jankowski, M., Bond, M., Martin, S., and Adamson, S. R. (1996) CHO DUKX cell lineages preadapted to growth in serum-free suspension culture enable rapid development of cell culture processes for the manufacture of recombinant proteins. *Biotechnol. Bioeng.* **52,** 518–528.
15. Zang, M., Trautmann, H., Gandor, C., Messi, F., Asselbergs, F., Leist, C, Fietcher, A., and Reiser, J. (1995) Production of recombinant proteins in Chinese hamster ovary cells using protein-free cell culture medium. *Biotechnology* **13,** 389–392.
16. Gandor, C., Leist, C., Feichter, A., and Asselbergs, F. (1995) Amplification and expression of recombinant genes in serum-independent Chinese hamster ovary cells. *FEBS Lett.* **377,** 290–294.
17. Kawamoto, T., Sato, J. D., Le, A., McClure, D. B., and Sato, G. H. (1983) Development of a serum-free medium for growth of NS-1 mouse myeloma cells and its application to the isolation of NS-1 hybridomas. *Anal. Biochem.* **130,** 445–453.
18. Kovar, J. and Franek, F. (1984) Serum-free medium for hybridoma and parental myeloma cell cultivation: A novel composition of growth-supporting substances. *Immunol. Lett.* **7,** 339–345.
19. Miyaji, H., Mizukami, T., Hosoi, S., Sato, S., Fujiyoshi, N., and Itoh, S. (1990) Expression of human beta-interferon in Namalwa KJM-1 which was adapted to serum-free medium. *Cytotechnology* **3,** 133–140.
20. Griffiths, J. B. and Racher, A. J. (1994) Cultural and physiological factors affecting expression of recombinant proteins. *Cytotechnology* **15,** 3–9.
21. Assoian, R. K. (1977) Anchorage-dependent cell cycle progression. *J. Cell. Biol.* **136(1),** 1–4.
22. Rouslahti, E. and Reed, J. C. (1994) Anchorage dependence, integrins and apoptosis. *Cell* **77,** 477,478.
23. Jeso, B. D., Ulianich, L., Racioppi, L., D'Armiento, F., Feliciello, A., Pacifico, F., Consiglio, E., and Fromisano, S. (1995) Serum withdrawal induced apoptotic cell

death in K1-Ras transformed but not normal differentiated thyroid cells. *Biochem. Biophys. Res. Commun.* **214,** 819–824.

24. Curling, E. M., Hayter, P. M., Baines, A. J., Bull, A. T., Gull, K., Strange, P. G., and Jenkins, N. (1990) Recombinant human interferon-γ: differences in glycosylation and proteolytic processing lead to heterogeneity in batch culture. *Biochem. J.* **272,** 333–337.
25. Watson, E., Shoh, B., Leiderman, L. Hsu, Y.-R., Lu, H. S., and Lin, F.-K. (1994) Comparison of N-linked oligosaccharides of recombinant human tissue kallikrein produced by Chinese hamster ovary cells in microcarriers and in serum-free suspension culture. *Biotechnol. Prog.* **10,** 39–44.
26. Chotigeat, W., Watanapokasin, Y., Mahler, S., and Gray, P. P. (1994) Role of environmental conditions on the expression levels, glycoform pattern and levels of sialyltransferase for hFSH produced by recombinant CHO cells. *Cytotechnology* **15,** 217–221.
27. Lifely, M. R., Hale, C., Boyce, S., Keen, M. J., and Phillips, J. (1995) Glycosylation and biological activity of CAMPATH-1H expressed in different cell lines and grown under different culture conditions. *Glycobiology* **5,** 813–822.
28. Jenkins, N., Parekh, R. B., and James, D. C. (1996) Getting the glycosylation right: implications for the biotechnology industry. *Nature Biotechnol.* **14,** 975–981.
29. Gawlitzek, M., Valley, U., Nimtz, M., Wagner, R., and Conradt, H. S. (1995) Characterization of changes in the glycosylation pattern of recombinant proteins from BHK-21 cells due to different culture conditions. *J. Biotechnol.* **42,** 117–131.
30. Ozturk, S. S. and Palsson, B. O. (1991) Physiological changes during the adaptation of hybridoma cells to low serum and serum-free media. *Biotechnol. Bioeng.* **37,** 35–46.
31. Barnes, D. and Sato, G. (1980) Methods for growth of cultured cells in serum-free medium. *Anal. Biochem.* **102,** 255–270.
32. Merten, Q.-W., Kierulff, J. V., Castignolles, N., and Perrin, P. (1994) Evaluation of the new serum-free medium (MDSS2) for the production of different biologicals: use of various cell lines. *Cytotechnology* **14,** 47–59.
33. Radford, K., Niloperbowo, W., Reid, S., and Greenfield, P. F. (1991) Weaning of three hybridoma cell lines to serum free low protein medium. *Cytotechnology* **6,** 65–78.
34. Qi, Y. M., Greenfield, P. F., and Reid, S. (1996) Evaluation of a simple protein free medium that supports high levels of monoclonal antibody production. *Cytotechnology* **21,** 95–109.
35. Chang, T. H., Steplewski, Z., and Koprowski, H. (1980) Production of monoclonal antibodies in serum-free medium. *J. Immunol. Methods* **39,** 369–375.
36. Merten, O.-W., Petres, S., and Couve, E. (1995) A simple serum-free freezing medium for serum-free cultured cells. *Biologicals* **23,** 185–189.

3

Viral Evaluation of Animal Cell Lines Used in Biotechnology

Alasdair J. Shepherd and Kenneth T. Smith

1. Introduction

Animal cell culture is being increasingly used for production of therapeutic reagents such as monoclonal antibodies, recombinant proteins, viral vaccines and replication incompetent viral vectors for gene therapy. Material derived from a variety of biological processes has been associated in the past with incidents involving the transmission of infectious agents, principally viruses *(1)*. Thus, virological evaluation of animal cell substrates for use in the manufacture of biologicals is essential to ensure the safety of products for pharmaceutical use.

The manufacture of safe products depends on development and safety testing of cell banks and processes according to stringent quality standards. A system of principles with emphasis on hierarchial testing and documentation, good manufacturing practice (GMP) *(2)*, has been developed for the manufacture of medicinal products, including biologicals. Safety testing should be performed according to good laboratory practice (GLP) *(3)*, a system designed to ensure the quality and validity of the data generated. Compliance with GMP and GLP requires monitoring by defined quality assurance staff operating independently of personnel involved in manufacture and testing.

In order to satisfy the requirement to demonstrate cell stock identity, safety, and quality, standard procedures have evolved for the selection, preparation, storage, and adventitious agent testing of animal cell cultures for use in the production process. The cell seed system involves the production and storage of designated (master, working, or postproduction) cell banks as a frozen stock in liquid nitrogen (*see* Chapter 8). Virological testing can thus be performed on samples from defined cell populations.

From: *Methods in Biotechnology, Vol. 8: Animal Cell Biotechnology*
Edited by: N. Jenkins © Humana Press Inc., Totowa, NJ

Guideline documents that provide a framework for consideration of virological testing of biopharmaceutical cells and processes have been issued by the appropriate regulatory authorities in Japan, the United States, and the European Union ***(4–8)***. Minor differences in the above-cited documents have been resolved through the International Committee on Harmonisation (ICH) document ***(9)***, which should provide a universally accepted regulatory approach to virological safety testing. Specific regulatory requirements also apply to particular product categories, such as human or veterinary vaccines ***(7,10)*** and viral vectors for gene therapy ***(11,12)***.

This chapter describes the testing strategy employed for viral characterization of cell banks and raw materials of animal origin. Most manufacturers and academic groups do not have the facilities to perform the wide range of assays required for full virological characterization of cell lines. There are, however, a number of commercial service companies that comply fully with GLP standards in the testing and control of these products. It is therefore inappropriate to provide detailed methods here for each of the many assays referred to below. The principal sources of contamination, organisms of concern, and the methodology of some of the assays used for viral detection are presented.

2. Cell Line Testing Strategy

The regulatory documents mentioned previously ***(4–8)*** present a common overall strategy to ensure virological safety of biopharmaceutical processes, based on the following:

1. Virological characterization of the cell banks from which the product is derived and other biological materials used in the production process.
2. In-process testing and testing of the final product.
3. Experimental studies to validate that the manufacturing process is capable of removal or inactivation of known or potential viral contaminants (virus validation).

In practice, the exact testing strategy employed may vary according to the nature of the product and the results of virological testing of cell banks or other source material. Where adequate data on viral validation of the process are available, as is generally the case for monoclonal antibody or recombinant protein products, virological testing of final product is not normally required. In the case of processes in which live viral vaccines or replication-incompetent viral vectors for gene therapy are the final product, downstream processing (and hence removal of potential contaminants) is minimal. Therefore, particular emphasis is placed on virological testing of the final product (viral vector or vaccine) for clinical use.

2.1. Sources of Contamination

Viral contamination of cell lines can arise from several sources, including the following:

1. Endogenous or persistent viral infections of the species from which the cell line was derived. Cell lines of murine (mouse, rat, hamster) origin may harbor an endogenous B- or C-type retrovirus *(Oncovirinae)* that may be infectious ***(13)***. Exogenous lentiviruses, spumaviruses, or oncornaviruses may also be present in cell lines from many species, including human and simian. Important nonretroviral contaminants include herpesviruses, arenavirus, and hantavirus.
2. Contamination of raw materials of animal origin, such as bovine serum or porcine trypsin, which may be used as part of the process. Bovine serum may be contaminated with several viruses, but bovine virus diarrhea virus (BVDV) is the most commonly encountered agent because of its high prevalence as a persistent infection of cattle ***(14)***. Similarly, porcine parvovirus (PPV) is highly prevalent in pigs and is a common contaminant of porcine trypsin because of its high resistance to physical or chemical inactivation ***(15,16)***. *See* Chapter 2 for alternatives.
3. Introduction of organisms from humans. Viruses of concern include agents that may be transmitted by aerosol or by contact, such as orthomyxo- or paramyxoviruses, adenoviruses, and enteroviruses.

3. Methods for Detection of Viruses

Viral infection can be detected by direct observation of cells using transmission electron microscopy, by specific assays for agents of concern, or by isolation of an infectious agent by inoculation of susceptible cell cultures and animals with cells or cell lysates. Viral infection of cells can be detected by observation of cytopathic effects on the cultures, involving lysis or morphological changes. In many cases, viruses are able to replicate in cells in the absence of a cytopathic effect. Such infection may be detected by a number of specific or general techniques, depending on the particular assay system. These include general techniques, such as hemadsorption and reverse transcriptase assay, and methods for detection of specific agents, such as polymerase chain reaction or immunofluorescence of infected cells using specific antisera.

The virological tests recommended by the International Committee on Harmonisation ***(9)*** to be performed on cell banks at each stage of the manufacturing process are summarized in **Table 1**. Selected agents of concern and the assays used for their detection are outlined in **Table 2**. Further information on the assays is presented in the following sections.

Characterization of the cell line is performed principally on the Master Cell Bank, where appropriate general and specific assays are conducted in order to detect contaminants. Particular emphasis is placed on assays for the detection and quantitation of endogenous or exogenous retroviruses.

Testing of the End of Production or Postproduction Cell Bank is performed primarily in order to detect contaminants that may have been introduced during the production process and to assess whether changes have occurred in the expression levels or tropism of endogenous viruses.

Table 1
Outline Testing Strategy for Cell Lines (According to International Committee for Harmonisation [ICH][*9*] Proposals)

Testing required	Master Cell Bank	Working Cell Bank	Extended Cell Bank[a]
General virus assays			
Electron microscopy	+	–	+
In vitro	+	–[b]	+
in vivo	+	–[b]	+
Retroviruses			
Reverse transcriptase assay	+	–	+
Infectivity assays for retroviruses	+	–	+
Species-specific viruses			
Bovine viruses	+	–	–
Porcine viruses	+	–	–
Murine viruses	+	–	–
Others[c]	+	–	–

[a]Cells at the limit of in vitro cell age used for production.
[b]Not required for the first working cell bank (WCB). For subsequent WCBs, a single in vitro and in vivo test can be performed either on the WCB or the resulting ECB.
[c]Specific tests for agents of concern in human, nonhuman primate, or other cell lines as appropriate.

3.1. Assays for Detection of Retroviruses

3.1.1. Electron Microscopy

Retroviral virions are enveloped particles approximately 100 nm in diameter with an internal spherical or conical core. Assembly of the capsid and envelope proteins into the virion occurs at budding from cell membranes. Classification of the retrovirus based on morphology and morphogenesis into A-, B-, C-, and D-type particles remains useful for viruses formerly in the Oncornavirus group, but identification must also take account of subsequently described lenti- and spumaviruses ***(17)***. Endogenous retroviral particle types encountered in commonly used cell lines in biotechnology are summarized in **Table 3**.

Samples of the pooled cell populations are prepared for transmission electron microscopy according to the following procedures:

1. Cells are fixed at 4°C for 2–4 h in 1.6% glutaraldehyde and 1.3% paraformaldehyde in 0.1 M sodium cacodylate buffer, pH 7.2–7.4.
2. The cells are washed in cacodylate rinse buffer, pH 7.2–7.4.
3. The cells are postfixed in 0.5% osmium tetroxide in Millonig's buffer for 1 h.
4. The cells are then dehydrated in a graded acetone series and embedded in Araldite, Agar Scientific, Stanstead, UK.

Table 2
Examples of Methods Used for Assay of Selected Agents of Concern

Virus family	Virus species	Assay[a]
Retroviridae	Murine leukemia virus	EM, RT, S^+L^-, XC, cocultivation
	Human immunodeficiency virus	EM, RT, PCR, cocultivation
Herpesviridae	Herpes simplex	In vitro, in vivo
	Cytomegalovirus	In vitro, PCR
	Epstein-Barr virus	PCR, IFA
	Human herpes virus types 6/7/8	PCR
	Bovine herpes virus	Bovine in vitro
Reoviridae	Reovirus types 1/2/3	In vitro, bovine in vitro
Parvoviridae	Porcine parvovirus	Porcine in vitro
	Minute virus of mice	In vitro, PCR, MAP
	Bovine parvovirus	Bovine in vitro
Adenoviridae	Human adenoviruses	In vitro
	Bovine adenoviruses	Bovine in vitro
Picornaviridae	Poliovirus	In vitro
	Encephalomyocarditis virus	In vitro, MAP
Paramyxoviridae	Parainfluenza types 1/2/3	In vivo, bovine in vitro
	Measles, mumps	In vivo, in vitro
Togaviridae	Bovine viral diarrhoea	Bovine in vitro
	Rubella	In vitro

[a]Abbreviations: EM—electron microscopy; RT—reverse transcriptase; S^+L^-/XC—infectivity assays for detection of xenotropic/ecotropic murine leukemia virus; PCR—polymerase chain reaction; IFA—immunofluorescent antibody test; MAP—mouse antibody production assay.

5. One-micrometer sections are cut and stained with Methylene Blue/Azure 2.
6. Thin sections are cut on a Reichert ultramicrotome, stained in uranyl acetate and lead citrate and examined in a transmission electron microscope.

Cells should be examined at both low and high magnifications and particular attention paid to locations such as the nucleus, the centrosomal area, and the cell periphery for the selective localization of contaminants.

3.1.2. Molecular Techniques

Reverse transcriptase (RT) assays are useful for nonspecific detection and quantitation of the retrovirus and may aid in the identification by determining the cation preference (Mg^{2+} or Mn^{2+}) of the enzyme. Samples are solubilized in disruption buffer and added to the reaction mixture containing Tris buffer, ^3H-labeled thymidine triphosphate (dTTP), and a [poly(rA)-oligo(dt)] template–primer to detect viral polymerase. Each sample is tested in individual reaction mixtures

Table 3
Endogenous Retroviruses Associated with Cell Lines Used in Production of Biologicals

Cell line	Particle type	Reverse transcriptase	Infectious virus[a]
Mouse	A/C type	Positive	E-/X-MLV
Rat	A/C type	Positive	E
Hamster (CHO)	C type	Positive	None
Hamster (BHK-21)	R type	Negative	None
Mouse/human	A/C type	Positive	X-MLV
Human	None	Negative	None
Monkey	None	Negative	None
Insect	None	Negative	None

[a]Abbreviations: E—ecotropic; X—xenotropic; MLV—murine leukemia virus.

containing either the Mn^{2+} or Mg^{2+} cation. The Mn^{2+} cation is preferred by the Type-C oncoviruses belonging to the murine leukemia virus (MLV) and feline leukemia virus group. The Mg^{2+} cation is preferred by type-C retroviruses such as bovine leukemia virus and human T-cell leukemia virus, type-B retroviruses such as mouse mammary tumor virus, and type-D retroviruses such as squirrel monkey retrovirus and the lentivirus group (such as HIV). In addition, samples are tested using a [poly(dA)-oligo(dT)] "cellular" template in place of the "viral" template–primer to detect endogenous levels of cellular DNA polymerase activity.

Polymerase-chain-reaction-enhanced reverse transcriptase (PERT) assays ***(18)*** have greater sensitivity than standard RT assays and may be useful for detection of RT in cell lines that do not harbor an endogenous retrovirus.

For specific detection of retroviral agents of concern in cell cultures, polymerase chain reaction (PCR) using oligonucleotide primers to conserve sequences of the viral genome is the method of choice.

3.1.2.1. Reverse Transcriptase Assay

Ten milliliters of test samples are clarified at 11,000*g* for 10 min. The samples are concentrated by centrifugation at approximately 100,000*g* for 1 h and each pellet is solubilized in 180 µL of disruption buffer (40 m*M* Tris pH 8.1; 50 m*M* KCl; 20 m*M* DTT (dithiothreitol); 0.2% NP40). An aliquot of 25 µL is then added to an equal volume of each of six reaction mixtures. Three of the mixtures contain 40 m*M* Tris pH 8.1, 50 m*M* KCl, 25 µCi [methyl-^{3}H] TTP (Amersham, Little Chalfont, UK) and 2 m*M* $MnCl_2$, plus either poly(rA) [0.05A_{260} units], oligo(dT) [0.05_{260} units] or 2 m*M* Tris pH 8.1/30 m*M* NaCl. Three further mixes contain 40 m*M* Tris pH 8.1, 50 m*M* KCl, 25 µCi [methyl-^{3}H] TTP (Amersham), and 20 m*M* $MgCl_2$ plus either poly(rA) (0.1 A_{260} units), oligo(dT) (0.1 A_{260} units), or 4 m*M* Tris pH 8.1/60 m*M* NaCl. The combined sample and reaction mixtures are incubated at 37°C for 1 h and the DNA or RNA templates precipitated by 10% trichloroacetic acid (TCA), 1% sodium pyrophosphate onto GFG filters (Whatman, Maidstone, UK). The ^{3}H-TTP incorporated into DNA or RNA templates and the background activity (no-template mixes) with either the manganese or magnesium cation is measured in Ecoscint in a scintillation counter (Beckman, High Wycombe, UK). Results are counted in disintegrations per minute (dpm). A positive result is concluded when the poly(rA) incorporation exceeds 2000 dmp with the poly(rA) incorporation at least twice the poly(dA) incorporation and at least four times the negative control poly(rA) incorporation. A negative result is concluded when the poly(rA) incorporation is less then 2000 dmp. Any other result is regarded as equivocal. Negative control (disruption buffer) and positive controls (ecotropic MLV for Mn^{2+}-dependent incorporation and bovine leukemia virus for Mg^{2+} incorporation) samples are tested in parallel.

3.1.3. In Vitro Assays

Mouse epithelial, fibroblast, hybridoma, myeloma, and plasmacytoma cell lines contain endogenous retroviruses and, in many cases, an infectious C-type MLV, which can be ecotropic (infectious for cell lines of murine origin or a closely related species), xenotropic (X-MLV) (infectious for species other than murine), or polytropic (recombinant viruses with both xenotropic and ecotropic properties). An endogenous C-type retrovirus is also released by hamster, rat, and porcine cell lines ***(13,19)***. Retroviral gene therapy products utilize recombinant retroviruses packaged within envelope gene products of amphotropic MLV, feline leukemia virus (FeLV), or gibbon-ape leukemia virus (GaLV) ***(20)***. There is potential for cell lines of human and simian origin (particularly human hybridoma cell lines) to be infected with exogenous type-C (human T-cell leukemia viruses), type-D (simian retrovirus types 1 and 2, squirrel monkey retrovirus), lentiviruses (human immunodeficiency viruses, simian immunodeficiency viruses), or spumaviruses (simian and human foamy viruses) ***(21–23)***.

An infectious retrovirus can be detected by inoculation of or cocultivation with susceptible cell lines. Most of the cell lines commonly used for in vitro assays are readily obtained from repositories such as the American Type Culture Collection (ATCC) and the European Collection of Animal Cell Cultures (ECACC). Cell lines used include mink S^+L^- (sarcoma positive, leukemogenic negative) cells or feline PG4 S^+L^- cells ***(24–26)*** for detection of xenotropic MLV, FeLV, or GaLV or amphotropic MLV, FG10 S^+L^-, or SC-1 cells for the detection of ecotropic MLV ***(27,28)*** or *Mus dunni* cells, which are susceptible to infection by xenotropic, amphotropic, mink cell focus (MCF) virus, and some ecotropic retroviruses ***(29)***. Cell lines that are of use for isolation of other exogenous retroviruses include Raji (simian retrovirus types 1 and 2, squirrel monkey retrovirus), H9 (human immunodeficiency viruses, human T-cell leukemia viruses), Molt 4 Clone 8 and 174XCEM (simian immunodeficiency viruses), and Vero (simian spumavirus) ***(30,31)***.

X-MLV from mouse hybridoma cell lines has been shown to infect human cells ***(32)***, and in vivo studies in which monkeys exposed to amphotropic MLV developed lymphomas ***(33)*** have heightened regulatory concern. The ability of endogenous X-MLV to infect human cells should therefore be determined. This is done by cocultivation of the cell line with susceptible cells such as the human fibroblast cell line, MRC-5, or the human rhabdosarcoma cell line (RD).

3.1.3.1. Assay Methods for Detection of Murine Leukemia Virus

Several assays (see Subheading 3.1.3.) can be used for detection of infectious ecotropic, xenotropic, or amphotrophic MLV in cell supernatants. Such assays can be performed in a direct and extended format. In direct assays, a given volume of sample is inoculated onto the cells in the presence of polybrene and the virus titer determined directly by enumeration of resulting foci of infection. Low levels of infection, which may not be detected by direct assays, can be detected by serial passage of inoculated cells ***(34)***. Hence, extended assays can be performed where the inoculated cultures are passaged five times to amplify low levels of MLV. Extended assays are not directly quantitative, but our unpublished data have shown them to be approximately one $\log_{10}$ more sensitive for detection of X-MLV than direct assays.

3.1.3.2. Mus Dunni Assay

In the direct Mus dunni assay, cells are seeded onto Labtek 2 chamber slides (NUNC Inc., Napaville, IL) at a concentration of 2×10^4 cells/well and inoculated the following day with 0.25 mL/well of test sample and 0.25 mL/well polybrene (effective concentration of 10 μg/mL) (Sigma, Poole, UK). Four days later, the cultures are fixed in cold acetone. Indirect immunofluorescence is performed using the broadly reactive murine anti-MLV monoclonal antibodies 34, R187, or 548 ***(35)*** and antimurine FITC conjugated antibody (Sigma) as the second antibody. Foci are identified and enumerated by fluorescent microscopy (Olympus, London, UK).

Negative control (culture medium) and positive control (amphotropic MLV) cultures are tested in parallel. In the extended assay, Mus dunni cells are seeded into duplicate 25-cm^2 tissue culture flasks at a concentration of 5×10^5/cells/flask and inoculated the following day with 0.5 mL/flask of test sample and 0.5 mL/flask polybrene (20 μg/mL). The cultures are passaged every 3–4 days for a total of five passages. At passage five, supernatant from the cultures is harvested and tested for MLV by direct mink S^+L^- mink assay. The negative control (culture medium) and positive control (amphotropic MLV) are tested in parallel.

3.1.3.3. Mink S^+L^- Assay

In the direct mink S^+L^- assay, cells are seeded onto duplicate 10-cm^2 dishes at a concentration of 2.5 × 10^5/cells/plate and inoculated the following day with 0.5 mL/plate of test sample and 0.5 mL/plate polybrene (8 µg/mL polybrene). The cells are examined daily for the formation of foci and the focus-forming units (ffu) counted. Samples with fewer than 10 ffu/mL are tested by the extended assay by passaging every 3–4 d for a total of five passages, and the foci were counted as before. The negative control (culture medium) and positive control (xenotropic MLV or FeLV) and tested in parallel.

3.1.3.4. PG4 S^+L^- Assay

In the PG4 assay, cells are seeded onto duplicate 10-cm^2 dishes at a concentration of 2 × 10^5/cells/plate and inoculated the following day with 0.5 mL/plate of test sample and 0.5 mL/plate polybrene (20 µg/mL polybrene). The cells are examined daily for focus formation and the ffu are counted (direct assay). The cells are then passaged every 3–4 d for a total of five passages and foci counted as before (extended assay). The negative control (culture medium) and positive control (FeLV or amphotropic MLV) are tested in parallel.

3.2. Other Viruses

The detection systems used to assay for viruses other than retroviruses vary according to the nature and culture history of the cell substrate and the tropisms and properties of the viruses of concern. Techniques used include direct examination of the cell substrate by transmission electron microscopy and by molecular techniques such as PCR, in vitro assays in cell culture, and in vivo assays in animals.

3.2.1. Direct Assays

Transmission electron microscopy is a relatively insensitive technique for detection of viral infection of cell lines, but it may be useful for detection of gross infection with agents that are noncytopathic in cell culture and nonpathogenic to animals used in in vivo assays. Assays by PCR for specific agents of concern may supplement general testing strategies in the case of certain cell lines, such as those derived from human B cells. Human hybridomas should be tested by PCR for hepatitis B, hepatitis C, Epstein–Barr, cytomegalovirus, and human herpes types 6, 7 and 8 viruses ***(4,5,9)***. Testing for human parvovirus B19 and hepatitis A viruses by PCR should also be considered for human cell lines.

3.2.2. In Vitro Assays

In vitro assays are based on standard techniques of clinical virology and as such rely foremost on isolation of virus by detection of cytopathic effect in susceptible cell cultures *(34–36)*. For cell lines producing biopharmaceuticals for human use, regulatory requirements *(5,8,9)* for in vitro assays for viral contaminants recommend that appropriate cell lines be inoculated with cell lysate and examined over a 14-d period for development of cytopathic effects, followed by hemadsorption assay.

For in vitro assays for detection of viruses in biological products for human use, US regulations *(5)* recommend that an appropriate volume of sample be inoculated into monolayer cultures of at least three types:

1. Monolayer cultures of the same species and tissue type as that used for production.
2. Monolayer cultures of a human diploid cell culture.
3 Monolayer culture of a monkey kidney cell culture.

Cell lines commonly used under **points 2** and **3** are the normal human fibroblast MRC-5 and African green monkey kidney (Vero). European *(8)* and ICH *(9)* regulations for monoclonal antibody producing cell lines recommend inoculation of cell cultures capable of detection of a wide range of murine, human, and bovine viruses.

Inoculated cultures should be observed for at least 14 d and tested by hemadsorption using human, guinea pig, and chicken erythrocytes *(8)*.

Specialized in vitro assays are also used extensively for detection of viral contaminants in cell lines used for veterinary vaccines. The choice of detector cell lines is dependent on the history of the cell line and the intended vaccine recipient and should, in addition, include a primary cell line of the species of origin and a cell line susceptible to pestiviruses *(10)*. Assays involve inoculation with cell lysates, followed by culture, blind passage, and hemadsorption. In addition, specified agents of concern are tested for at the end of the observation period by indirect immunofluorescence using specific antisera *(10)*.

Specific in vitro assays may also be used for detection of potential contaminants of animal origin. Such assays may be employed to test cell substrates or specific lots of raw material, such as bovine serum or porcine trypsin. As above, such assays involve inoculation of susceptible cell lines, followed by culture, passage, hemadsorption, and indirect immunofluorescence for specific agents *(37)*.

3.2.3. In Vivo Assays

General safety tests in animals are performed in order to detect viruses that may be pathogenic to vertebrates but may, nevertheless, be undetectable by other assays. These assays involve intracerebral, intramuscular and intraperitoneal inoculation of suckling and adult mice, intramuscular inoculation of guinea pigs, and inoculation of embryonated eggs via the allantoic cavity, the amniotic cavity, and the yolk sac. Animals and eggs are observed for signs of morbidity and death and hemagglutination assays performed on allantoic fluid of the eggs at the conclusion of the observation period. Suckling mice are fatally susceptible to a wide range of viruses, including many arthropod-borne viruses (alpha-, flavi-, and bunyaviruses), picornaviruses (poliovirus, coxsackievirus, encephalomyocarditis, and echovirus), Herpes simplex, rabies and some arenaviruses ***(40,41)***. Adult mice are also susceptible to some of these agents. Embryonated eggs are susceptible to poxviruses, orthomyxo- and paramyxoviruses, and Herpes simplex ***(42)***. Guinea pigs are susceptible to a number of viral agents, including paramyxovirus, reovirus, filoviruses, and some arenaviruses ***(14,41)***.

Antibody production assays are of use for detection of viruses of mouse, rat, or hamster origin which may be present as contaminants in cultured cell lines. These tests involve inoculation of susceptible adult mice, rats, or hamsters with lysate of the appropriate cell line. Animals are held for 28 days, whereupon they are bled and the serum tested for antibody to viruses of concern. Antibody tests for up to 16 viruses are routinely performed on serum, but viruses of particular concern in rodent cell lines include those which may cause human disease, such as lymphocytic choriomeningitis and Hantaan and encephalomyocarditis viruses ***(14,43,44)***.

References

1. Petricciani, J. C. (1991) Regulatory philosophy and acceptibility of cells for the production of biologicals. *Develop. Biol. Standard* **75,** 9–15.
2. Medicines Control Agency. (1993). *Rules and Guidance for Pharmaceutical Manufacturers*, Her Majesty's Stationery Office, London.
3. Department of Health. (1987) *The Good Laboratory Practice Regulations 1997.* Department of Health, London.
4. Center for Biologics Evaluation and Research. (1993) *Points to Consider in The Characterization of Cell Lines Used to Produce Biologicals*, US Food and Drug Administration, Washington, DC.
5. Center for Biologics Evaluation and Research. (1994) *Draft Points to Consider in the Manufacture and Testing of Monoclonal Antibody Products for Human Use*, US Food and Drug Administration, Washington, DC.
6. Committee for Proprietary Medicinal Products. (1995) *Virus Validation Studies* (Revised within CPMP BWP 3-4 July 1995) CPMP document CPMP/268/95, European Commission, Brussels.
7. World Health Organisation. (1995) *Draft Requirements for Cell Substrates Used for Biological Production.* WHO Expert Committee on Biological Standardization 17–24 October 1995, Document BS/95. 1792, World Health Organization, Geneva.

8. Committee for Proprietary Medicinal Products. (1995) *Production and Quality Control of Monoclonal Antibodies*. CPMP Approved 13 December 1994. Document III/5271/94, European Commission, Brussels.
9. Committee for Proprietary Medicinal Products. (1995) *Notes for Guidance on Quality of Biotechnological Products: Viral Safety Evaluation of Biotechnology Products Derived from Cell Lines of Human or Animal Origin* (CPMP/ICH/295/95) Step 2, Draft, 1,12,95, European Commission, Brussels.
10. Committee for Veterinary Medicinal Products. (1995) *The Rules Governing Medicinal Products in the European Union. Volume VII; Guidelines for Testing of Veterinary Medicinal Products*, European Commission, Brussels.
11. Center for Biologics Evaluation and Research. (1991) *Points to Consider in the Human Somatic Cell Therapy and Gene Therapy*, US Food and Drug Administration, Washington, DC.
12. Center for Biologics Evaluation and Research. (1996) *Addendum to the Points to Consider in the Human Somatic Cell Therapy and Gene Therapy*, US Food and Drug Administration, Washington, DC.
13. Kozak, C. A. and Ruscetti, S. (1992) Retroviruses in rodents, in *The Retroviridae* (Levy, J. A., ed.), Plenum, New York, vol. 1, pp. 405–481.
14. Andrewes, C. A. (1989) *Andrewes' Viruses of Vertebrates*, 5th ed. (Porterfield, J. S., eds,), Baillière Tindall, London.
15. Mengeling, W. L. (1992) Porcine parvovirus, in *Diseases of Swine*, 7th ed. (Leman, A. D., Straw, B. E., Mengeling, W. L., D'Allaire, S., and Taylor, D. J., eds.), Wolfe Publishing Ltd. London, pp. 299–311.
16. Croghan, D. L., Matchett, A., and Koski, T. A. (1973) Isolation of porcine parvovirus from commercial trypsin. *Appl. Microbiol.* **26,** 431.
17. Coffin, J. M. (1992) Structure and classification of retroviruses, in *The Retroviridae* (Levy, J. A., ed.), Plenum, New York, vol. 1, pp. 19–50.
18. Silver, J., Maudru, T., Fujita, K., and Repaske, R. (1993) An RT-PCR assay for the enzyme activity of reverse transcriptase capable of detecting single virions. *Nucleic Acids Res.* **21,** 3593–3594.
19. Armstrong, J. A., Porterfield, J. S., and Teresa de Madrid, A. (1971) C type virus particles in pig kidney cell lines. *J. Gen. Virol.* **10,** 195–198.
20. Smith, K. T., Shepherd, A. J., Boyd, J. E., and Lees, G. M. (1996) Gene delivery systems for use in gene therapy: an overview of quality assurance and safety issues. *Gene Therapy* **3,** 190–200.
21. Shepherd, A. J., Boyd, J. E., Hogg, C. E., Aw, D., and James, K. (1992) Susceptibility of human monoclonal antibody producing B cell lines to infection by human immunodeficiency virus. *Hum. Antibody Hybridomas* **3,** 168–176.
22. Popovic, M., Kalyanaraman, V. S., Reitz, M. S., and Sarngadharan, M. G. (1982) Identification of the RPMI 8226 retrovirus and its dissemination as a significant contaminant in some widely used human and marmoset cell lines. *Int. J. Cancer* **30,** 93–100.
23. Hooks, J. J. and Gibbs, C. J. (1975) The foamy viruses. *Bacteriol. Rev.* **39,** 169–190.

24. Peebles, P. T. (1975) An *in vitro* focus induction assay for xenotropic murine leukaemia virus, feline leukaemia virus C and the feline primate virus RD-114 CCC/M7. *Virology* **67,** 288–291.
25. Bassin, R. H., Ruscetti, S., Ali, I., Haapala, D. K., and Reins, A. (1982) Normal DBA/2 mouse cells synthesize a glycoprotein which interferes with MCF virus infection. *Virology* **123,** 139–151.
26. Haapala, D. K., Robey, W. G., Oroszlan, S. D., and Tsai, W. P. (1985) Isolation from cats of an endogenous type C virus with a novel envelope glycoprotein. *J. Virol.* **53,** 827–833.
27. Bassin, R. H., Tuttle, N., and Fischinger, P. J. (1971) Rapid cell culture assay technique for murine leukaemia virus. *Nature* **229,** 564–566.
28. Klement, V., Rowe, W. P., Hartley, J. W., and Pugh, W. E. (1969) Mixed culture cytopathogenicity: a new test for growth of murine leukaemia virus in tissue culture. *Proc. Natl. Acad. Sci. USA* **63,** 753–758.
29. Lander, M. R. and Chattopadhyay, S. K. (1984). A *Mus dunni* cell line that lacks sequences closely related to endogenous murine leukaemia viruses and can be infected by ecotropic, amphotropic, xenotropic and mink cell focus-forming viruses. *J. Virol.* **52,** 695–698.
30. Popovic, M., Sarngadharan, M. G., Read, E., and Gallo, R. C. (1984) Detection, isolation and continuous production of cytopathic retroviruses (HTLV III) from patients with AIDS and pre-AIDS. *Science* **224,** 497–500.
31. Sommerfelt, M. A. and Weiss, R. A. (1990) Receptor interference groups of 20 retroviruses plating on human cells. *Virology* **176,** 58–69.
32. Weiss, R. A. (1982) Retroviruses produced by hybridomas. *N. Engl. J. Med.* **307,** 1587.
33. Donahue, R. E., Kessler, S. W., Bodine, D., McDonagh, K., Dunbar, C., Goodman, S., Agricola, B., Byrne, E., Raffeld, M., Moen, R., Bacher, J., Zsebo, K. M., and Niehuia, A. W. (1992) Helper virus induced T cell lymphoma in nonhuman primates after retroviral mediated gene transfer. *J. Exp. Med.* **176,** 1125–1135.
34. Bhatt, P. N., Jacoby, R. O., Morse, H. C., III, and New, A. E. (1986) Murine leukaemia viruses, in *Methods in Virology*, Academic, New York, pp. 349–388.
35. Chesesbro, B., Britt, W., Evans, L., Wehrly, K., Nishio, J., and Cloyd, M. (1983) Characterization of monoclonal antibodies reactive with murine leukemia viruses: Use in analysis of strains of Friend MCF and Friend ecotropic murine leukemia virus. *Virology* **127,** 134–148.
36. McIntosh, K. (1990) Diagnostic virology, in *Virology*, 2nd ed. (Fields, B. N. and Knipe, D. M., eds.), Raven, New York, pp. 411–440.
37. Schmidt, N. J. (1989) Cell culture procedures for diagnostic virology, in *Diagnostic Procedures for Viral, Rickettsial and Chllamydial Infections,* 6th ed., (Schmidt, N. J. and Emmons, R. W., eds.), American Public Health Association, Washington, DC, pp. 51–100.
38. Morgan-Caper, P. and Pattison, J. R. (1985) Techniques in clinical virology, in *Virology. A Practical Approach* (Mahy, B. W. J., ed.), IRL, Oxford, pp. 237–258.

39. Animal and Plant Health Inspection Service. United States Department of Agriculture (1995). *Code of Federal Regulations 9. Animals and Animal Products.* 1995; Part 113. 53. 8. US Government Printing Office, Washington, DC.
40. Gould, E. A. and Clegg, J. C. S. (1985) Growth, assay and purification of alphaviruses and flaviviruses, in *Virology. A Practical Approach* (Mahy, B. W. J., ed.), IRL, Oxford, pp. 43–78.
41. Shepherd, A. J. (1988) Viral hemorrhagic fever: laboratory diagnosis, in *Handbook of Viral and Rickettsial Hemorrhagic Fever* (Gear, J. H. S., ed.), CRC, Boca Raton, FL, pp. 241–250.
42. Schmidt, N. J. and Emmons, R. W. (1989) General principles of laboratory diagnostic methods for viral, rickettsial and chlamydial infections, in *Diagnostic Procedures for Viral, Rickettsial and Chlamydial Infections,* 6th ed., (Schmidt, N. J. and Emmons, R. W., eds.), American Public Health Association, Washington, DC, pp. 1–36.
43. Carthew, P. (1986) Is rodent virus contamination of monoclonal antibody preparations for use in human therapy a hazard? *J. Gen. Virol.* **67,** 963–974.
44. Mahy, B. W. J., Dykewicz, C., Fisher-Hoch, S., Ostroff, S., Tipple, M., and Sanchez, A. (1991) Virus zoonoses and their potential for contamination of cell cultures. *Develop. Biol. Standard* **75,** 183–189.

II

Molecular Methods

4

Optimizing Gene Expression in Mammalian Cells

Lucia Monaco

1. Introduction

Important post-translational modifications affect both the physicochemical identity and the biological activity of many recombinant polypeptides. Mammalian cells are essential to perform such modifications correctly, and efficient methods have been established to transduce or transfect them, based on either viral or synthetic vectors, or combinations of both. Several parameters affect the efficiency of expression of the transgene, such as the nature and molecular design of the vector, the strength and identity of transcription regulatory signals, and the transfection procedure. Each of these parameters can be optimized to increase gene expression. An approach based on transfection of mammalian cells with DNA consisting of multiple copies of the gene of interest is described in this chapter.

2. Increasing Gene Dosage

One of the parameters that can be optimized to increase production of the recombinant protein is the gene copy number. This can be achieved by high copy number viral vectors or by exploiting viral replication mechanisms, e.g., in the COS cell expression system *(1)*. High-level transient expression of the transgene is usually obtained in this way. To increase gene dosage in permanently transfected cells, amplification of the integrated transgene is obtained following cotransfection with an amplifiable selectable marker.

One of the most commonly used expression systems relying on amplification mechanisms employs cotransfection of the dihydrofolate reductase (*dhfr*) gene with the gene of interest into one of the Chinese hamster ovary (CHO) cell lines deficient for *dhfr*, obtained by Urlaub and Chasin *(2)*. Following selection of clones stably expressing *dhfr* by exposure to selective medium

From: *Methods in Biotechnology, Vol. 8: Animal Cell Biotechnology*
Edited by: N. Jenkins © Humana Press Inc., Totowa, NJ

lacking nucleosides, cells are treated stepwise with increasing concentrations of methotrexate (MTX), an inhibitor of dhfr. This results in amplification of the DNA region where the *dhfr* gene and often the gene of interest have been integrated (up to several hundred copies) and consequently in overproduction of the recombinant protein ***(3)***. A disadvantage of this approach is in the tedious procedure, requiring several months to isolate clones producing high amounts of the desired product. Moreover, the selection mechanism is a recessive one, requiring dhfr-deficient host cells. Methods for the *dhfr*/MTX system are described in *Current Protocols for Molecular Biology* ***(4)***, Unit 16.14.

2.1. In Vitro Amplification

A method allowing transfection with long DNA molecules, constructed using several complete expression units of the gene of interest linked to the selectable marker gene, has been developed in our laboratory ***(5)*** and is shown in **Fig. 1**. The method has been named in vitro amplification, because the multimerization process takes place in the test tube (in vitro), as opposed to the in vivo amplification taking place inside the cell, as mentioned above for the *dhfr*/MTX system in CHO cells.

2.2. Plasmids for In Vitro Amplification

The key feature of plasmid vectors for in vitro amplification is the presence of *Sfi*I sites flanking the expression unit, which includes suitable promoter/enhancer and polyadenylation sequences placed upstream and downstream of a multiple cloning site, respectively. *Sfi*I is a rare cutter enzyme recognizing an 8-bp sequence. The sticky ends generated by *Sfi*I cleavage are nonpalindromic, yielding forced head-to-tail orientation on ligation. The *Sfi*I sequence 5'-GGCCA/AAAAGGCC-3' has been inserted into vectors pSfiSV19 ***(5)*** and pCISfiT (L. Monaco, unpublished results), carrying promoter regions from simian virus 40 (SV40) and human cytomegalovirus (CMV), respectively. Moreover, a transcription terminator sequence from the human gastrin gene has been inserted in pCISfiT. Plasmids pSfiSVneo ***(5)*** and pSfiSVdhfr ***(6)***, carrying suitable selectable markers, have been similarly constructed (**Fig. 2**). pSfiSV19, pSfiSVneo, and pSfiSVdhfr contain two *Sfi*I sites flanking the expression unit, while a unique *Sfi*I site is present in pCISfiT. In principle, any mammalian expression vector can be adapted to in vitro amplification by insertion of one *Sfi*I site in a convenient region of the plasmid. Other restriction sites with properties similar to *Sfi*I can be employed to build comparable plasmid series, essentially constituted by a vector providing a multiple coloning site and a companion plasmid for the selectable marker. This would be needed to express coding regions naturally containing *Sfi*I sites. For example, an *Rsr*II plasmid series has been built in our laboratory.

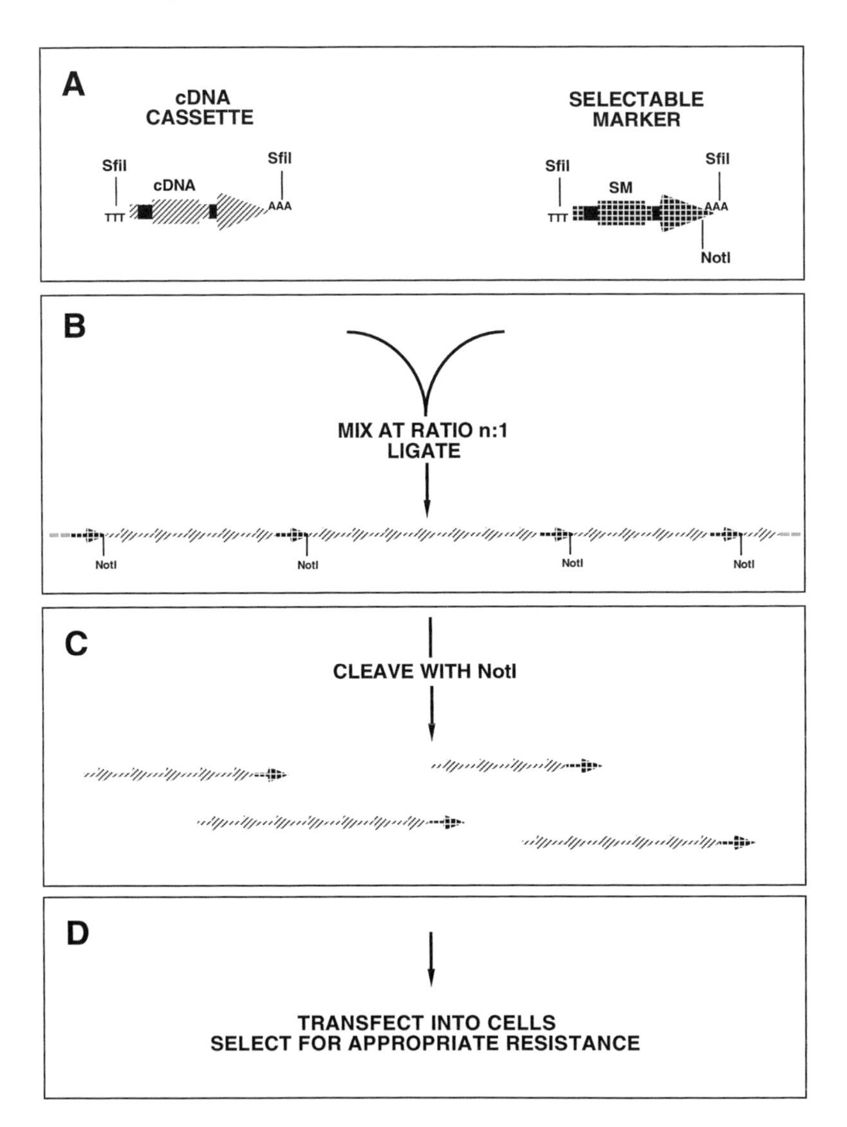

Fig. 1. Schematic drawing of the in vitro amplification method. **(A)** Expression units for the cDNA to be expressed (stripes) and the selectable marker (SM, squares) consist of the coding region (large boxes) and suitable promoter and polyadenylation regions (black boxes). The units are flanked by *Sfi*I nonpalindromic cohesive ends. **(B)** The cDNA and selectable marker units are combined at a molar excess of cDNA over selectable marker. Ligation at high DNA concentration allows formation of DNA concatenamers consisting of several head-to-tail units. **(C)** Cleavage with *Not*I yields DNA concatenamers terminating with one selectable marker unit. **(D)** DNA concatenamers are transfected into cells by either calcium phosphate coprecipitation or cytofection.

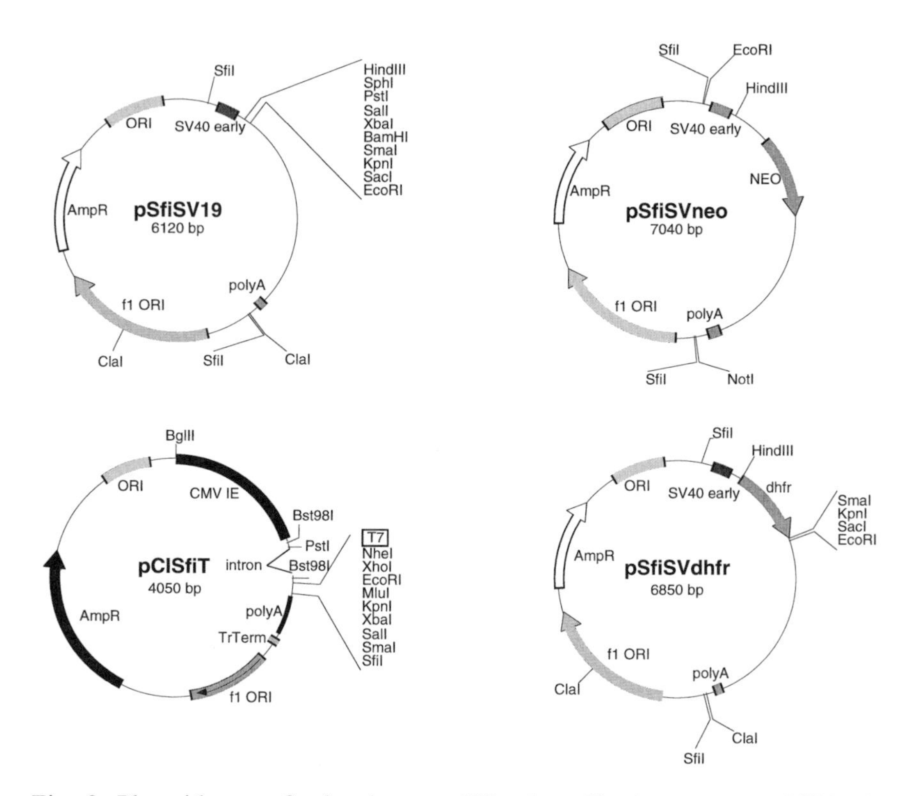

Fig. 2. Plasmid maps for in vitro amplification. Cloning vectors pSfiSV19 and pCISfiT carry multiple cloning sites downstream of the SV40 early and CMV immediate-early (IE) promoters, respectively. Selectable marker plasmids pSfiSVneo and pSfiSVdhfr are both under the SV40 early promoter and carry the neomycin resistance (NEO) and the dihydrofolate reductase (dhfr) genes, respectively. Polyadenylation sites are either from SV40 early (pSfiSV19, pSfiSVneo, pSfiSVdhfr) or SV40 late (pCISfiT) regions. A transcription terminator sequence from the human gastrin gene has been inserted in pCISfiT.

2.3. Steps of In Vitro Amplification

Once the cDNA of interest has been inserted into the desired plasmid vector, *Sfi*I is used to generate linear DNA molecules carrying the complete expression unit of either the cDNA of interest or the selectable marker (**Fig. 1A**). These are then mixed at the desired molar ratio, usually at a molar excess of the cDNA of interest over the selectable marker, and ligated (**Fig. 1B**). Ligation conditions have been established to favor the formation of intermolecular rather than intramolecular linkages, to obtain DNA concatenamers. The length distribution of the ligation products is controlled by pulsed-field electrophoresis and typically reaches 200 kb or longer (**Fig. 3**).

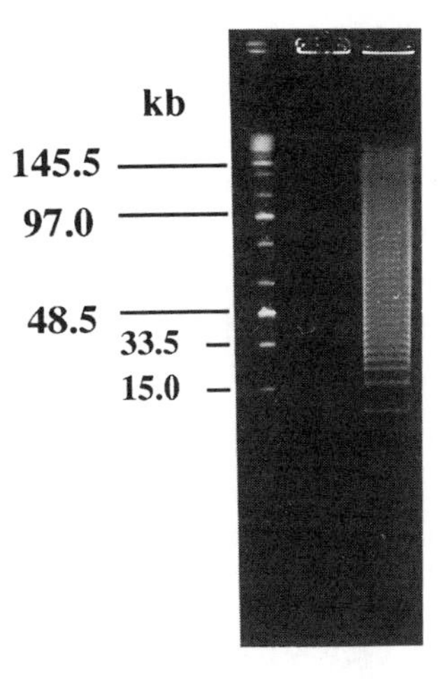

Fig. 3. Pulsed-field electrophoresis of ligation products. One microgram of DNA concatenamers obtained by ligation of *Sfi*I-flanked fragments (as detailed in **Subheading 3.3.**) were loaded on a 1.2% agarose gel and separated by pulsed-field electrophoresis, according to methods described in **Subheading 3.5.**

To obtain concatenamers containing one terminal selectable marker unit, concatenamers can then be truncated by cleavage at the unique *Not*I site of the selectable marker unit (**Fig. 1C**). This is preferred when concatenamers containing a low number of expression units are desired and a low cDNA/selectable marker molar ratio is employed, since the presence of more than one copy of the selectable marker gene would not be optimal. In fact, production of the recombinant product of interest should not be favored under this circumstance. Conversely, when desiring concatenamers containing as many gene copies as possible, the cDNA/selectable marker molar ratio is increased, and *Not*I cleavage is not needed. For example, a DNA concatenamer of 100 kb average length would consist of 20 copies of 5-kb monomeric units. By mixing the cDNA unit in a 19:1 ratio with the selectable marker unit, each concatenamer should on the average contain one copy of the selectable marker. Since ligation products are heterogeneous in size distribution, a certain proportion of the shorter polymers will not contain any selectable marker unit, whereas longer polymers might contain more than one. As a rule of thumb, we suggest choosing the cDNA/selectable marker molar ratio as follows: 5:1 plus *Not*I cleavage when overexpression is not required, or 10:1 to 20:1 without *Not*I cleavage when high-level expression is desirable. Given an average length of DNA concatenamers of 100 kb, the appropriate ratio will be chosen on the basis of the length of the monomeric unit: the shorter the monomeric unit, the higher the ratio.

Finally, the polymerized DNA is used to transfect adherent mammalian cells, typically by either calcium phosphate co-precipitation or cytofection (**Fig. 1D**). Following the appropriate selection procedure, production levels of the recombinant product by isolated colonies are assayed.

Expected results of in vitro amplification are discussed in **Note 1**. The possibility to combine in vitro and in vivo amplification is described in **Note 2**. Known limitations of the method are discussed in **Note 6**.

3. Methods

Standard molecular biology techniques are required, for which excellent manuals are available *(4,7)*. Only a few steps need particular stringency, as discussed below. Different commercial sources of enzymes proved satisfactory for performing both DNA cleavage and ligation (*see* **Note 4** for comments on transfection procedures; **Note 5** for time considerations).

3.1. DNA Purification

Purity of DNA proved essential to allow large-size ligation products. Plasmid DNA purification by commercially available chromatography-based methods did not yield reproducibly satisfactory DNA in our hands. Conversely, DNA purified by CsCl gradient ultracentrifugation (*see* Section 1.40 in **ref.** *7*) consistently yielded large-size ligation products.

3.2. Cleavage with SfiI

Expression cassettes for the cDNA of interest and for the selectable marker are obtained from the parent plasmids by cleavage with *Sfi*I. To ensure complete *Sfi*I cleavages, 50 µg of purified plasmid DNA is incubated with 100 U of enzyme at 50°C overnight.

Following *Sfi*I cleavage, DNA is extracted with phenol/cloroform and chloroform and ethanol-precipitated according to standard procedures (*see* Section E3 in **ref.** *7*).

3.3. Preparation of DNA Concatenamers for In Vitro Amplification

cDNA and selectable marker cassettes are mixed at the desired molar ratio and ligated in 50 m*M* Tris-Cl, pH 8.0, 10 m*M* $MgCl_2$, 1 m*M* DTT, 0.5 m*M* ATP for 4–16 h at room temperature, using 2 U of T4 DNA ligase/µg DNA. The choice of the appropriate molar ratio has been discussed in **Subheading 2.3.** DNA concentration in the ligation mixture is 200 ng/µL, favoring intermolecular ligation events, as opposed to DNA recircularization. Ligations with 20–50 µg DNA are prepared (*see* **Note 3** for handling DNA concatenamers).

Following ligation, DNA is extracted with phenol/cloroform, chloroform, and diethyl ether. Ethanol precipitation is not recommended at this stage, as it might subsequently prove difficult to redissolve the pellet.

3.4. Truncation of DNA Concatenamers with NotI

Plasmid pSfiSVneo contains a unique *Not*I site at the 3'-end of the expression cassette, immediately upstream of the *Sfi*I site (**Fig. 2**). This allows DNA concatenamers to be obtained containing only one terminal selectable marker unit by cleaving the ligation products with *Not*I. The appropriateness of performing this step has been discussed in **Subheading 2.3.** Extractions are finally performed as in **Subheading 3.3.**

3.5. Pulsed-Field Electrophoresis

The quality of the ligation products is evaluated by pulsed-field electrophoresis of 1 μg of the ligated DNA on 1.2% agarose gel in 1× TBE buffer (0.09 *M* Tris-borate, 0.002 *M* EDTA), using a Bio-Rad Chef DR III apparatus, Bio-Rad, Hercules, CA. Electrophoresis is performed at 4.8 V/cm, applying 2.5–7.5-s pulses over 16 h at 14°C. A smear or a ladder of discrete bands should be visible after ethidium bromide staining of the gel, extending up to at least 200 kb (**Fig. 3**).

4. Notes

1. Results of in vitro amplification. In vitro amplification has been used to express various genes in different cell lines *(5,6,8)* as summarized in **Table 1**. By comparing results with those from parallel transfections, in which nonligated DNA of exactly the same composition had been employed, we observed that maximal production levels in clones from the in vitro amplification protocol were higher or as high as those from the control cotransfection, employing physically distinct DNA molecules. The number of clones displaying high production levels was always higher for the in vitro amplification. This means that it is possible to reduce the number of clones to be screened for high productivity.

   Quantitative Southern blot analysis of the pattern of integration of the DNA concatenamers into the genome of the host cells revealed that intact multimers consisting of several (3–30) copies of the monomeric expression unit had been stably integrated. A broad, nonlinear correlation between number of integrated expression units and expression levels could be observed. This is expected, since gene dosage is only one of the parameters affecting expression levels, whereas other factors such as the influence of the specific integration site are not controlled in the present approach.
2. Combining in vitro and in vivo amplification. In one application of the in vitro amplification method, DNA concatenamers were prepared containing the cDNA for human granulocyte colony-stimulating factor (G-CSF) and two selectable markers (coding for neomycin resistance and dhfr) in a 10:1:1 molar ratio *(6)*. Following transfection into CHO dhfr$^-$ cells, neomycin-resistant clones were isolated and G-CSF secretion measured. Placing cells from individual clones in medium lacking nucleosides and containing either 10 or 100 μ*M* MTX increased

Table 1
Examples of Transfections Employing the in vitro Amplification Expression Method[a]

Gene	Cell	Transfection method	Yield	Comparison with nonligated DNA	Ref.
CAT	HeLa	Calcium phosphate ***(4)***	100% CAT activity in all clones	Variable (0–100%) CAT activity	***(5)***
ppET-1	hVSMC	DOTMA/ DOPE ***(11)***	1–3 ng/mL ET-1/24 h in 100% clones	Undetectable	***(5)***
G-CSF	CHO dhfr⁻	DOTMA/ DOPE	max. production: 1.5 μg/mL G-CSF/24 h; 90th percentile: 0.6 μg/mL G-CSF/24 h	Max. production: 1.3 μg/mL G-CSF/24 h; 90th percentile: 0.1 μg/mL G-CSF/24 h	***(6)***
α2,6–ST	CHO-320	DOTMA/ DOPE	70% clones with >80% α2,6–ST-positive cells	Not performed	***(8)***
IFN-γ	CHO dhfr⁻	Calcium phosphate, DOGS ***(12)***	14–42 ng/mL IFN-γ/24 h before MTX amplification; 2–6 μg/mL IFN-γ/24 h after MTX	Not performed	L. Monaco, unpublished data

[a]*CAT,* cDNA coding for bacterial chloramphenicol acetyl transferase: *ppET-1,* cDNA coding for human preproendothelin-1; *G-CSF,* cDNA coding for human granulocyte colony-stimulating factor; *α2,6-ST,* cDNA coding for rat α2,6-sialyltransferase; *IFN-γ*, cDNA coding for human interferon-γ; hVSMC, human vascular smooth muscle cells; CHO 320, CHO cells stably producing human interferon-γ. Comparison is made to parallel transfections using the same DNA composition, but with nonligated DNA.

G-CSF production by a factor of up to 30-fold. This proved the possibility of including more than two different expression units in the DNA concatenamers, a useful property in view of co-expressing multi-subunit proteins. Moreover, application of only one step of MTX amplification allowed a substantial enhancement of production levels within a reasonable time span.

3. Handling DNA concatenamers. DNA solution following ligation is viscous. Different handling procedures have been compared at this stage, to minimize breakage of the long DNA molecules. Analysis of ligated DNA by pulsed-field electrophoresis (*see* **Subheading 3.5.**) revealed that neither pipetting with normal pipet tips, compared to wide-bore ones, nor storing DNA at –20°C, compared with 4°C, reduced the length distribution of the ligation products.
4. Transfection. Polymerized DNA is transfected into cells by calcium phosphate or cytofection (as described in Sections 9.1.1 and 9.4.2, respectively of **ref. *4***). The optimal transfection method for a particular cell line needs to be determined empirically by test transfections. Transfections with the selectable marker gene only can be used to optimize the method for the highest number of colonies surviving the appropriate selection conditions.

Care must be taken to ensure that mixing DNA polymers to the transfection reagent does not produce large precipitates, which would reduce transfection efficiency. This occurred in our hands when preparing complexes with polyethylenimine ***(9)*** and, to a minor extent, with lipofectin ***(10)*** or DOGS (dioctadecylamidoglycylspermine) ***(11)***. As a general rule, diluted solutions of both DNA and reagent are mixed by addition of small aliquots of the reagent into DNA. From time to time, calcium phosphate coprecipitation yielded somewhat larger precipitates than usual.

Cells are placed in selective conditions 48 h after transfection. No particular requirements are needed at this step. Transfection efficiency, in terms of number of resistant colonies, is generally comparable to that obtained in parallel transfections employing standard plasmid DNA. *Current Protocols for Molecular Biology* ***(4)*** offers a basic protocol for selection of stably transfected colonies in Unit 9.5.1 and an overview of the available selectable marker genes in Unit 9.5.2.

5. Time considerations. Applying the in vitro amplification method requires a relatively short time, apart from the time needed to obtain the suitable DNA constructs. *Sfi*I cleavages to generate linear expression units and ligation can be performed in 1 day; pulsed-field electrophoresis to check the length of DNA concatenamers is performed overnight. Overall, this is an acceptable time span, if the length of the whole procedure is considered, including selection of stable clones.
6. Known Limitations. Transfecting cells with the DNA concatenamers by electroporation did not prove advantageous when compared with standard cotransfection ***(6)***. Neither maximal expression levels nor distribution of expression levels significantly differed among the two procedures. We speculate that entry of the long DNA molecules into the cells would not be favored, in this case, due to the large size of DNA, which is not compacted by any synthetic carrier.

Acknowledgments

The author is grateful to Mathias Uhlèn, with whom she started her work on the in vitro amplification system, and to Marco Soria, Gianfranco Distefano, and all the people in the Biotechnology Unit at Dibit who contributed to a continuous improvement of the methodology.

References

1. Warren, T. G. and Shields, D. (1984) Expression of preprosomatostatin in heterologous cells: biosynthesis, posttranslational processing, and secretion of mature somatostatin. *Cell* **39,** 547–555.
2. Urlaub, G. and Chasin, L. A. (1980) Isolation of Chinese hamster cell mutants deficient in dihydrofolate reductase activity. *Proc. Natl. Acad. Sci. USA* **77,** 4216–4220.
3. Kaufman, R. J. and Sharp, P. A. (1982) Amplification and expression of sequences cotransfected with a modular dihydrofolate reductase complementary DNA gene. *J. Mol. Biol.* **159,** 601–621.
4. Ausbel, F. M., Brent, R., Kingston, R. E., Moore, D. D., Seidman, J. G., Smith, J. A. and Struhl, K., eds. (1996) *Current Protocols in Molecular Biology.* John Wiley & Sons, New York.

5. Monaco, L., Tagliabue, R., Soria, M. R., and Uhlen, M. (1994) An *in vitro* amplification approach for the expression of recombinant proteins in mammalian cells. *Biotechnol. Appl. Biochem.* **20,** 157–171.
6. Monaco, L., Tagliabue, R., Giovanazzi, S., Bragonzi, A., and Soria, M. R. (1996) Expression of recombinant human granulocyte colony-stimulating factor in CHO dhfr$^-$ cells—new insights into the *in vitro* amplification expression system. *Gene* **180,** 145–150.
7. Sambrook, J., Fritsch, E. F., and Maniatis, T. (1989) Molecular Cloning: A Laboratory Manual, 2nd ed. Cold Spring Harbor Laboratory Press, Cold Spring Harbor, NY.
8. Monaco, L., Marc, A., Eon-Duval, A., Acerbis, G., Distefano, G., Lamotte, D., Engasser, J. M., Soria, M. R., and Jenkins, N. (1996) Genetic engineering of α2,6-sialyltransferase in recombinant CHO cells and its effects on the sialylation of recombinant interferon-γ. *Cytotechnology* **22,** 197–203.
9. Boussif, O., Lezoualc'h, F., Zanta, M. A., Mergny, M. D., Scherman, D., Demeneix, B., and Behr, J. P. (1995) A versatile vector for gene and oligonucleotide transfer into cells in culture and *in vivo*: polyethylenimine. *Proc. Natl. Acad. Sci. USA* **92,** 7297–7301.
10. Felgner, P. L., Gadek, T. R., Holm, M., Roman, R., Chan, H. W., Wenz, M., Northrop, J. P., Ringold, G. M., and Danielsen, M. (1987) Lipofection: a highly efficient, lipid-mediated DNA-transfection procedure. *Proc. Natl. Acad. Sci. USA* **84,** 7413–7417.
11. Behr, J. P., Demeneix, B., Loeffler, J. P., and Perez-Mutul, J. (1989) Efficient gene transfer into mammalian primary endocrine cells with lipopolyamine-coated DNA. *Proc. Natl. Acad. Sci. USA* **86,** 6982–6986

5

Cytogenetic Characterization of Recombinant Cells

Florian M. Wurm and Dietmar Schiffmann

1. Introduction

The use of mammalian cell systems for recombinant protein production has become the dominant method to produce high-value proteins for therapeutic purposes. Recombinant cell lines are often generated by transfection with specific DNA; therefore a selectable system is desirable to discriminate cells that have integrated the foreign DNA. Chinese hamster ovary (CHO) cells are most commonly used because of the availability of both a dihydrofolate reductase (DHFR)-deficient mutant cell line (CHO-DUKX *(1)* and DHFR expression vectors. Since DHFR expression vectors can be linked to DNA coding for desired proteins, co-integration of the DHFR expression vectors and the associated DNA often occurs. Integration of the DHFR expression vector into the CHO genome enables the cells to grow in a medium lacking the components glycine, hypoxanthine, and thymidine. Also, when multiple copies of the DHFR expression vector have been integrated, cellular resistance to methotrexate (MTX), a folate analog, is mediated. MTX can therefore be used to select for subclones that have amplified the transfected plasmid sequences to very high copy numbers.

In recombinant CHO cell lines, as with other recombinant mammalian cells, the efficiency of protein production is dependent, among other factors, on the fraction of cells carrying the inserted gene(s) in a functional and non-rearranged form. Even though a particular recombinant cell line is cloned, subtle pressures applied by continuous growth in culture, cell handling, and media manipulations may affect the stability of the inserted sequence in some cells.

Most recombinant CHO cells have, like the K1 parent line and the various $DHFR^-$ mutant lines, a modal, quasidiploid chromosome number of 21. This num-

From: *Methods in Biotechnology, Vol. 8: Animal Cell Biotechnology*
Edited by: N. Jenkins © Humana Press Inc., Totowa, NJ

ber may fluctuate slightly during the course of subcultivation, a phenomenon, that is observed in all immortalized cell lines. In general, CHO cells are considered to have a quite stable karyotype. It must be noted however, that CHO cell populations selected for MTX resistance and subcultivated in the presence of a drug for extended periods often contain substantial subpopulations of cells with increased chromosome numbers. Up to 30% of cells may contain a nearly tetraploid chromosome number of 40 or more (Wurm, unpublished data).

Verification and copy number estimations on transfected sequences in recombinant cells can be performed using dot-blot and Southern hybridizations. These methods use total DNA extracted from cells as a substrate for analysis. They do not give insights into the chromosomal localization of the recombinant sequences, nor do they give information about possible heterogeneities with respect to copy number among individual cells of a population. (For the use of filter-based hybridization techniques with extracted DNA *see* **ref.** ***2***).

The method of fluorescence in situ hybridization (FISH; **refs.** ***3,4***) allows characterization of recombinant CHO cells despite possible chromosomal variabilities. In particular, this method can be used to distinguish independent recombinant cell lines of the same origin (before transfection) but carrying different recombinant sequences. It may even permit the distinction of "clonal" cell lines obtained after transfection with one particular DNA.

FISH involves the hybridization of appropriate DNA probes with the DNA in chromosomes of the cell host. It allows the visualization of the location and extent of amplification of recombinant and nonrecombinant sequences on chromosomes in mammalian cells ***(5,6)***. Even single-copy genes can be localized on chromosomes of metaphase spreads with this method when optimized ***(7,8)***. The application of this method for the characterization of cells (CHO and others) carrying low copy or amplified recombinant sequences is based on the following features:

1. Hybridization is a powerful method capable of verifying the presence of a specific type of DNA sequence in a large excess of unrelated sequences.
2. Integration of plasmid sequences into the chromosomal DNA of CHO cells occurs at random sites ***(6,9–11)***. Therefore, an independent clonal cell line should exhibit a unique hybridization pattern when probed with corresponding plasmid sequences.
3. After treatment with MTX the transfected plasmid sequences are often amplified to high copy numbers. Even though large stretches of host DNA are coordinately amplified with accompanying recombinant sequences, the amplicon units are usually localized within narrowly defined chromosomal regions.
4. Frequently (at least with highly amplified expression vector DNA sequences), because of the high intensity of the hybridization signal, it is possible to screen hundreds of metaphase cells rapidly on a microscope slide. This allows statistical analysis to be performed on the resultant data.

1.1. Limitations of FISH for Analysis of Recombinant CHO Cell Populations

In clonal cell lines, cultivated in the absence of MTX for at least 6 weeks, unique identifiable hybridization signals or "master integration" patterns have been found over distinct chromosomes in these cell lines, using FISH ***(5,6)***. Intensity, location, substructure, and length of the hybridized motifs seem to be reproducible and can serve as suitable markers for the characterization of such lines. When the same cell lines were cultivated in the presence of MTX, however, the hybridization signals observed were more heterogeneous. Thus the MTX-induced heterogeneity may sometimes make it more difficult to distinguish independently developed cell lines containing the same type of transfected plasmid sequences. However, FISH would still allow verification of the presence of a specific type (gene/sequence) of transfected DNA in the chromosomes and would give an indication of the degree of amplification.

2. Materials

1. Nick-translation kit (for example, by Promega, Madison, WI).
2. Stop solution for nick translation: 200 m*M* EDTA.
3. Tris-EDTA buffer (TE): 10 m*M* Tris-Cl, pH 8.0, 1 m*M* EDTA, pH 8.0.
4. Colcemid solution for metaphase spreads: 0.1 µg/mL (Sigma, St. Louis, MO).
5. 2× standard saline citrate (SSC), pH 7.0. A large quantity of 2× SSC, pH 7.0, is needed for all the wash steps involved, so prepare at least 5 L in a plastic tank (*see* **item 15**).
6. Hybridization solution: for each slide 20 µL of hybridization mixture is needed.
7. Master mix: 5 mL formamide, 1 mL 20× SSC, 1 g dextran sulfate, H_2O up to a volume of 7 mL; adjust pH to 7.0; mix thoroughly and store in small aliquots at –20°C.
8. Hybridization solution (for five slides): 70 µL master mix (see **item 7**), 2.5 µL salmon sperm DNA (1 mg/mL), 4 µL biotinylated probe (40 ng), 23.5 µL H_2O.
9. PN buffer: quantities given here are for 1 L; prepare 4–8 L in a larger plastic tank. Dissolve 14.19 g Na_2HPO_4 (dibasic) in 700 mL H_2O; dissolve 345 g NaH_2HPO_4 (monobasic) in 250 mL of H_2O, add slowly to dibasic buffer, until pH is 8.0, then add H_2O up to 1 L.
10. PMN buffer: Add nonfat milk powder up to 5% and sodium azide up to 0.02% (w/v) to PN buffer. Stir for 30 min. Centrifuge for 30 min at 1000–1200*g* to clarify. Transfer supernatant to fresh tube. Adjust pH to 8.0 if necessary. Store at –20°C. Upon storage more precipitate may appear. Reclarify if necessary.
11. Biotinylated antiavidin antibody: 5% (v/v) in PMN buffer.
12. Avidin-fluorescein isothiocyanate (FITC) solution: 5 µg/mL avidin-FITC DCS grade (Vector, Burlingame, CA) in PMN buffer.
13. Antifade solutio: 100 mg phenylenediamine dihydrochloride in 10 mL phosphate-buffered saline (PBS). (Filter through 0.22-µm filter, adjust to pH 9.0 with 0.5 *M* NaOH].) Add 90 mL glycerol. Store in the dark at –20°C.
14. Coplin jars and Ektachrome 400 ASA color slide film.

15. SSC: 0.3 *M* trisodium citrate × $2H_2O$), pH 7.0, with HCl. Prepare fresh.
16. Paraformaldehyde 4%: add 2 g paraformaldehyde to 25 mL distilled water, and add 2 droplets of 1 *M* NaOH. Incubate at 70°C for 15 min. Add 25 mL 2× PBS, 100 m*M* $MgCl_2$. Filter through 0.22-μm filter.

2.1. Materials for CY3-dUTP Labeling

1. Polymarase chain reaction (PCR) mix:
   a. 20× Tfl buffer (Tris, pH 9.0): 25 μL, to 50 m*M* final concentration (f.c.).
   b. 100 m*M* MgCl2: 12.5 μL, to 25 m*M* f.c.
   c. 30 μ*M* primer (to be selected): 40 μL, to 2.5 μ*M* f.c.
   d. 2.5 m*M each* dNTP (A, C, G): 40 μL, to 200 μ*M* f.c.
   e. 2.5 m*M* dTTP: 40 μL, to 200 μ*M* f.c.
   f. 5% NP-40: 3.75 μL, to 0.38% f.c.
   g. 5% Tween-20: 3.75 μL, to 0.38% f.c.
   h. DNA 500 ng/μL): 1.0 μL, to 500 ng.
   i. Tfl polymerase (1.0 U/μL): 5.0 μL, to 5 U.
   j. H_2O bidest.: 9.5 μL (ad 50 μL total volume).
2. Nick translation mix:
   a. 10× Nick Translation (GIBCO-BRL) buffer: 50 μL.
   b. 2.5 m*M each* dNTP (A, C, G): 1.2 μL, to 60 μ*M* f.c.
   c. 500 μ*M* dTTP: 1.2 μL, to 16 μ*M* f.c.
   d. 1.0 m*M* Cy3-dUTP: 0.8 μL, to 16 μ*M* f.c.
   e. PCR mix: 25.0 μL.
   f. DNA polymerase/DNAse I (GIBCO-BRL) 0.4 U + 40 pg/μL: 5.0 μL, to 2 U + 200 pg.
   g. H_2O bidest.: 11.8 μL (ad 50 μL total volume).

The nick translation mix is incubated for 4 h at 16°C, and the reaction is stopped by addition of 5 μL 200 m*M* EDTA.

See also **Note 2**.

3. Methods

3.1. *Fluorescence* In Situ *Hybridization*

See also **Notes 4–12**.

FISH for visualization of recombinant sequences on chromosomes is a modification of an older method used for chromosome analysis ***(12)***. The use of biotinylated instead of radiolabeled probes, and the option for significant enhancement of the primary detection signal through a "sandwich" procedure containing both avidin and biotinylated antibodies directed against avidin are the main differences from previous methods. A biotinylated probe is hybridized to the corresponding sequences on the chromosomes of the recombinant cells. After extensive wash steps, fluoresceinated avidin is bound to the biotinylated probe, followed by the binding of biotinylated antibodies to the avidin molecules on the chromosomes. Incubation with fluoresceinated avidin enhances the primary signal, which can be further enhanced by additional rounds of incubation.

3.1.1. Preparation of Indirectly Labeled DNA Probes by Nick Translation (see also **Note 1***)*

1. In a 1.5-mL reaction tube (Eppendorf), kept on ice, add sequentially:
   a. 26 µL H_2O.
   b. 10 µL dNTP mix minus dTTP (mix individual nucleotides 1:1:1).
   c. 5 µL 10× nick translation buffer.
   d. 2 µL plasmid or DNA fragment (0.2–2 µg/µL).
   e. 2 µL biotin dUTP (0.4 m*M*; Bethesda Research Laboratories, Gaithersburg, MD).
   f. 5 µL enzyme mix (DNase I, DNA polymerase I). (*See also* **Note 3**.)
2. Vortex, centrifuge for 10 s in an Eppendorf centrifuge or equivalent for collection of droplets, and incubate at 15°C for 2 h.
3. Add 5 µL stop solution (EDTA) and 3 µL 5 mg/mL *Escherichia coli* tRNA, mix, and transfer reaction tube into ice.
4. Add sequentially:
   a. 200 µL H_2O.
   b. 125 µL 7.5 *M* NH_4OAc.
   c. 750 µL ethanol.

   Vortex and incubate in dry ice (CO_2) for 7 min.
5. Centrifuge at maximum speed in an Eppendorf centrifuge (10,000*g*) at 4°C for 7 min.
6. Aspirate supernatant cautiously (without touching the invisible pellet with tip of pipet).
7. Rinse pellet once with 70% ethanol, centrifuge as above, and remove supernatant. Dry under vacuum.
8. Dissolve pellet at a concentration of 10–20 ng/µL in TE buffer or water (assume about 50% loss of original DNA during preparation of probe). Store probe in aliquots of 20 µL at –20°C.

3.1.2. Preparation of Metaphase Chromosome Spreads on Microscope Slides

1. Cells should be grown in 75-cm^2 or 150-cm^2 flasks. Prior to treatment with colcemid, cells should be between 65 and 90% confluent, preferably 65%. Tap the flask (if cells are adherent) and pour off medium (shake-off technique for collection of cells in mitosis). Add fresh medium (approx 7 mL to a 75-cm^2 or 15 mL to a 150-cm^2 flask) with colcemid.
2. Incubate at 37°C for about 4 h. Cells that have been arrested in metaphase will be spherical.
3. Tap the flask once or twice to detach cells that have been arrested in metaphase. Collect by decanting with medium into a 15-mL conical centrifuge tube and centrifuge at 100*g* for 2 min at 4°C (do not trypsinize; cells that are not in mitosis will remain attached).
4. Decant medium and resuspend the pellet in the remaining volume (no greater than 300 mL) by gently vortexing for 5–10 s. If necessary, repeat several times.

5. Quickly add 5 mL 75 m*M* KCl and incubate at 37°C for 13 min.
6. With a Pasteur pipet, slowly add about 1 mL fresh (prepared the same day) Carnoy's fixative (3:1 methanol/glacial acetic acid). Centrifuge cells as in **step 3**. (This is a prefix step.)
7. Gently resuspend the pellet in the remaining volume. (If the cells are not treated gently, cell membranes will rupture and scatter the chromosomes.)
8. Add approx 5 mL fresh Carnoy's fixative to the cells.
9. Centrifuge at 100*g*, for 2 min at 4°C. Resuspend pellet in remaining volume.
10. Repeat **steps 8** and **9** twice.
11. Add fresh Carnoy's fixative until the solution becomes clear.
12. Permit one droplet of each cell suspension cells to fall onto ethanol-cleaned, dry slide. (Drop cells with a Pasteur pipet from a height of 6–8 in. (20–25 cm). Circle the area on the slide that contains the attached chromosomes immediately with a diamond marker. Check density of metaphases using a phase contrast microscope at ×100–400. The metaphase spread should not be too close together in order to allow clean analysis. If necessary, dilute sample in more fixative.
13. Store slides at –20°C under nitrogen for >5 days (aging).

3.1.2.1 RNase and Proteinase Treatment of Metaphase Spreads

1. For RNase treatment remove metaphase slides from –20°C storage and incubate at 37°C for 1 h in a solution of 100 μg/mL RNase A (Boehringer Mannheim) in 2× SSC.
2. Wash slides in 2× SSC for 5 min.
3. For proteinase K treatment, incubate slides at 37°C for 7.5 min in 20 m*M* Tris-HCl, pH 7.6, 2 m*M* $CaCl_2$ containing 0.5 μg/mL proteinase K (Boehringer Mannheim).
4. For paraformaldehyde fixation treat slides at room temperature for 10 min in 4% paraformaldehyde.
5. Wash slides three times in 2× SSC.

3.1.2.2. Hybridization Procedure

Caution: A number of chemicals used during this procedure are toxic, carcinogenic or mutagenic. Most noteworthy are MTX, colcemid, formamide, and paraformaldehyde. Gloves should be worn at all times, and inhalation of fumes should be avoided. A chemical fume hood is strongly advised.

1. Denature metaphase chromosomes on slides in 70% formamide, 2× SSC, pH 7.0, at 70°C for 5 min.
2. Transfer slides quickly to 70% ethanol at 4°C for 1 min, quickly transfer to cold 80%, 90%, and finally 100% ethanol, and incubate for 1 min in each.
3. Air-dry slides at room temperature.
4. Prepare hybridization solution. For each slide 20 μL of hybridization mixture is needed.
5. Denature the hybridization solution by incubation at 70°C for 5 min. Cool quickly in ice water.
6. Apply 20 μL of chilled hybridization solution per slide. Place a cover slip over the mixture, avoiding trapping air bubbles, and cover the perimeter of the cover

slip with a film of rubber cement. Incubate the slides overnight at 37°C in a dark humidified slide chamber.

The next day:

7. In three Coplin staining jars, warm 50% formamide, 2× SSC to 40°C (the water bath may have to be at a higher temperature).
8. Remove the rubber cement from the slides. Place slides (with coverslips) into the first staining jar. The cover slips should then slide off, but may need encouragement. Allow the slides to sit for 3 min. Transfer into a second jar. Again, allow them to sit for 3 min and then transfer the slides into a final staining jar for another 3 min.
9. Transfer the slides into PN buffer.
10. Rinse the slides twice for 2 min in fresh PN buffer.
11. Place a droplet (50 µL) of avidin-FITC solution onto each slide. Cover slides with a strip of parafilm, and then cover all slides with aluminum foil. Incubate slides at room temperature for 20 min.
12. Rinse the slides as in **step 10**.
13. Place a droplet of biotinylated antiavidin antibody solution (goat; Vector, Burlingame, CA) onto the slides and cover them as above. Incubate at room temperature for 20 min.
14. Rinse slides twice in fresh PN buffer.
15. Repeat **step 11**.
16. Rinse slides twice in fresh PN buffer.
17. Place 10 µL of antifade solution containing the DNA counterstain propidium-iodide at a concentration of 1 µg/mL (Sigma) on each slide. Cover with cover slip, avoiding the trapping of air bubbles.
18. View slides with a fluorescence microscope at ×400 magnification. Simultaneous observation of the fluorescence by fluorescein and propidium-iodide is possible when slides are excited at a wavelength range of 450–490 nm. The green fluorescein emission will be rendered yellow on the counterstaining background of the red-fluorescing propidium-iodide. Ektachrome ASA 400 color slide film is useful for color images. Black and white pictures can be taken with Tri-X (ASA 400) film. Storage of hybridized and "enhanced" slides should be done for limited periods only in a dark chamber at 4°C.

3.1.3. Preparation of Directly Labeled DNA Probes

Alternatively, using PCR-based techniques, probes can be generated directly from genomic DNA, cell hybrids or *in situ* ***(13–15)***. These methods use fluorescent dyes covalently attached to the nucleic acid probe so that the respective hybridized probe can be visualized microscopically immediately following hybridization. For example, fluorescein-labeled uridine can replace thymidine in the probe. Following hybridization, the hybridized complex can be detected by microscopy using the appropriate excitation and emission wavelengths to detect the green fluorescein dye. A number of direct fluorochrome-labeled nucleotides (covalently bound to DNA) have become available ***(16)***.

3.1.3.1. Preparation of CY3-dUTP-Labeled DNA Probes

The DNA sequences to be used as probes are amplified using a standard PCR procedure ***(17)***. Subsequently, Cy3-dUTP (cyano chromophore) is introduced via nick translation.

3.1.3.2. Hybridization Procedure

For hybridization with directly labeled probes a modification of the standard procedure of Trask and Pinkel ***(18)*** is applied: The reaction mixture contains 50% formamide, 10% dextrane sulfate, 1× SSC, 1 µg sheared herring sperm DNA, 20–100 ng labeled probe DNA, and deionized H_2O as required to a volume of 10 µL. The probe is denatured at 70°C for 5 min and then chilled rapidly on ice slush. The hybridization mix is placed on the warm slides and a #2 cover slip is sealed onto the slides with rubber cement. The slides are then hybridized overnight in a humidified chamber at 37°C.

3.1.3.3. Fluorescence Signal Detection

Slides hybridized with probes directly labeled with Cy3™ are washed three times in 60% formamide/2× SSC at 45°C for 5 min each and then rinsed in PN buffer. Subsequently, the slides are counterstained with 4,6-diamidino-2-phenylindol (DAPI) at 2.5 µg/mL. The fluorescence signals can be visualized in the metaphase preparations using a bandpass filter (Omega, Brattleboro, VT; Chroma, Brattleboro, VT) for simultaneous visualization of DAPI (blue) and Cy3™ (red) fluorescent signals.

4. Notes

1. For the preparation of biotinylated DNA probes, nick-translation or oligolabeling methods can be used (for details of these methods, *see* **ref. *2***). The difference from the methods given here is the use of [^{32}P]phosphate or otherwise radiolabeled DNA precursors instead of biotinylated deoxynucleotides. The nick-translation procedure is given as a reliable method for which there are also convenient kits available. (Promega, Madison, WI). Biotinylated probes can be stored indefinitely at –20°C. A probe useful for many different cell lines (dihydrofolate reductase [DHFR] vector transfected) is made with either the whole DHFR expression plasmid or with a gel-purified fragment containing the DHFR cassette.
2. A trace of [^{3}H]dATP can be added in a pilot nick-translation experiment to verify ad optimize incorporation of deoxynucleotides during the reaction. Minimum incorporation of 20% should be possible.
3. This 50-µL reaction should give enough biotinylated DNA for hybridizations with about 50–200 slides. If more DNA is needed, set up a larger reaction mixture.
4. Addition of too much Carnoy's fixative will dilute the cells on the slide; addition of too little fixative will clump cells together.
5. For highly amplified sequences generally one round of the labeling procedure is sufficient; for cells with low-copy-number integrations, more than two additional

rounds can be implemented (**Subheading 3.1.2.2., steps 13–15**). Bear in mind that the noise signals will be amplified as well. Three layers of avidin result in approximately 5^3 fluorescein molecules/detected biotin. If every biotin molecule were detected, 200 bp of bound probe would result in approx 1000 fluorescein molecules.

6. Photobleaching will occur after extended exposure of the slides to intensive light. Therefore, it is advisable to take pictures immediately after preparation of the slides and also limit their exposure to daylight.
7. FISH, in common with Southern hybridization ***(19)***, makes it possible to investigate whether a cell line contains transfected DNA sequences as integrated amplicons within (specific) chromosomes. As it deals with DNA isolated from the totality of a population of cells this method does not give any information on the heterogeneity of amplified sequences within that population. However it can to a certain degree, give information with respect to possible rearrangements or disturbances within those sequences visualized by the probe used. In contrast, FISH gives information on the degree of heterogeneity from cell to cell with respect to integrated transfected sequences.
8. A comparison of the indirect versus the direct method reveals that the direct method has the advantage that hybridizations require less time to perform as fewer detection steps are needed. The resolution of the signals is also better in many cases. However, the signals often tend not to be as strong as amplified signals obtained using the indirect method. It is important to consider these factors in particular if low copy numbers need to be detected.
9. Successful application of FISH is dependent on a number of factors, e.g., the quality of the metaphase chromosomes fixed on the microscope slide.
10. A rapidly growing, healthy cell population is required, and the timing of the treatment with colchicine is critical. Depending on the population doubling time (usually between 16 and 30 h in CHO cells) a treatment schedule shorter or longer than 4 h with colchicine may be optimal. One should be careful, however, not to treat cells for more than 10 h, as endo-reduplication without separation of cells may occur, giving a nonrepresentative, misleading result.
11. Another set of parameters deals with the probe length (50–200 nucleotides being optimal), probe concentration, amount of nonspecific carrier DNA, and hybridization conditions (temperature, time, salt concentration during hybridization and wash). These parameters should be optimized using a cell line with a high copy number of recombinant sequences, facilitating the detection of signal with only one round of enhancement.
12. Finally, the wash procedure and the number of rounds for enhancement of the primary signal are critical with respect to background staining. Here a good balance between a strong positive signal and the inevitable noise signal has to be found.

The sensitivity, specificity, and speed of performance have made FISH the method of choice to address a wide variety of experimental questions. Despite the obvious advantages of FISH compared with classical techniques, a number of improvements would further enhance the specificity and sensitivity of the method: Cooled charge-coupled device cameras with high sensitivity in combination with residual light enhancement systems are available today to detect hybridization signals not detectable with conventional techniques; and brighter fluorochromes with greater resistance to fading upon UV exposure may be developed. This would facilitate detection of shorter DNA sequences of hybridized DNA and of lower integrated DNA copy numbers, allowing a highly detailed analysis of chromosomes. Thus the advantages of FISH assure its continuous application in the future.

For fluorescent figures, *see* **ref. *20***.

References

1. Urlaub, G. and Charin, L. A. (1980) Isolation of Chinese hamster cell mutants deficient in dihydrofolate reductase activity. *Proc. Natl. Acad. Sci. USA* **77,** 4216–4220.
2. Ausubel, F. M., Brent, R., Kingston, R. E., Moore, D. D., Seidman, J. G., Smith, J. A., and Struhl, K. (eds.) (1988) *Current Protocols in Molecular Biology,* vols. 1 and 2, Wiley Interscience, New York.
3. Pinkel, D., Straume, T. , and Gray, J. W. (1986) Cytogenetic analysis, using quantitative, high sensitivity fluorescence hybridization. *Proc. Natl. Acad. Sci. USA* **83,** 2934–2938.
4. Swiger, R. R. and Tucker, J. D. (1996) Fluorescence in situ hybridization: a brief review. E*nviron. Mol. Mutagen.* **27,** 245–254.
5. Pallavicini, M. G., DeTeresa, P. S., Rosette, C., Gray, J. W., and Wurm, F. M. (1990) Effects of methotrexate (MTX) on transfected DNA stability in mammalian cells. *Mol. Cell. Biol.* **10,** 401–404.
6. Wurm, F. M., Bajaj, V., Tanaka, W., Fung, V., Smiley, A., Johnson, A., Pallavicini, M., and Arathoon, R. (1990) Methotrexate and CHO cells: productivity and genetics of amplified expression vector sequences, in *Production of Biologicals from Animal Cells in Culture,* Spier, R. E., Griffiths, J. B., and Meignier, B., eds., Butterworths, Oxford, pp. 316–326.
7. Lawrence, J. B., Villnave, C. A., and Singer, R. H. (1988) Sensitive, high resolution chromatin and chromosome mapping in situ : presence and orientation of two closely integrated copies of EBV in a Iymphoma line. *Cell* **52,** 51–61.
8. Lichter, P., Chang Tang, C.-J., Call, K., Hermanson, G., Evans, G. A., Housman, D., and Ward, D. C. (1990) High resolution mapping of human chromosome 11 by in situ hybridization with cosmid clones. *Science* **247,** 64–69.
9. Kaufman, R. J., Wasley, L. C., Spiliotes, S. J., Gossels, S. D., Latt, S. A., Larsen, G. R., and Kay, R. M. (1985) Coamplification and coexpression of hum an tissue-type plasminogen activator and murine dihydrofolate reductase sequences in Chinese hamster ovary cells. *Mol. Cell. Biol.* **5,** 1750–1759.
10. Ruiz, J. C. and Wahl, J. M. (1990) Chromosomal destabilization during gene amplification. *Mol. Cell. Biol.* **10,** 3056–3066.
11. Wurm, F. M. (1990) Integration, amplification and stability of plasmid sequences in CHO cell cultures. *Biologicals* **18,** 159–164.

12. Gall, J. G. and Pardue, M. L. (1971) Nucleic acid hybridization in cytological preparations. *Methods Enzymol.* **38,** 470–480.
13. Narayanan, S. (1992) Overview of principles and current uses of DNA probes in clinical laboratory medicine. *Ann. Clin. Lab. Sci.* **22,** 353–376.
14. Dunham, I., Lengauer, C., Cremer, T., and Featherstone, T. (1992) Rapid generation of chromosome-specific alphoid DNA probes using the polymerase chain reaction. *Hum. Genet.* **88,** 457–462.
15. Hindkjaer, J., Koch, J., Tekelsen, C., Brandt, C. A., Kolvraa, S., and Bolund, L. (1994) Fast, sensitive mulicolored detection of nucleic acids in situ by primed in situ labeling (PRINS). *Cytogenet. Cell Genet.* **66,** 152–154.
16. Wiegant, J., Wiesmeijer, C. C., Hoovers, J. M. N., Schuuring, E., d'Azzo, A., Vroliijk, J., Tanke, H. J., and Raap, A. K. (1993) Multiple and sensitive fluorescence in situ hybridization with rhodamine-, fluorescein- and coumarin-labeled DNAs. *Cytogenet. Cell. Genet.* **63,** 73–76.
17. Saiki, R. K., Gelfand, D. H., Stoffel, S., Scharf, S. J., Higuchi, R., Horn, G. T., Mullis, K. B., and Ehrlich, H. A. (1988) Primer-directed enzymatic amplification of DNA with a thermostable polymerase. *Science* **239,** 487–491.
18. Trask, B. and Pinkel, D. (1990) Fluorescence in situ hybridization with DNA probes. *Methods Cell Biol.* **33,** 383–400.
19. Southern, E. M. (1975) : Detection of specific sequences among DNA fragments separated by gel electrophoresis. *J. Mol. Biol.* **98,** 503–517.
20. Eastmond, D. A. and Pinkel, D. (1990) Detection of aneuploidy and aneuploidy-inducing agents in human lymphocytes using fluorescence *in situ* hybridization with chromosome-specific DNA probes. *Mutat. Res.* **234,** 303–318.

6

Immortalization Strategies for Mammalian Cells

John Stephen McLean

1. Introduction

Mammalian cells normally have a limited life span and after having undergone approximately 30 population doublings stop dividing; this is known as senescence. However, when the control mechanisms which limit life span are modified cells can become immortalized i.e. have an unlimited growth potential *(1)*. One such way of subverting these control mechanisms is to express an immortalizing oncogene in normal cells. These proteins override the normal control points and one of the most studied and best understood of this class of oncogenes is large T antigen which is contained within the genome of the monkey virus SV40 *(2)*. SV40 large T antigen binds a number of proteins, but the most important are p53 and Rb which, when functioning normally, are negative regulators of cell division. Upon being bound by SV40 large T antigen, p53 and Rb are inactivated and the cells have an unlimited life span. There are a number of good reviews on immortalization *(3–5)* that deal in greater detail with biology involved in the process which is beyond the scope of this review.

1.1. Choice of Oncogene

SV40 large T antigen is the oncogene used most often in immortalization studies. This is because it is has strong immortalizing capabilities and has been shown to successfully immortalize a wide range of cells from a number of species. It is recommended therefore that a SV40 large T antigen be used therefore as a first choice when immortalizing animal cells. Other oncogenes are suitable for immortalization; for example, E1a from the adenovirus has been use to create cell lines, although it has much weaker immortalizing activity than SV40 large T antigen. Also, it has been proposed that the Polyoma large T antigen is more effective at immortalizing human cells than SV40 large T antigen *(6)*. However,

From: *Methods in Biotechnology, Vol. 8: Animal Cell Biotechnology*
Edited by: N. Jenkins © Humana Press Inc., Totowa, NJ

there is only a limited database on the use of this oncogene. Upon expression of SV40 large T antigen (and other oncogenes) in human cells, initially, lifespan is only extended and immortalization *per se* involves secondary genetic events.

1.2. Temperature-Sensitive or Wild-Type Large T Antigen

Although SV40 large T antigen has strong immortalizing capabilities, it also has the disadvantage in that has weak transforming capabilities (i.e., although it creates cells with an extended replicative capacity, its expression can also lead to dedifferentiation of cells). For this reason scientists active in this field often use a temperature-sensitive mutant of SV40 large T antigen (tsTag) which is functional at 33°C (i.e., it causes immortalization). This is known as the "permissive temperature," and when cells immortalized with tsTAg are grown under these conditions, they would grow but be dedifferentiated. At what is known as the "nonpermissive temperature" (usually about 39°C, the immortalized cells will not proliferate but would be differentiated *(5)*. At the should be noted that when tsTag is used to immortalize human cells incubating at nonpermissive temperature, in cell death occurs and so this approach is of little benefit. It is recommended, therefore, that a plasmid expressing SV40 large T antigen be used for immortalization of mammalian cells.

1.3. Types of Cell-Immortalized

There is a now a significant database on immortalization of cells by SV40 large T antigen, and cell types (with some important exceptions; see below) from all mammalian species can be immortalized using this oncogene. **Table 1** gives some examples of the wide range of immortalized lines made from a number of species through expression of SV40 large T antigen. This list is by no means exhaustive. However, interestingly, a number of cell types have proven refractory to immortalization by SV40 large T and many other oncogenes. Specifically, these are B and T lymphocytes and they can only be immortalized through use of specific methods such as EBV (Epstein-Barr virus) and HIV (human immunodeficiency virus) infection, respectively *(3)*.

2. Practical Considerations

Immortalization of mammalian cells depends on three techniques (**Table 2**):

1. Construction of a plasmid containing the oncogene
2. Method of gene transfer
3. Target cells to be immortalized

2.1. Plasmid Containing Oncogene

A plasmid containing the oncogene must be obtained or produced in-house for immortalization. The quickest and easiest route is to obtain a plasmid containing the

Table 1
Types of Differentiated Immortal Cell Lines from Transfection with SV40 Large T Antigen

Tissue	Species
Adrenocortex	Bovine
Pancreas	Hamster
Brain	Mouse
Endothelial cells	Mouse
Endothelial cells	Human
Endothelial cells	Bovine
Glial cells	Mouse
Hepatocytes	Mouse
Hepatocytes	Rat
Hepatocytes	Human
Hepatocytes	Monkey
Hypothalamic cells	Mouse
Macrophages	Mouse
Mammary epithelium	Rabbit
Endometrium	Rat
Heart	Rat
Myoblasts	Rat
Schwann cells	Rat
Smooth-muscle cells	Rat

oncogene from an investigator who has successfully immortalized cells. This area of biotechnology has matured sufficiently such that most scientists will make plasmids available subject to the conditions normally applied for release of material.

However, if a plasmid cannot be obtained, then it may be essential to generate one in-house. There are now a number of good commercially available mammalian expression vectors that could be used for immortalization studies. The following vectors are available from Invitrogen (**Table 3**). The most suitable is pcDNAneo3, which uses the strong CMV (cytomegalovirus) promoter to drive expression and contains a multiple-cloning site and the marker neomycin resistance (allowing selection of stable clones). Other useful vectors are pZeoSV2, which uses the zeocin resistance gene and the family of episomal vectors pCEP and pREP which contain the hygromycin resistance gene. The availability of multiple resistance genes means that more than one plasmid can be stably expressed in cells.

Ensure that you do not use a vector containing the SV40 origin of replication to express SV40 large T antigen, as the oncogene can act via this origin and so result in runaway replication of the vector and will lead to cell death ***(4)***.

Table 2
Checklist for Immortalization of Mammalian Cells

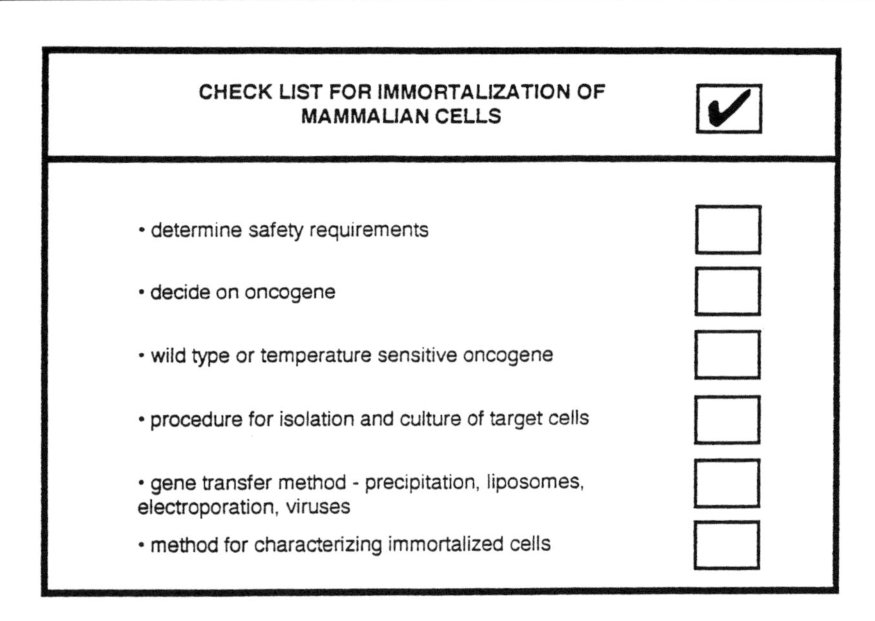

CHECK LIST FOR IMMORTALIZATION OF MAMMALIAN CELLS	✔
• determine safety requirements	☐
• decide on oncogene	☐
• wild type or temperature sensitive oncogene	☐
• procedure for isolation and culture of target cells	☐
• gene transfer method - precipitation, liposomes, electroporation, viruses	☐
• method for characterizing immortalized cells	☐

Another very important consideration is the quality of the purified plasmid DNA. It is important that the DNA be free from chromosomal DNA, RNA, and endotoxins which compromise gene transfer. The commercially available purification kits such as Qiagen (http://qiagen.com) give high-quality DNA quickly that can be used directly for transfer into mammalian cells. Some investigators prefer to use cesium chloride gradient centrifugation for preparing DNA. However, in our laboratory, commercially available purification kits work well, and we believe the evidence in favor of cesium chloride gradient centrifugation to be anecdotal. Upon isolating the plasmid from bacteria the DNA should be maintained in sterile conditions.

2.2. Method of Gene Transfer

As well as choice of oncogene, the method of gene transfer is vital if successful immortalization is to take place. These are discussed briefly and summarized in **Table 3**. Commercial sources of kits for carrying out gene transfer experiments are also available. **Figure 1** illustrates the parameters involved in immortalization that would affect the method of gene transfer chosen.

2.2.1. Precipitation Methods

The most commonly used approach for gene tenser is calcium phosphate precipitation. However, it has the very serious drawback of being toxic to spe-

Table 3
Methods of Introducing Immortalizing Oncogene

Technique	Advantages	Disadvantages	Supplier (WWW site address)
Calcium phosphate precipitation	Cheap, robust	Toxic to some specialized cell types, low transfer efficiencies	Pharmacia (http://www.biotech.pharmacia.se/) Life Technologies (http://www.lifetech.com/) Promega (http://www.promega .com/)
Strontium phosphate precipitation	Cheap, robust, low toxicity	Low transfer efficiencies	Biovations
Liposomes	Low toxicity, good transfer efficiencies	Expensive, requires cell line to cell line optimization	Promega (http://www.promega .com/) Invitrogen (http://www.invitrogen.com/) Life Technologies (http://www.lifetech.com/)
Electroporation	Works with suspension cells	Poor transfer efficiencies, needs large numbers of cells	Biorad (http://www.biorad.com)
DEAE dextran	Good transfer efficiencies	Some toxicity	Promega (http://www.promega.com/)
Microinjection	Excellent transfer efficiencies	Needs complex and expensive equipment, time-consuming safety considerations	
Retroviral vectors	Very high transfer efficiencies	Specialized facilities, possible replication-competent retrovirus, cannot infect nondividing cells	
Adenoviral vectors	Exceptionally high transfer efficiencies, can infect nondividing cells	Specialized facilities, no tsTag construct available	

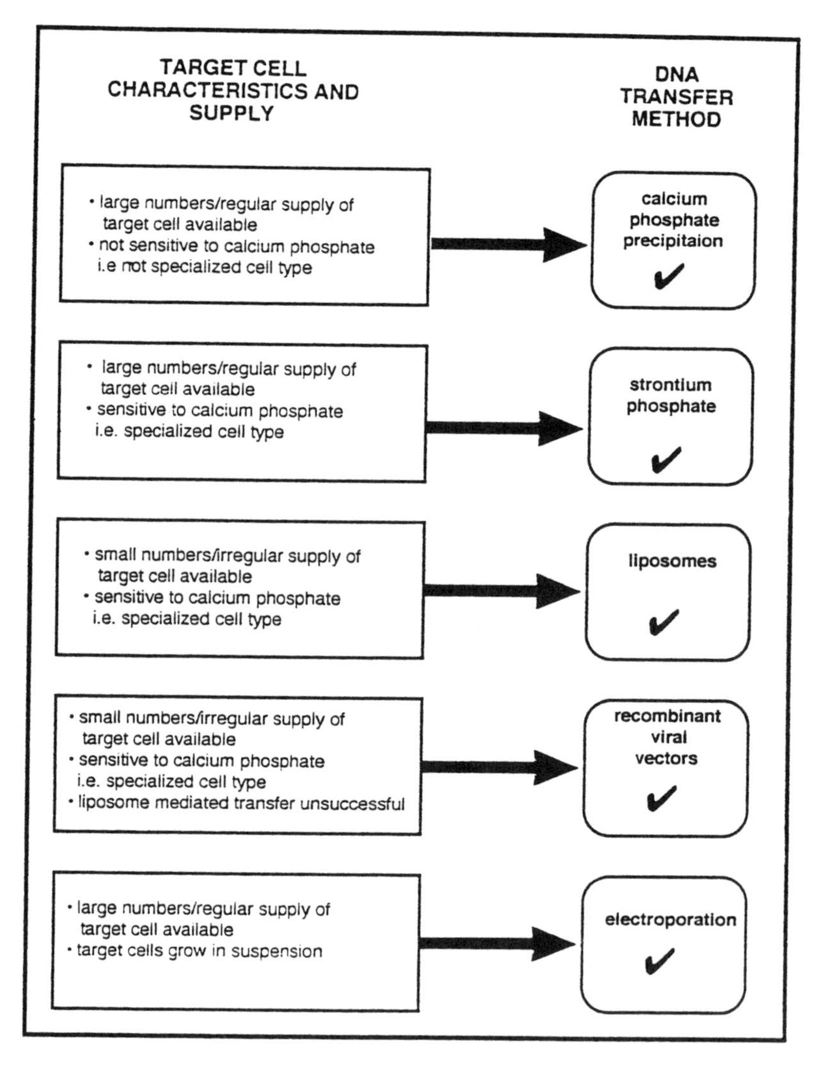

Fig. 1. Factors affecting choice of gene transfer technique.

cialized cell types which are often targets for immortalization (e.g., endothelia, hepatocytes, neurones, and epithelia) due to the high concentrations of calcium. Also, gene transfer efficiencies are low (approximately 0.001%). Nonetheless, calcium phosphate precipitation is a robust technique that works well even in the hands of inexperienced investigators and is inexpensive; therefore, it is often the first choice in gene transfer studies *(4)*.

An alternative to this is to replace the calcium with strontium. No toxicity is observed, and gene transfer efficiencies, albeit low, are maintained at low cost.

Strontium phosphate precipitation has been used to mediate gene transfer into and/or immortalize a range of cell types ***(7)*** and it is envisaged that its usage will increase, especially with the advent of commercially available kits.

2.2.2. Lipid Mediated Gene Transfer

DNA can be encapsulated in liposomes and this can circumvent the toxicity observed with calcium phosphate precipitation. Liposomes also have another major advantage over the phosphate-precipitation-based approaches in that when conditions are properly optimized, very high transfer efficiencies can be obtained (approaching 50%). However, there are problems with cost, optimizing conditions, and the poor performance of some lipids in serum containing media which reduce their utility. Finding the optimal conditions and best liposome for the target cell type can be expensive, and synthesis of liposomes involves complex chemistry which can make them costly. Also, there is a wide range of liposomes now available for gene transfer studies and their efficiency varies from cell type to cell type. Consequently, a significant amount of time and resources can be spent in optimizing conditions. Obviously, if the cells to be immortalized are in only limited supply, this can also render this technique inappropriate. Nonetheless, if time is spent on optimizing conditions, this results in transfection efficiencies significantly higher than other techniques and so this approach may be the one of choice if only small numbers of target cells are available.

2.2.3. Electrical Techniques

By applying an electrical voltage across the membrane, cells become permeabilized and any DNA present in the solution is Absorbed. This is known as electroporation. However, the gene transfer efficiencies attained through this approach are the lowest for all the techniques described and the high voltages are often toxic to the target cells. Also, the technique requires an electroporator which is not inexpensive (BioRad, http://biorad.com). However, for cells which normally grow in suspension, electroporation is a method often used in gene transfer especially if large cell numbers are available.

2.2.4. Viral Approaches

Recombinant viruses give the highest levels of gene transfer, with reports of 100% efficiency for retroviral- and adenoviral-mediated transfer. There have been numerous publications (e.g. ***10***) on the use of retroviral vectors in immortalization studies, but there are some drawbacks in the use of this approach. The major problem is that retroviruses only mediate genetic transfer into non-dividing cells, and because many target cells proliferate poorly or not all, retroviruses have limited uses. Another consideration is the possibility that replication-competent retroviruses may be produced due to recombination events between the recombinant retrovirus and endogenous retroviral sequences.

An adenovirus can infect nondividing cells and can be obtained at very high virus titer levels (a million-fold better than retrovirus) and there are fewer safety considerations. Therefore, they have distinct advantages over retroviral vectors. A recombinant adenovirus that expresses SV40 large T antigen is available and has been used to successfully immortalize a number of cell types ***(5)***. However, there have been no reports of an adenoviral vector expressing tsTag as of yet. The availability of such a tool would be invaluable in immortalization.

2.2.5. Microinjection

DNA (or RNA) can be introduced into cells by microinjection. This is where small volumes of liquid are inserted into cells through using very fine glass micropipets and the application of slight pressure. This can give rise to very high frequencies of transfer but has a number of drawbacks mainly the cost and expertise needed to carry out this technique. Also, the use of glass in conjunction with oncogenic DNA is at best undesirable.

2.2.6. Cell Fusion

A specialized approach to cell immortalization is the use of cell fusion. This is most common in the production of hybridomas through the fusion of B cells with myelomas for the manufacture of monoclonal antibodies. It has been used as an approach in basic cancer research to identify chromosomes which contain oncogenes, but it is not a technique which can be used routinely to immortalize cells ***(4)***.

2.3. Isolation and Preparation of Cells for Immortalization

Other important considerations in immortalization are the target cells and the culture conditions. It is recommended that some that the culture conditions needed for growth be fully investigated before starting any immortalization work. A number of publications are now available which explain in detail the techniques most suitable for culturing even the most difficult cell types ***(8,9)***.

It is unlikely that the culture which contains the cells to be immortalize consists of only that cell type. It is more probable that the primary cell culture contains a mixture of cells, some of which may overgrow the cell of interest. The most common contaminant are fibroblasts, and unless precautions are taken, they will overgrow most cultures. It is therefore essential that the cultures contain as high a proportion of the cells of interest as possible. Specialized techniques for culturing the target cells prior to immortalization are therefore recommended. Alternatively, if there are mechanisms for separating the immortalized target cells from contaminating cells, then pure populations can be obtained after immortalization. FACS sorting is a very useful tool in this situation (*see* Chapter 11), although it is not available to every investigator.

As immortalization is completely dependent on transfer of the oncogene into the target cell, criteria that are essential for successful gene transfer must be met. Specifically the cells must be actively dividing when exposed to the DNA if it is to be taken up and stably expressed in the genome.

3. Materials

1. Water. The quality of water used is very important in determining the efficiency of gene transfer. Purchasing water from a commercial source may incur extra cost but will ensure quality. Sigma (http://www.sigma.sial.com) and Life Technologies sell cell-culture-grade water.
2. 10X Strontium chloride (2*M*). 10.66 *g* of $SrCl_2 \cdot 6H_2O$ dissolved in 20 mL of H_2O and filter sterilize before storing at –20°C.
3. 10X Calcium chloride (2*M*). 5.88 g of $CaCl_2 \cdot 2H_2O$ dissolved in 20 mL of H_2O and filter sterilize before storing at –20°C.
4. 2X HBSS. Dissolve 1.6 g of NaCl, 0.074 g of KCl, 0.027 g of $Na_2HPO_4 \cdot 2H_2O$, 0.2 g of dextrose, and 1.0 g of HEPES in ~90 mL of H_2O before adjusting the pH to 7.05 using NaOH and diluting to 100 mL. Filter sterilize and store at 4°C.
5. 0.5 g/mL G418. G418 is sold by LTI under the name Geneticin. Dissolve 5 g of G418 in 100 mL of H_2O. Filter sterilize and store at 4°C.

4. Methods

4.1. Preparation of DNA for Immortalization

It is recommended that the plasmid be linearized and made sterile for transfer into the target cells. Following linearization, it is essential to remove any proteins by phenol/chloroform extraction, otherwise transfer efficiencies will be compromised. As a general rule of thumb, the more DNA present, the better the chances of gene transfer and so immortalization. However, with about 20 μg of DNA, saturation is achieved and no further increase in efficiency is observed with further increases. For this reason normally 5–20 μg of plasmid are used in immortalization work.

1. Linearized DNA should be made up to 500 μL with H_2O and to this, 500 μL of phenol–chloroform added. Gently mix by inversion.
2. Spin in a microfuge at maximum speed for 2 min. Remove top (aqueous) layer and place in a fresh microcentrifuge tube.
3. Add 500 μL of chloroform and gently mix by inversion.
4. Spin in a microfuge at maximum speed for 2 min. Remove top (aqueous) layer and place in a fresh microcentrifuge tube.
5. Add 1 mL of ice-cold 100% ethanol and 50 μL of $NaAc \cdot 3H_2O$ and mix gently by inversion. Spin at maximum speed in a centrifuge for 5 min.
6. Gently pour off liquid and wash the precipitated DNA pellet with ice-cold 70% ethanol. Spin at maximum speed in a centrifuge for 5 min.
7. Gently pour off liquid and resuspend DNA in 50 μL of H_2O.

4.2. Preparing Cell Cultures for Immortalization

The quality of the target cells to be immortalized is of vital importance if successful results are to be obtained and there are three main criteria to be considered. First, cell viability should be as high as possible prior to transfection and a viability of <80% (as determined by Trypan Blue staining) is recommended. However, in some instances, obtaining cultures with such a high level of viability is problematical, and gene transfer will not occur unless cells are actively dividing on the addition of DNA. To ensure a high percentage of cells are undergoing cell division, fresh media should be added to the cells a few hours prior to transfection. Alternatively, enough cells should be used so that cultures are 30–50% confluent at time of transfection.

Remember to set up cultures in culture dishes that you can easily pick immortalized colonies of cells from. Failure to do this may mean losing is successfully immortalized cells.

1. (a) Determine viability. If viability is low (<80%), then it may be worthwhile to postpone transfection until cell viability improves. Add sufficient cells to a petri dish so that on the following day, the culture is 30–50% confluent. Incubate overnight.
   (b) If culture has already been established then 3–4 h prior to transfection, add fresh *prewarmed* media (37°C).
2. Immediately prior to transfection, examine cultures to ensure they are suitable for transfection (i.e., they are 30–50% confluent).

4.3. Introduction of Plasmid into Cells Using Precipitation Methods

All the following steps should be carried out under appropriate conditions so as to ensure maintenance of sterility. The following protocol is for transfer cells in a 35-mm petri dish. Volumes should be changed *pro rata* on the basis of the surface area of the cell culture flask or petri dish.

1. Polypropylene tubes are recommended for the following steps. Mix 20 μL of plasmid (5–20 μg) with 30 μL of 2 *M* $SrCl_2$ or $CaCl_2$ (10X solution). Add 200 μL of water (final volume of 250 μL).
2. The DNA–strontium/calcium chloride solution is then added dropwise over a 30-s period to a second tube containing 250 μL of 2X HBSS. Gentle agitation of the second tube during addition of the DNA–strontium/calcium chloride solution results in a finer precipitate which increase gene transfer efficiencies.
3. Incubate at room temperature for 15–20 min. The solution should appear "milky." This is due to the formation of the precipitate.
4. Mix the precipitate to ensure resuspension and add to the cells dropwise ensuring addition to all of plate.
5. Incubate at 37°C. As a result of the lack of toxicity observed with strontium phosphate precipitation, cells can be incubated for up to 24 h to ensure maximal gene transfer. However, by approximately 6 h, gene transfer will have begun and be completed following exposure for approximately 16 h.

6. Gently wash the cells two to three times with prewarmed medium to remove the precipitate. Add fresh medium and return to incubator.
7. Allow cells to recover for 3 days following transfection. If cultures have grown and flasks are confluent, trypsinize cells and add to fresh culture dishes. If the culture dish is still sub-confluent, then it is possible to carry out the selection in the same culture dish.
8. When distinct colonies are visible and contain approximately 1000 cells, they can be picked and expanded for characterization.
9 To this medium, add an appropriate amount of selection agent (e.g., G418 [neomycin resistance]). A concentration of 750 μg/mL of this agent will be toxic to the vast majority of cells. However, if this concentration is not lethal, then it can be increased to 1 mg/mL, although this has cost implications. If cells are expressing the neomycin resistance gene, they will can grow well in concentrations as high as 2 mg/mL of G418. These criteria are common to all selection strategies. The appearance of colonies resistant to the selection, and thus expressing the gene of interest, will vary depending on the cell type and toxic agent but should occur within 14–28 days. It is advisable to add fresh prewarmed medium containing the selection agent every 4–5 days. For some cells, conditioned medium (medium that has already been used to grow cells and which contains growth-promoting agents produced by them) is essential for growth and should still be used during selection of immortalized cells.

4.4. Modifications for Improving Calcium-Phosphate-Precipitation Transfer Efficiencies

A variety of modifications for improving calcium-phosphate-precipitation efficiencies have been described *(4)*. These include the addition of glycerol or dimethyl sulfoxide (DMSO) following removal of the phosphate–DNA precipitate. These treatments have been reported to improve DNA transfer efficiencies but can also be toxic to more specialized cell types.

4.5. Introduction of Plasmid into Cells Using Liposomes

The exact details for carrying out liposome-mediated DNA transfer vary depending on the lipid used. The description given is for Lipofectin (Life Technologies), one of the most commonly used lipids.

1. Ensure that cells are 3–50% confluent.
2. Make up 10–20 μg of plasmid DNA to 50 μL in H_2O.
3. In a separate tube, add 10 μL of H_2O to 40 μL of Lipofectin and mix.
4. Add the DNA mix dropwise to the Lipofectin and allow to stand at room temperature for 15 min.
5. Meanwhile, remove media from cells and wash twice with serum-free medium (no fetal calf serum). Add 5 mL of serum-free medium.
6. Add DNA–Lipofectin mix and incubate for 6 h at 37°C.
7. Remove medium and add complete growth medium (containing fetal calf serum).
8. Allow cells to recover for 3 days following transfection. Carry out selection as in DNA precipitation-based approaches (**Subheading 4.3.**).

4.6. Introduction of Plasmid into Cells Using Electroporation

The exact conditions for electroporation will vary from cell type to cell type. These will have to be optimized accordingly. The following conditions represent a good starting point for gene transfer by this technique into human cells.

1. Harvest 5–106 cells. The cells should be in log-phase when harvested.
2. Spin cells down in bench centrifuge at 4°C for 5 min at approximately 200*g*.
3. Resuspend in 1 mL of medium of choice and transfer to electroporation cuvette (BioRad) which contains plasmid DNA (40–50 µg). Incubate on ice for 10 min.
4. Deliver a single pulse to the cells (0.26 kV, 960 µF) using an electroporator (BioRad).
5. Cool on ice for a further 10 min.
6. Add about 20 mL of complete growth medium.
7. Allow cells to recover for 3 days following transfection. Carry out selection as in DNA precipitation based approaches (**Subheading 4.3.**).

5. Safety Considerations

As in all scientific activities, the most important consideration must be safety. The following rules are most pertinent to immortalization studies.

1. Follow local and national rules on the use of oncogenic sequences.
2. Never attempt to immortalize cells from an investigator in the laboratory where the work is being performed. These immortalized cells cannot be recognized as foreign by the donor and may be able to colonize if accidentally infected.
3. Fetal material does not display the full cell-surface antigens involved in the immune response and so may not be recognized as foreign if an investigator becomes infected with these cells.

References

1. Glover, D. M. and Hames B. D. (eds.) (1989) *Oncogenes.* Oxford University Press, Oxford.
2. Fanning, E. and Knippers, R. (1992) *Annu. Rev. Biochem* **61,** 55–85.
3. MacDonald, C. (1991) *Mammalian Cell Biotechnology* (Butler, M., ed.), Oxford University Press, Oxford.
4. MacDonald, C. (1990) *Crit. Rev. Biotechnol.* **10,** 155–178.
5. McLean, J. S. (1993) *Trends Biotechnol.* **11,** 232–238.
6. Strauss, M., Hering, S., Lubbe, L., and Griffin, B. E. (1990) *Oncogene* **5,** 1223–1229.
7. Brash, D. E., Reddel, R. R., Quanrad, M., Yang, K., Farrell, M. P., Harris, C. C. (1987) *Molec. Cell. Biol.* **7,** 2031–2034.
8. Doyle, A., Griffiths, J. B., and Newell, D. G. (eds.) (1995) *Cell and Tissue Culture: Laboratory Procedures.* John Wiley & Sons, Chichester.
9. Freshney, R. I. (1994) *Culture of Animal Cells: A Manual of Basic Techniques*, 3rd ed., Wiley–Liss, New York.
10. Ito, M. and Kedes, L. (1997) *Human Gene Therapy.* **8,** 57–63.

7

Mouse Monoclonal Antibodies

Immunization and Production Protocols

Margaret Goodall

1. Introduction

The main principle on which the production of monoclonal antibodies-depends is that a single antibody producing cell produces only one type of antibody of a single specificity. As immunization results in a polyclonal response with the production of many antibodies with different specificities and isotypes, then to obtain a monoclonal antibody, a single antibody-producing cell must be isolated. However, these cells have a short life-span; thus, they must be fused with a nonproducing myeloma cell line so that the resulting hybrid cell will be both immortal and antibody producing.

Monoclonal antibodies (MAb) were made first in 1975 by Kohler and Milstein who immortalised spleen cells from an immunised mouse by fusing them with mouse myeloma cells. The resulting cell line proliferated indefinately with the continuous production of antibody of a single specificity. They used inactivated Sendai virus as the fusion agent ***(1,2)***, but polyethylene glycol (PEG) subsequently became the method of choice ***(3)***. It was not appreciated at the time just how MAbs would revolutionise research in science and in medicine, in particular, but the contribution of Kohler and Milstein was recognized, along with Jerne, with the award of the Nobel Prize in 1984. Detailed procedure protocols using PEG as the inducer fusion were published by Galfre and Milstein in 1981 ***(4)***. The theory, production, and application of MAbs has been detailed in **ref.** ***5***.

The methodology presented here uses PEG as the inducer of fusion. It is inexpensive, easy, and, provided there is attention to detail, the fusion rate can be high; so much so, that a very efficient primary screening assay needs to be

From: *Methods in Biotechnology, Vol. 8: Animal Cell Biotechnology*
Edited by: N. Jenkins © Humana Press Inc., Totowa, NJ

in place to cope with the numbers of supernatants to be assessed for MAb content. It is advisable to set up a timetable for immunizations, growth of the myeloma cells, final boost, and fusion date especially if more than one spleen are to be fused. It takes about 7 days before the supernatant is assayed, although it may be as long as 20 days, with the slow-starting colonies being possibly the most rewarding.

The fusion of a spleen cell and myeloma cell results in a hybridoma cell which now has a complete set of enzymes to enable it to survive in selective medium containing hypoxanthine, aminopterin and thymidine (HAT). Aminopterin blocks the biosynthetic pathway for guanosine and unless the cells possess the enzyme hypoxanthine guanine phosphoribosyltransferase (HGPRT), which offers an alternative salvage pathway using hypoxanthine, the cells will die. Spleen cells do not survive for long in tissue culture under these conditions but they do possess HGPRT. Thus, only the spleen cell–myeloma fusions will be able to grow in HAT medium.

1.1. Choice of Fusion Partner and Lymphocyte Donor

The choice of lymphocyte donor and the fusion partner must be considered carefully to reduce the risk of incompatibility, especially if it is necessary to grow the cells in vivo. The source of the lymphocytes is usually the spleen, although lymph nodes may be used, in which case, the numbers of cells available will be reduced and the difficulty in isolating the lymphoid tissue aseptically will be increased. Cells from the rat can be used, especially if anti-mouse MAbs are required, but the usual donor of cells is the Balb/c mouse. In certain instances, other strains of mouse will be necessary, but, historically, the Balb/c mouse was used because the hybridoma cells could then be grown in inbred Balb/c mice and the MAb harvested as ascitic fluid (*see* **Subheading 11.2.**). This method of production may not be acceptable ethically and the protocol may be severely restricted in some countries and illegal in others. However, growth of cells intraperitoneally may be necessary on a very small scale to salvage an important clone if it is in danger of being lost due to infection or difficulty in bulking up the cells in vitro in the early stages of securing the clone. Balb/c mice are usually the least expensive and most available strain because of their rapid breeding time and acknowledged use in MAb production. Animals must be obtained from reputable breeders. It is not unknown for inbred mice not to be as inbred as claimed. This will result in rejection of the hybrid cells due to histocompatibility differences if they need to be grown in vivo.

1.2. Choice of Fusion Partner and Myeloma Cells

The fusion partner must be immortal and compatible with the strain of animal from which the lymphocytes are taken or there will be a graft rejection reaction to the hybrid cells should they need to be grown in vivo. The original fusion partners (NS-1) produced their own immunoglobulins either as whole molecules or free light chains. This resulted in the secretion of a mixture of hybrid molecules such as lymphocyte donor heavy chain with myeloma cell light chain, although these hybrid immunoglobulins performed in assays very satisfactorily; after all, the clones had been selected based on the ability of their product to perform well. Currently, there are many mouse myeloma cell lines of Balb/c origin suitable for fusion that do not synthesize immunoglobulins or their fragments. A necessity for these cells is that they die in selective media containing HAT.

Recommended conditions are as follows:

NS0 mouse myelomas which do not express IgG (a subclone of NS-1 which is a nonsecreting variant of P3X63Ag8), available from
European Collection of Animal Cell Cultures (ECACC)
Department of Cell Resources
Centre for Applied Microbiology and Research
Porton Down,
Salisbury, SP4 0JG, UK

or Sp2/0 - Ag14 mouse nonsecreting myeloma (does not synthesize IgG or Kappa light chains), available from ECACC or
American Type Culture Collection
12301 Parklawn Drive,
Rockville, MD 20852, USA

If tempted to obtain cells from other sources, especially "in house," then it is strongly recommended that the cells be tested to prove that they are free from Mycoplasma contamination (*see* **Subheading 12.**).

2. Immunization

Contrary to common belief, it is not always necessary to prepare the antigen with adjuvant for injection. However, it is necessary for any adjuvant used to be obtained from a reputable supplier. Even then there may be batch variation which could influence the production of antibody and the isotype. Also, the adjuvant may have quite severe effects at the site of injection, which are unacceptable and unnecessary. Freund's Complete Adjuvant (FCA) can be obtained from a number of suppliers but there can be great variation in quality. FCA is composed of a suspension of *Mycobacteria tuberculosis* in a purified oil. Not all oils are suitable and injection may lead to necrosis of the surrounding tissues. In some products, *M. tuberculosis* may be substituted with other species of Mycobacteria and these may not be effective at inducing a good antibody response with the minimum of distress to the animal. Freund's Incomplete Adjuvant (FIA) does not contain the bacterial component. The selection of the correct adjuvant, its dose, and application are crucial. For detailed information on the use and choice of adjuvants, *see* **refs. *6–8***.

It is possible to carry out in vitro immunization, but this does not eliminate the use of animals (required for spleen and thymus cells), although it may be of value if the antigen is in extremely short supply ***(9)***. The MAbs produced tend to be of the IgM class. This may not be desirable, as they may be unstable in long-term storage. They tend to be of low affinity and it can be difficult to purify them from large volumes of supernatant by euglobulin precipitation and further derivitization with labels such as peroxidase may interfere with their performance in assays.

The commonest and most effective method of producing antibodies is by immunization of the lymphocyte donor with the antigen emulsified in FCA. However, whole cells act as adjuvants themselves and thus, in this case, an additional adjuvant may not be necessary. Precipitation of the antigen may be effective, as this will contribute to the retention and slow release of the antigen from the site of inoculation. Increasing the amount of whole antigen, because the specific epitope is at a very low concentration, or ultrapurification may not be successful in producing antibodies. This may induce tolerance or immune paralysis, especially if the specific epitope is poorly immunogenic but the whole antigen is highly immunogenic. However, it may be beneficial to conjugate the antigen of interest to a more immunogenic protein such as IgGFc, BSA (bovine serum albumin), tetanus toxoid, or KLH (keyhole limpet haemocyanin).

Note that in most countries, immunization of live animals is a regulated procedure (i.e., it is controlled by government legislation). In the United Kingdom, for example, the appropriate licenses, both personal and project, must be obtained from the Home Office before any animal work can commence.

The three alternatives are immunization with antigen emulsified in adjuvant, precipitates of antigen, or cells either whole or membrane preparations.

2.1. Using Freund's Adjuvant

2.1.1. Materials

1. Freund's Complete Adjuvant (FCA) (Gibco BRL Life Technologies Ltd, Paisley, UK).
2. Freund's Incomplete Adjuvant (FIA) (Gibco BRL Life Technologies Ltd).
3. 1- or 2-mL glass syringe with luer lock (*see* **Note 1**).
4. Luer lock disposable needles: 21G (0.8 × 40-mm) and 25G (0.5 × 16-mm)
5. Immunogen diluted in phosphate buffered saline (PBS) (*see* **Note 2**).
6. Disposable gloves and safety goggles (*see* **Note 3**).
7. Young adult mice of 6–8 weeks old, preferably female (*see* **Note 4**).

2.1.2. Method

1. Calculate how much emulsion is needed: based on 0.2 mL per animal and allow for loss during preparation. The emulsion is made up of equal parts of adjuvant and antigen.
2. Warm the vial of FCA to 37°C so that the oil becomes less viscous. Shake vigorously to disperse the bacteria.
3. Dilute the immunogen in PBS to between 10 and 500 mg/mL and dispense the required volume in a small glass bottle, e.g., bijou.
4. With the glass syringe and 21-gage needle, remove an equal volume of adjuvant and add it to the diluted antigen.
5. Using the same syringe and needle, withdraw the aqueous phase first (beneath the oil) and expel into the oil. Mix to give a thick water-in-oil emulsion by withdrawing and expelling the mixture rapidly (*see* **Note 5**).
6. Withdraw the emulsion into the syringe and, with care, change to a 25-gage needle.
7. Shave an area of approx 2 cm^2 on the back of each mouse towards the tail.
8. Expel a small amount of emulsion from the needle and wipe clear. Inject no more than 0.1 mL subcutaneously (sc) into two sites. Before withdrawing the needle, wait at least 5 s. for the pressure to dissipate and thus minimise seepage from the wound. Wipe away any seepage. The emulsion should be visible beneath the skin as a discrete globule.
9. Leave the animals for approx 4 weeks before "boosting."
10. Booster injections are usually given in FIA (prepare as in **steps 1–6** above). Injections are subcutaneous with a maximum of 0.2 mL per mouse (*see* **Note 6**). Alternatively, the booster may be the antigen only (1–50 mg) in PBS, and up to 1 mL can be injected intraperitoneally (ip) (*see* **Note 7**).
11. A series of booster injections may be given with an interval of 3–4 weeks in between. It may be advantageous to take a sample of blood (one drop will suffice) from the tail and perform a preliminary assay to acertain the production of polyclonal antibodies to the antigen under investigation (*see* **Note 8**).
12. When the antibody is deemed to be at a good titre (*see* **Note 9**), the final booster injection is given 60–70 h before the fusion, preferably ip in PBS without adjuvant.

Notes

1. Difficulties may arise during mixing if using plastic syringes. Glass is preferred because there is less likelihood of the plunger jamming in the barrel. High pressures are needed to mix the emulsion; thus, the needle must be fixed with a luer lock or it may be forced off.
2. The immunogen should nor contain any preservative (e.g., azide), detergents, or organic solvents.
3. Safety is important: Wear gloves, eye protection mask and coverall. Accidental injection or spraying of the emulsion can occur occasionally because of the high pressures involved. Hypersensitivity reactions to Mycobacteria can be severe. If an accident happens, seek help immediately and/or render the area safe.

4. The age of the mice is important especially if the immunization regime is likely to be lengthy. Old mice do not give good responses. Male mice may fight if kept together for some time and the resultant stress affects the immune response. Mice kept singly are also under stress; therefore keep young female mice in groups and identify individuals by tail marking at the base of the tail with a nontoxic marker pen.
5. As the emulsion thickens, it becomes more difficult to expel. To test if the emulsion is stable, allow two drops to fall on the suface of water. The first drop may disperse, but the second should remain as a globule.
6. Booster injections must not contain Freund's Complete Adjuvant, as the animals are likely to be hypersensitive to the Mycobacteria and develope severe delayed hypersensitivity lesions.
7. Have the inoculum warm so that the animals do not suffer from a temperature shock.
8. If the production of polyclonal antibodies cannot be detected, it may not be worthwhile to continue boosting. It could be more advantageous to begin again with new animals, use another preparation of antigen, or try a different concentration, or even consider another mode of immunization.
9. It should be possible to dilute the test serum to at least 1/100 and still demonstrate specific binding to antigen in the appropriate assay.

2.2. Alum Precipitates of Antigen, With or Without B. pertussis

Alum-precipitated antigen without *Bordetella pertussis* may be as effective as that with *B. pertussis*. However, with *B. pertussis*, there is an increased primary response (which is of little consequence when boosting several times) with perhaps the isotypes of the antibodies being in a different proportion. This may be important when wanting to raise MAbs with specific properties.

2.2.1. Materials

1. Sterile antigen dissolved in deionized water at 1 mg/mL.
2. Sterile 9% w/v aluminum potassium sulfate (alum) in deionized water.
3. Sterile phosphate-buffered saline pH 7.4 (PBS).
4. Heat killed *B. pertussis* (Lederle Labs., Gosport, UK).
5. 10 *M* NaOH
6. Female mice 6–8 weeeks old (*see* **Note 4** above).

2.2.2. Method

1. Add an equal volume of alum to an aliquot of antigen (e.g., 1 mg protein in 1 mL water plus 1 mL of alum solution).
2. Adjust pH to 6.5 by adding approx 50 µL of 10 *M* NaOH.
3. Leave 30 min to allow maximum amount of precipitate to form.
4. Centrifuge at 2000 rpm for 5 min.
5. Wash twice with PBS. The pH should be 7.4 after washing.
6. Resuspend to desired concentration (e.g., in 4 mL PBS to give 0.25 mg antigen per mL and add *B. pertussis*. One dose should contain approximately 0.2 mL

with 50 mg of antigen and 10^9 *B. pertussis* cells. Inject subcutaneously (sc) or intraperitoneally (ip).
7. Booster injections of 1–50 mg of antigen can be given ip at 3–4 week intervals. If necessary, acertain the production of polyclonal antibodies by taking one drop of blood from the tail. Test in the assay which will be used for assessing MAb production in the supernatants.
8. A final booster injection is given ip in PBS, 60–70 h prior to the fusion.

2.3. Cells as Antigen

Whole cells (alive or dead or cell membranes) do not require an adjuvant usually and thus can be injected ip. The cells should be washed and suspended in 1 mL PBS (warm). The cells should be washed free of growth medium or fixatives, as they will prove antigenic or toxic. Phenol red is toxic and results in liver damage and jaundice. A dose of between 10^6 and 10^8 cells should suffice for primary immunization and boosters. Boost every 4 weeks with the cells; and boost 60–70 h before the fusion.

3. Media

The tissue culture medium is usually RPMI 1640 containing antibiotics [penicillin at 200 U/mL, streptomycin at 100 mg/mL and gentamycin at 5 mg/mL (optional)] and heat inactivated (56°C for 56 min) fetal bovine serum (HIFBS). Media and the additives are obtainable from many suppliers. The purity of the water is important: Use deionized water. Glutamine is unstable in solution and its concentration may vary in commercially prepared media ***(10)***; thus, it is advisable to add it from frozen stock.

Some batches of fetal bovine serum (FBS) do not support cell growth well and may even be toxic. Test several batches from different suppliers for the ability to support cell growth. If possible, take two representative clones, one of which must be the myeloma fusion partner. Test by setting up duplicates in 2 mL of RPMI 1640 with glutamine and 10% HIFBS at a seeding density of 5×10^3 cells/mL. Record cell numbers and their viability every 48 h by resuspending the cells and taking 200 µL for Trypan Blue assay (*see* **Subheading 4.**). Choose the serum that supports the highest number of viable cells for the longest period, which may not be that which gives the steepest log-growth phase. This may be important for maximum production levels. For many cell lines, optimum MAb production occurs as the cells enter the stationary phase, with little secretion during the early or midlog phases.

Store media, additives, and HIFBS in clear containers so that any contamination or precipitation can be seen clearly and keep out of light as much as possible to minimize the formation of toxic products.

All sterile procedures must be performed in a Class II laminar flow cabinet. All incubations of cells are at 37°C in an incubator with high humidity and gassed with 5% CO_2 in air.

3.1. Materials

1. RPMI 1640 without glutamine (Gibco BRL Life Technologies Ltd., Paisley, Scotland).
2. Glutamine stock solution (100×): 150 mg in 5 mL dist. H_2O, filter sterilised (0.2-mm filter) and stored in 5 mL aliquots at –20°C. It will require warming to redissolve (Gibco BRL Life Technologies Ltd., Paisley, UK or Sigma–Aldrich Company Ltd., Poole, UK).
3. Antibiotics: penicillin 500× stock solution containing 100,000 U/mL in deionized water; streptomycin 500× stock solution containing 50 mg/mL in deionized water.These two antibiotics may be combined in one stock solution and stored in 1 mL ampoules at – 20°C. They are unstable in solution if unfrozen. gentamycin 1000 x stock solution containing 5 mg/mL in deionised water (optional) (Gibco BRL Life Technologies Ltd., Paisley, UK or Sigma–Aldrich Company Ltd., Poole, UK).
4. HIFBS aliquoted into 50 mL volumes and stored at 4°C and the excess stored at –20°C.

3.2. Method

1. To 500 mL of RPMI 1640 single strength, add 1 mL stock penicillin/streptomycin and 100 µL stock gentamycin (if used). Divide into separate batches of 100 mL to minimize cross-contamination. Use different batches for separate fusions.
2. Add the appropriate volume of heat-inactivated FBS: 22 mL to make 20% v/v for the fusion; 11 mL to make 10% v/v for growing the myeloma partner cells prior to the fusion; 5 mL to make 5% v/v for maintaining the myeloma partner cells.
3. Aliquot 5 mL of the medium and incubate for at least 48 h at 37°C as a sterility control to check for the absence of microbial growth and precipitate. Store the remainder at 4°C .
4. Add 1 mL of stock glutamine per 100 mL medium immediately before use.

4. Trypan Blue Exclusion Test for Viability

This is a quick and easy assay to give some indication of cell viability. It relies on the property of cell membranes to be semipermeable when viable but become permeable to Trypan Blue when dead. This property is not absolute and some cells which are no longer capable of division may still exclude the dye. Upon experience, it can be seen that these cells are not as "bright and shiny" as truly viable cells and appear dull on adjusting the focus of the microscope.

4.1. Materials

1. Sterile trypan blue: 0.4% w/v in saline (Gibco BRL Life Technologies Ltd., Paisley, Scotland). Aliquot into 5-mL volumes as required. Store at room temperature out of direct sunlight.
2. Hemocytometer and microscope.
3. Clean, but not necessarily sterile, 96-well plate.

4. Pipeter and sterile tips for 25-µL volumes.
5. Pipeter and sterile tips or Pasteur pipets for 200-µL volumes.

4.2. Method

1. Take approx 200 µL of the cell suspension and place in a well of the 96-well plate. Sterility of this sample is now not important.
2. Take 25 µL of Trypan Blue asceptically and add to the cells. Contamination of the Trypan Blue may give rise to impressions of contamination in the cultures.
3. Mix and pipet a small volume under the cover slip of the hemocytometer.
4. Check for contamination, count at least 200 cells alive and dead, and calculate the percentage viability.

5. Myeloma Cells Fusion Partner

Cells should be shown to be free of Mycoplasma (*see* **Subheading 12.**). They should be maintained in 5% v/v FBS/RPMI at greater than 98% viability. At least a week before the fusion, cells should be grown in the presence of 2×10^{-5} *M* thioguanine to eliminate any revertant cells containing normal levels of the enzyme hypoxanthine guanine phosphoribosyltransferase (HGPRT).

5.1. Materials

1. Thioguanine 100× stock solution of 2 m*M*. 33.44 mg (anhydrous) thioguanine is dissolved in 100 mL deionised water. 1 *N* NaOH is added as necessary, to dissolve the thioguanine. The pH is adjusted to 9.5 with acetic acid. Sterilise by filtration (0.2-mm filter), aliquot in 5- or 10-mL volumes and store at –20°C.
2. Sterile conical centrifuge tubes: 20 mL and 50 mL.

5.2. Prepreparation of the Cells

1. Add sterile thioguanine stock (100×) to 10% v/v FBS/RPMI at 1% v/v just before seeding the cells.
2. After about 2 ds, perform a trypan blue exclusion test (*see* **Subheading 4.**) to acertain cell viability.
3. If the viability is greater than 96%, then continue to expand the cells in medium containing thioguanine,until there are at least 10^7 for each fusion.
4. If there is significant cell death, then culture the cells in medium containing thioguanine until 98% viability is achieved before expanding them in readiness for the fusion. If there are not enough cells, delay giving the pre-fusion booster injection.

5.3. Harvesting the Cells for the Fusion

1. Acertain the viability of the cells in each flask by trypan blue exclusion (*see* **Subheading 4.**).
2. Centrifuge the cells at 1800 rpm on a bench centrifuge for 5 min. Pool the cell pellets and resuspend in 20 mL of 10% v/v FBS/RPMI without thioguanine. Keep at room temperature within the cabinet and count the cells.

6. Spleen Cells

6.1. Materials for Dissecting the Spleen Aseptically

1. Prepare three sterile sets of dissecting scissors and forceps, each in a separate container or bag.
2. Approximately 50 mL alcohol or alcohol spray.
3. 5 mL sterile RPMI in 20-mL sterile universal container.

6.2. Materials for Preparing Spleen Cell Suspension

1. Prepare 2 × 10-mL sterile syringes with 25G (0.5 × 16-mm) needles and each containing 10 mL RPMI without serum. Replace the needle shields. Keep at 37°C
2. Sterile Petri dish: approx 90-mm diameter.
3. Trypan Blue viability test.
4. Sterile conical centrifuge tube, 20-mL.
5. Sterile Pasteur pipets.

6.3. Method

1. Cull the animal and flood the skin with alcohol after laying it out on its back.
2. Aseptically section the skin and peel away using the first set of scissors and forceps.
3. Using the second set of scissors and forceps, cut and peel away the abdominal muscle layer to reveal the spleen.
4. With the third set of sterile instruments, remove the spleen, keeping it whole if possible. Place it in the 5 mL of sterile RPMI.

Proceed by performing the following procedures aseptically in a sterile cabinet under good tissue culture practice.

5. Pour off and discard the medium in which the spleen has been placed.
6. Tip the spleen into the sterile Petri dish.
7. Puncture the spleen several times with the needle of one of the filled syringes.
8. Skewer one end of the spleen with one needle/syringe and gently inject the medium of the other syringe. The punctured spleen sac acts as a sieve and the cells are flushed out into the Petri dish.
9. Skewer the other end of the spleen with the needle of the now empty syringe and flush out the remaining cells with the medium of the second syringe. It will become apparent when the spleen is empty. If necessary, take up some of the cell suspension and flush out again.
10. Pipet the spleen cell suspension (20 mL) into a sterile centrifuge tube. Hold at room temperature within the Class II cabinet.
11. Sample (200 μL), count the number of lymphocytes and assess their viability by the Trypan Blue exclusion test. Ideally at least 5×10^7 viable spleen cells are required with a viability of 80%.

7. Fusion Procedure

Ideally, a ratio of 5 fusion partner myeloma cells to 1 viable spleen cell is sought. Adjust the number of myeloma partner cells according to the spleen cell yield.

7.1. Materials

1. Pre-prepare sterile 40% w/v polyethylene glycol (PEG) 1500 (Sigma, Poole, UK) in RPMI as follows: weigh 2 g PEG in a small bottle (it is solid at room temperature) and add 3 mL RPMI. Either autoclave to sterilise (it will go into solution) or warm to 37°C to dissolve and filter sterilize. This must be freshly prepared (*see* **Note 1**). The 40% PEG will be slightly alkaline which is more likely to give a successful fusion.
2. Sterile 1 mL syringe with a 25-gage (0.5 × 16-mm) needle in its guard. Fill with 0.8 mL 40% PEG/RPMI and keep in the incubator.
3. Six 96-well tissue culture plates (Sarstedt, Leicester, UK or Gibco BRL Life Technologies Ltd., Paisley, Scotland).
4. Sterile Pasteur pipets.
5. 8-channel multi-pipeter and sterile tips for 50- to 150-μL volumes.
6. Sterile media boats (aluminum pie dishes are excellent). Autoclave in bags.
7. 21 mL sterile RPMI held at 37°C but not in the CO_2 incubator so that it goes slightly alkaline.
8. 24 mL 20% HIFBS/RPMI held at 37°C.
9. Stock solution (×1000) of 2-mercaptoethanol (2 ME) at 50 m*M* (Gibco BRL Life Technologies Ltd., Paisley, Scotland).
10. Stock hypoxanthine and thymidine (100×): Hypoxanthine is prepared from 408 mg in 100 mL distilled water on a stirrer. Add 1 *M* NaOH slowly until dissolved. Thymidine is prepared by dissolving 114 mg in 100 mL distilled water. These two solutions are combined and made up to 300 mL with distilled water. The pH is adjusted to 10 with 1 *M* HCl. Filter sterilize and aliquot into 5 mL and store at –20°C (*see* **Note 2**).
11. Methotrexate can be used to replace aminopterin (Sigma-Aldrich Company Ltd., Poole, UK) A 100× stock solution in distilled water is prepared at 5×10^{-5} *M*. Adjust the pH to 7.5 and filter sterilize. Aliquot in 5-mL volumes and store at –20°C.
12. Minute timer.
13. 20% HIFBS/RPMI.
14. Complete HAT/RPMI: 20% HIFBS/RPMI containing H, A, and T and 2 ME.
15. Aspirating pipet: heat 2 cm from the end of a Pasteur pipet. When the glass has softened, pull and bend at 90°, with forceps. Allow to cool. Cut off to give a bent end of approximately 0.5 cm. Check that the lumen is open. Store sterile.

Notes

1. There is great variability in the effectiveness/toxicity of PEG. If fusions fail repeatedly, then try a different batch/supplier of PEG.
2. Hypoxanthine, aminopterin, and thymidine can be purchased as stock solutions as HAT or HT from Gibco BRL Technologies Ltd., Paisley, Scotland. Medium already containing HAT or HT can be purchased from the same company but it is better to have the flexibility of adding the selecting reagents as required and using fresh media for optimum growth and selection conditions.

7.2. Method (Fig. 1)

1. Spin both types of cells separately for 5 min at 2,000 rpm in a bench centrifuge.
2. Discard the supernatant carefully by aspiration.
3. Resuspend each pellet in 10 mL RPMI (without serum) at room temperature.
4. Mix the cells by pipeting together in one of the centrifuge tubes.
5. Spin the cells at 1850*g* for 7 min.
6. Aspirate the supernatant.
7. Resuspend the cells by flicking the base of the tube.
8. Warm the pellet by holding the tube in the hand or 37°C water bath for 1 min.
9. Add 0.8 mL 40% PEG quickly using the syringe. Rotate the tube so that the cells coat the wall and keep warm by holding in the hands or at 37°C in the water bath for 1 min.
9. Add 1 mL RPMI (from the 21-mL prewarmed to 37°C) dropwise from a Pasteur pipet over 1 min, while rotating very gently in the hand.
10. Add remaining warmed 20 mL RPMI over 5 min, while rotating gently to mix.
11. Spin for 15 min at 1500 rpm, the cells are very fragile at this stage.
12. Aspirate supernatant.
13. Resuspend cells in 24 mL 20% FBS in RPMI containing 2 ME (50 μ*M*).
14. Add one drop of cell suspension using a Pasteur pipet to each well of the 96-well tissue culture plates; 6–7 plates will be needed. Label and incubate for 24 h.
15. Add 50 μL complete HAT/RPMI: dispense medium into a medium boat and pipet from this with a multichannel pipeter. Incubate for 24 h.
16. Add a further 100 μL complete HAT/RPMI. Incubate for 48 h.
17. Carefully aspirate 100 μL from each well using an aspirating pipet. Add 150 mL fresh complete HAT/RPMI.
18. Incubate and observe daily for (a) medium turning acid (yellow) and(b) score on the plate lid for single clones.

8. Screening Assays

8.1. Specificity Assays

If possible, it is necessary to have a fast, efficient assay which can cope with large numbers of supernatants requiring screening. In the first instance, there will be only 50–100 μL of supernatant available. The titer of the antibody is not important at this stage; what is to be determined is whether any antibody is being produced by the clones observed and which are of the desired specificity. Either an enzyme-linked immunosorbent assay (ELISA) or hemagglutination assay (HA) or immunohistology using anti-mouse antibody can be useful in the first instance. Care must be taken to check that the antimouse antibody recognizes all classes of mouse antibody. A one-well assay can be carried out for each well of the fusion plate with the assay plate mimicking the culture layout.

8.1.1. Materials

1. Multichannel pipeter with sterile tips for 100mL.
2. Complete HAT/RPMI and complete HT/RPMI.

3. Sterile media boats.
4. Prepared ELISA plates or 96-well plates which need not be sterile.
5. 24-well tissue culture plates (Gibco BRL Life Technologies Ltd., Paisley, Scotland).

8.1.2. Method

1. Remove 100 µL from each well and place in similar well of an empty 96-well plate and use for assaying or place directly in a prepared ELISA plate (*see* **Note 1**).
2. Add fresh complete HAT/RPMI to the culture wells and re-incubate.
3. Perform the assay with the supernatants. Wells that give positive supernatants should be highlighted by marking the lid (*see* **Note 2**).
4. When the supernatant turns acidic (yellow) within 24 h, transfer the cells to a well of a 24-well plate into 1mL of complete HT medium.
5. After 1–3 d, add a further 1 mL of medium and leave for the medium to begin to turn acidic.
6. Freeze down most of the cells in liquid nitrogen (*see* **Subheading 10.**) and label as uncloned. Use the supernatant for further assays.
7. Grow he cells for a further 24 h before recloning (*see* **Subheading 9.**).
8. Continue assaying the original plate for slow-growing clones.

Notes

1. Use a new sterile tip for each well, as there may be carryover of MAb, which will give a false positive in the next well.
2. The assay should be repeated at least once with fresh supernatant to verify each result. Subsequent samples of supernatant can be tested for titer, which gives some indication of the productivity of the clone.

8.2. Determination of Isotype

This may be necessary when determining the method of purification or for certain assays especially when using labeled antibodies, the feasability of further derivitization, and defining the hybridoma as completely as possible. It can be done by ELISA or precipitation in gel using polyclonal antisera specific for the different classes of mouse immunoglobulin.

8.2.1. Materials

1. Sheep polyclonal antisera specific for each class/subclass of mouse antibody: anti-IgG1, anti-IgG2a, anti-IgG2b, anti-IgG3, anti-IgM, and anti-IgA (The Binding Site, Birmingham, UK). The Binding Site also sells monoclonal antibody-typing kits to establish the isotypes of MAbs being produced by hybridoma cells.
2. Clean grease-free glass slides or plates (*see* **Note 1**).
3. Leveling board.
4. 1.4% agar (ICN Biomedicals Ltd., Thame, Oxon., UK) in barbitone buffer pH 8.6 (High resolution buffer: Pall Gelman Scinces Ltd., Northampton, UK) containing 6% w/v PEG 6000 (Sigma–Aldrich Company Ltd., Poole, UK) (*see* **Note 2**).

Remove spleen asceptically.

⇓

Perfuse out cells with 2 × 10 mL RPMI (no serum) with syringes (25-gage needles)
Need at least 5×10^7 live spleen cells.

⇓

Harvest and count NS0 cells (or 653)
Need 10^7 live cells.

⇓

Spin each cell type separately at 2000*g*.

⇓

Discard supernatants by aspiration with a sterile Pasteur pipet.

⇓

Resuspend each cell pellet in 10 mL RPMI (no serum) at room temperature.

⇓

Mix cells in one tube.

⇓

Spin cells for 7 min at 1850*g*

⇓

Discard supernatant by aspiration.

⇓

Resuspend cells.

⇓

Warm cells at 37°C.

⇓

Add 0.8 mL 40% PEG 1500 in RPMI prewarmed to 37°C.

⇓

Keep at 37°C for 1 min.

⇓

Add 1 mL RPMI (prewarmed to 37°C) over 1 min.

⇓

Add 20 mL RPMI (prewarmed to 37°C) over 5 min.

⇓

Spin for 15 min at 1500 rpm.

⇓

Discard supernatant. Resuspend cells in 25 mL RPMI containing 20% HIFBS + 2 ME

⇓

Add one drop suspension to each well of 96-well tissue culture plate—requires 6–7 plates.

⇓

Label and incubate plates at 37°C in 5% CO_2 for 24 h.

⇓

Fig. 1. Flow diagram for fusion protocol.

Add 50 µL complete HAT medium to each well and incubate for a further 24 h.

⇓

Add 100 µL complete HAT medium and leave a further 24 h.

⇓

Change the medium every 48 h by aspiration of the old medium and adding 200 µL fresh media to each well until clones are obvious.

⇓

Score for single clones.

⇓

Assay supernatants for antibody production.

⇓

Isolate and grow on cells from those wells which have a single clone and produce antibody of the required specificity.

Fig. 1 *(continued)*. Flow diagram for fusion protocol.

Boil together in a water bath until dissolved. Aliquot into 10 mL in glass containers and cap tightly. Store at 4°C (*see* **Note 3**).

5. Humidity box (sandwich box with damp tissue in the bottom).
6. Graduated pipets.
7. Hole punches 2-mm and 8-mm (size 4 bung borer) in diameter.
8. Needle.
9. Boiling water bath.

8.2.2. Method

1. Melt an aliquot of agar/PEG in a boiling water bath.
2. Warm the levelling board (place under running hot water) and align the glass slides on it.
3. Warm the graduated pipette in the hot agar before dispensing agar to cover the glass slide. Use the pipette to guide the agar almost to the edges.Allow to set and store in a humidity box at 4°C.
4. Punch holes in the agar in the following pattern: the central well is 8 mm diam., surround with six holes of 3-mm in diameter placed equidistant on the circumference of an imaginary circle of 2 cm in diameter.
5. With the needle lift out the agar from the punched holes.
6. Fill the central well with 50 µL supernatant containing the MAb under test.
7. Fill the 3-mm wells with 5 µL of each of the 6 anti mouse isotype antibodies.
8. Incubate in the level humidity box at room temperature or 4°C until lines of precipitation can be seen between the 3-mm well and 8-mm well.
9. In any one pattern, there should be only one line indicating the isotype of the MAb.
10. It is desirable to set up controls with MAbs of known isotype (*see* **Note 4**).

Notes

1. A plate 8 cm × 8 cm will hold 10 mL of agar. Slides 2.5 cm × 7.5 cm will take 3 mL.
2. Both agarose and PEG can result in cloudy gels after cooling and setting; thus, the quality must be as high as possible. If the clarity is not satisfactory, change the batch of either or both reagents.

3. These gels will give cloudy preparations after storage of a few weeks.
4. MAb sourced from ascitic fluid cannot be used for controls as the mouse's own immunoglobulins will give positive reactions.

8.2.3. ELISA Determination of Isotype

8.2.3.1. Materials

1. ELISA plates or strips. If appropriate, the ELISA plate can mimic the tissue culture plate from which the supernatants are being taken.
2. Sheep polyclonal antisera specific for mouse isotypes conjugated to HRP (The Binding Site, Birmingham, UK).
3. Appropriate antigen.
4. Supernatant containing MAb and controls of known isotype (*see* **Note 4** above).
5. Pipettors with tips for 100 μL.
6. Coating buffer: 0.05 *M* carbonate buffer pH 9.6.
7. Washing/diluting/blocking buffer: PBS containing 0.05% Tween 20.
8. Substrate buffer: 0.15 *M* citrate/phosphate buffer pH 5.0. Store at 4°C and check for growth of organisms especially fungus.
9. Substrate: 35 mg o-PD in 100 mL citrate/phosphate buffer plus 35 μL hydrogen peroxide. Make up fresh.
10. Stopping solution: 12.5% v/v sulfuric acid (H_2SO_4).

8.2.3.2. Method

1. Fill each well with 100 μL of antigen at 5 μg/mL in coating buffer. For each supernatant to be tested, six wells will be required; one for each anti-mouse isotype (*see* **Note 1**).
2. Incubate at 37°C for 2 h in a humidity box or at 4°C overnight.
3. Tip out the excess antigen and wash each well three times with the washing buffer.
4. Fill each well with the washing buffer and leave for 10 mins for the unused sites to be blocked.
5. Tip out the washing/blocking buffer and bang out as much liquid as possible onto absorbant tissue.
6. Add 100 μL of supernatant to each of 6 wells. Include 6 control wells with 100 μL diluting buffer only. Incubate for 1 h at 37°C in a humidity box.
7. Tip out the excess supernatant and wash three times with the washing buffer. Bang out excess onto absorbant tissue.
8. To each well, add 100 mL sheep antiserum conjugate, (diluted 1/5,000 in washing/diluting buffer) in such a manner that each supernatant is tested with each anti-isotype. Incubate at 37°C for 1 h in a humidity box.
9. Wash six times with the washing buffer and bang dry.
10. Develope by adding 100 μL of the substrate and stop the reaction with 50 μL sulfuric acid after 5 min or when the color has changed to bright yellow. Negative controls should remain colorless and positive controls should be specific.
11. Monoclonal supernatants should be positive with only one antiserum, indicating the isotype.

Notes

1. It can be costly testing for each isotype. Assay for IgG1 first, as most clones will be of this isotype, especially if the immunization protocol has been lengthy with a protein antigen. Those that are not IgG1 test in the following order: IgM, Ig2a, Ig3, Ig2b, IgA. Note that this method will not reveal wells with more than one clone of different isotypes.

9. Cloning

9.1. Macrophage Monolayers

In some instances, single cells may only grow with the help of factors produced by other cells such as macrophages. In this case, a monolayer of macrophages is cultured for 24–48 h before adding the cells. The medium is not changed, as it will contain growth-promoting factors.

9.1.1. Materials

1. Genetically compatible mice (there may not be enough cells from one mouse).
2. Autoclaved 1% suspension of starch in saline.
3. Sterile 1-mL syringe with 25-gage (0.5 × 16-mm) needle.
4. RPMI without serum
5. Sterile syringe (10 mL) with 21-gage (0.8 × 40-mm) needle.
6. Sterile 20-mL container.
7. 96-well tissue culture plates (Gibco BRL Life Technologies Ltd., Paisley, Scotland).
8. RPMI with 10% HIFBS.
9. Ice bath.

9.1.2. Method

1. Inject the mice ip with 1 mL (25-gage needle) of the starch suspension 48 h before harvesting the macrophages.
2. Cull the mouse, flood with alcohol and immediately inject ip with 10 μL RPMI but do not remove the syringe and needle (21-gage). Gently knead the peritoneum and slowly withdraw the peritoneal fluid. Dispel the cell suspension into the sterile 20-mL container and keep in the ice bath. Alternatively, after injecting and kneading, remove the syringe and allow the peritoneal fluid to drip from the needle (21-gage) into the 20 μL sterile container. Keep the cells cold so that the macrophages do not adhere to the internal surfaces of the container.
3. Spin the cells, resuspend in cold 10% HIFBS/RPMI, count the macrophages and adjust the cell number to 10^4 cells/mL. Seed 96-well tissue culture plates with 100 μL and incubate for 24–48 h before use.
4. Cells to be cloned should be suspended in 10% FBS/RPMI to give one cell in one drop from a Pasteur pipet.
5. Add one drop hybridoma cells to each well of macrophages. Incubate and monitor until the medium begins to turn yellow.

6. Score for single clones and assay the supernatant for specific antibody production. It is possible to detect nonspecific antibody production in this technique; thus, assays to detect antibody production with polyclonal antiserum should be reviewed carefully.
7. When the clones are well established, transfer them to a 24-well plate without macrophages. After changing the supernatants twice, they can be assayed for product and the cells can be frozen.

9.2. Soft Agar Method

The aim is to prepare a basal layer of 5% agar diluted 1/10 in complete medium and to add cells in a top layer of 0.3% agar. The cells should be dispersed but fixed in the agar. Clones will grow so that they can be seen with the naked eye or eyepiece and picked off.

9.2.1. Materials

1. Water bath at 56°C.
2. 3% w/v and 5% w/v agar (ICN Biomedicals Ltd., Thame, Oxon., UK) (*see* **Note 1**) in saline heated to 100°C in a water bath to dissolve. Keep at 56°C.
3. Small Petri dishes approx 5cm in diameter; calculate the number required for plating out each cell suspension in duplicate and at three dilutions.
4. Warm graduated pipets or glass ones that can be heated in a flame as required.
5. Complete HAT/RPMI kept at 56°C (*see* **Subheading 7.1.**).
6. Sterile tubes 1 cm × 10 cm with loose caps kept at 37°C.

9.2.2. Method

1. Calculate how much agar is required and make up the appropriate volume by diluting the 3% and 5% agar in complete HAT/RPMI. Keep at 56°C.
2. Dispense 5 mL into dishes using a warmed pipette and leave to set in the cabinet.
3. Aliquot into warm sterile tubes, 5 mL 0.3% agar in complete HAT/RPMI. Keep at 56°C.
4. Take 6 of these tubes and put in the 37°C incubator. Time is now important as the agar will set at about 45°C.
5. Take the clone of cells to be recloned into 1 ml complete HAT/RPMI at 37°C.
6. Add to two tubes of 0.3% agar , 50 μL cell suspension, mix quickly by rolling between the hands and pour onto each of two plates containing 0.5% agar. Leave to set. Repeat with 2 × 100 μL and 2 × 300 μL of cell suspension.
7. Incubate at 37°C and monitor daily.
8. When individual clones can be seen, pick out and grow in 96-well tissue culture plates in 150 μL complete HAT/RPMI.
9. Monitor daily, and when the medium begins to turn acid, assay for antibody production.

Notes

1. Some batches of agar do not support growth well. The agar must be of the best quality available. If problems are still experienced, change the batch/supplier.

9.3. Limiting Dilutions

This technique can be used to (1) enrich a cell population for antibody producing cells or (2) obtain clonality of an antibody cell line.

9.3.1. Enrichment

Antibody-producing cells will probably grow more slowly than nonproducing cells. Thus, in a mixed population, the productive cells will soon be overgrown. To isolate the producing cells, it may be necessary to first enrich a population by a series of dilution/growing up procedures whereby the highest producing wells, as indicated by the titer compared with cell number, are used for second or third round of dilution/growing up/assessment of production procedures.

9.3.1.1. Materials

1. Complete HAT/RPMI (*see* **Subheading 7.1.**).
2. Five 96-well plates (Gibco BRL Life Technologies Ltd., Paisley, Scotland).
3. Sterile medium boats (aluminum pie dishes).
4. Eight-channel multipipeter with sterile tips to take 100 µL.
5. Sterile Pasteur pipets.

9.3.1.2. Method

1. Suspend cells in 1 mL complete HAT and assay viability and cell number by Trypan Blue (*see* **Subheading 4.**).
2. Dilute cell suspension in complete HAT to give 3×10^4 cells/mL.
3. Mark the lid into two halves.
4. With the multipipeter, fill the wells of column 1 and column 7 with 200 µL complete HAT; the remainder fill with 100 µL complete HAT.
5. With a Pasteur pipet, add one drop of cell suspension.
6. With the multipipeter, remove 100 µL from the wells of column 1 and place them in the wells of column 2. Mix and remove 100 µL from the wells of column 2 and place in the wells of column 3. Continue doubly diluting the cells until column 6 discarding the remaining 100 µL.
7. Repeat **step 6** in the wells of columns 7–12.
8. Assay for antibody production as early as possible when cell division is obvious, probably within 48 h for some wells. Put the cells from the highest producing wells through another round of enrichment until nearly all wells give high titers before cloning out for single cells.

9.3.2. Cloning for Single Cell Derived Colonies

The procedure is the same as that for enrichment except fewer cells are needed in the first wells; as few as 5. Some cells such as J588L or Chinese Hamster Ovary (CHO) cells are more robust than NSO and thus may be easier to obtain single cell clones from fewer starting cells. NSO may require an initial dilution of at least 10 cells per well.

9.3.2.1. Materials

Materials are as in **Subheading 9.3.1.1.**

9.3.2.2. Method

1. Suspend cells in 1 mL complete HAT and assay viability and cell number by Trypan Blue.
2. Dilute cells with complete HAT to give 1000 cells/mL.
3. With multipipeter, fill wells with column 1 of a 96-well plate with 200 μL complete HAT; fill remainder with 100 mL complete HAT.
4. With Pasteur pipet add 1 drop of the cell suspension of each well of column 1.
5. Doubly dilute as in **step 6** above across the whole plate.
6. This dilution series should result in single cells within the 96 wells of each plate used. Monitor daily for wells with obvious single colonies and highlight by marking the lid.
7. Every two days, replenish the medium by aspirating the old and adding fresh complete HAT. Use a fresh tip for each well to avoid any carry over of cells, and take care not to disturb the colony.
8. Transfer the cells to 24-well plate once the colonies cover more than half the well. Grow up as **Subheading 11**. It may be necessary to trypsinise strongly adherent cells (*see* Chapter 16).

10. Storage

10.1 Cells

A liquid nitrogen storage system is required to hold cells at –196°C. Cells need to be taken down in temperature slowly and, upon access, they need to be thawed quickly. If an automated freezing down facility is not available, then slow freezing can be achieved by making a freezing box (*see* **Note 1**).

10.1.1. Materials

1. Ice bath.
2. Dimethylsulfoxide (DMSO) (Sigma–Aldrich Company, Fancy Road, Poole.UK).
3. Fetal bovine serum or 10% v/v HIFBS/RPMI.
4. Sterile cryovials: 1.8 mL Nunc brand with external thread (Gibco BRL LifeTechnologiesLtd., Paisley. Scotland).
5. Freezing down box kept at –70°C.
6. Sterile Pasteur pipets.
7. Sterile absorbant tissue.

10.1.2. Method

1. Prepare sterile 10% v/v DMSO in either FBS or 10% FBS/RPMI. Excess freeze mixture can be stored frozen in aliquots of 1–5 mL and the aliquot in use can be stored at 4°C.
2. Place an aliquot of 10% DMSO in the ice bath.

3. Harvest the cells to be frozen and place in the ice bath.
4. Spin the cells at 1800*g* for 5 min in a bench centrifuge.
5. Label the cryovials with permanent marker and record the details.
6. Aspirate the cell supernatant if not required or pour off and leave the container inverted on sterile tissue to drain.
7. Take approx 300 μL freezing mixture with a Pasteur pipet and suspend the cells in it.
8. Transfer the suspension to the cryovial and place it in the ice bath.
9. Place the cryovials in the freezing box at —70°C and leave to freeze for 2–24 h.
10. Transfer the cryovials to the liquid nitrogen banks and record their location.

Notes

1. Line the sides and base and inside of the lid of a card box (approx 10 cm × 10 cm × 10 cm) with expanded polystyrene sheet approximately 2 cms thick. Fill the centre of the box with polystyrene in which 6–8 holes have been bored which are just large enough to take a freezing down vial.

10.2. Supernatants

Supernatants can be stored frozen at –20°C or below. For short term storage, they may be stored sterile or with the addition of preservatives such as 0.1% w/v sodium azide. Be aware that azide might interfere with ELISAs. Also of the possibility of deterioration of the product from enzymes or contaminating micro-organisms in the supernatants.

11. Production

The production of MAb must be monitored frequently for specificity and titre. It is less likely for the isotype to change, but it is possible for a clone which was deemed productive in the first instance, to "switch off" or to become overgrown with nonproducing cells.

11.1. In Vitro

The method of in vitro production depends on how much material is required and how much the clone is producing. Once the clone is deemed to be from a single cell-producing antibody of the specificity desired, then the cells must be expanded with the product being continually monitored both for specificity and titer. It is possible for cells to stop producing their product, for the clone to be lost through mishap such as contamination, or for the clone not to have been from a single cell as originally thought. Thus, cells should be frozen down as soon as there are enough for this to be a viable proposition so that the cells can be reaccessed and recloned if necessary.

For initial characterization of the MAb, a few milliliters may be sufficient. For larger amounts, then the cells may have to be grown in many tissue culture flasks of larger volume. There are a number of methodologies available commercially; from flasks of various shapes and sizes with high surface area or membranes separating cells from the bulk of the medium to roller bottles to bioreactors such as stirred tanks and fermentors. For small-scale laboratory requirements, a hollow-fiber bioreactor is recommended which gives large amounts of product at a high concentration relative to the normal tissue culture supernatant. An inexpensive, efficient and successful methodology is described in **ref. *13***.

11.1.1. Materials

1. 24-well tissue culture plates (Sarstedt Ltd., Leicester, UK).
2. Tissue culture flasks: 50 mL and 250 mL (Sarstedt Ltd., Leicester UK).
3. Sterile Pasteur pipets.
4. Complete HAT/RPMI and complete HT/RPMI medium (*see* **Subheading 7.1.**).

11.1.2. Method

1. The cells are transferred from the 96-well plate to a 24-well tissue culture plate where each well holds 2 mL of medium.
2. When the medium begins to turn yellow, harvest and freeze the cells and retain the supernatant for further characterization such as determining the isotype of the product or its performance in other assays.
3. Add 2 mL of fresh complete HAT/RPMI to the remaining cells.
4. Allow the cells to grow up again before transferring them into 10 mL of complete HAT/RPMI in a small tissue culture flask holding 50 mL. When the medium begins to go acid, add a further 10 mL of complete HAT/RPMI.
5. Harvest most of the cells when the medium begins to turn acid, freeze down and retain the supernatant. Add 20 mL fresh complete HT/RPMI medium to the remaining cells.
6. Grow and freeze cells as required. Monitor the titer of the antibody: If it falls, and the cells appear healthy, then the cells may require to be recloned because a subpopulation has changed its genotype and is no longer producing antibody. These cells tend to grow more quickly and will soon dominate the population.

11.2. In Vivo

This methodology is for emergency use only, although it has been used extensively in the past. It is considered unethical to use animals for the production of MAbs as it can cause severe distress. However, it may be the only method available to "rescue" a cell line especially if the only source of cells now available are frozen down as ascitic cells. By the in vivo method, it may be possible to grow cells within the peritoneal cavity of a mouse and to harvest the cells aseptically and adapt them to tissue culture.

11.2.1. Materials

1. Pristane (2,6,10,14 tetramethylpentadecane) (Sigma-Aldrich Company Ltd., Poole, UK).

2. Glass syringe (1 mL) with 25-gage × 16-mm (orange) needle.
3. Cell suspension of 2×10^6 viable cells/mL saline.
4. Disposable 1 mL syringe with 25-gage × 16-mm needle.
5. Young adult genetically compatible mice (e.g., Balb/c) two or three animals for each clone to be "rescued."
6. Sterile 20-mL containers with 100 μL heparin.
7. Sterile 20-mL containers with 10 mL RPMI.
8. Disposable sterile 5-mL syringes with 21-gage (0.8 × 40-mm) needle.
9. Tissue culture flasks or 24-well plate (Gibco BRL Life Technologies Ltd., Paisley, Scotland).
10. Alcohol.

11.2.1. Method

1. Inject ip, 0.2 mL pristane into each mouse 7–14 d before required. Use the glass syringe to minimize jamming and orange needle to minimize seepage from the wound.
2. Inject 1 mL cell suspension ip and monitor daily.
3. When swelling of the abdomen is very evident, cull the mouse and flood with alcohol.
4. Either (a) insert the green needle ip and gently withdraw the ascitic fluid. Dispense into the heparin bottle or (b) inject 5 mL RPMI ip and gently withdraw fluid and dispense into the heparin bottle or (c) penetrate the abdominal wall with the green needle and collect the drops of fluid directly into the heparin bottle or into the RPMI medium. Harvest into more than one bottle to circumvent contamination problems. Continue under good culture practice.
5. Remove aliquots of cell suspension and place in tissue culture to grow.
6. Excess cell suspension/ascitic fluid can be centrifuged at 2000 rpm for 5 min the fluid removed and assayed and/or stored and the cells frozen down as in **Subheading 10.1.**

12. Mycoplasma Contamination

Mycoplasma spp. are ubiquitous parasites of man and animals and are implicated in a number of infections, although most people are symptom free but may be carriers. The presence of Mycoplasma organisms, of which there are a number of different species/strains, will interfere with growth and/or production of the hybridoma cells. Cells from national cell banks will be supplied as Mycoplasma-free. However, it is good practice to test all hybridomas for the presence of the micro-organisms on a regular basis, especially when bulking up for production and if culturing more than one cell line, any of which may be of human origin or have been received from other laboratories. It is not important to know the specific identity of the Mycoplasma, only whether it is present or not.

12.1. Detection

The specific antibody kits can only use the first reagent on mouse monoclonal antibody-producing hybridomas. The second reagent is a goat antimouse antibody and will bind to the hybridoma cells as well as the reagent 1 bound to the Mycoplasma. Not all strains of Mycoplasma-like organisms are detected with this method, but it is a rapid test. The Hoechst stain binds specifically to DNA. Uninfected cultures under fluorescent microscopy are seen as cell nuclei with a negative background. Infected cultures are seen with extranuclear mycoplasmal DNA surrounding the cell nuclei, in the cytoplasm, and in the surrounding areas. All infections will be detected by this method.

12.1.1. Use of Specific Antibody Kits

12.1.1.1. Materials

1. Immu-Mark Myco-Test kit from ICN Biomedicals Ltd., Thame, UK.
2. Grease free multispot microscope slides (12-spot) PTFE coated (C. A. Hendley (Essex) Ltd., Loughton, UK).
3. Humidity chamber.
4. PBS.
5. Cold ethanol (keep at –20°C).
6. Microscope with filter system for fluoroscein and ×400 to ×600 magnification.
7. Warm fan or warming block at 50°C.
8. Pasteur pipets.
9. Cover slips.

12.1.1.2. Method

1. Allow cells to grow to high density without changing the medium.
2. Onto each of two spots of a slide, place one drop of cell suspension and leave to dry, preferably under a warm fan.
3. Place the slide in the –20°C freezer and flood with the cold ethanol for 1–2 min.
4. Pour off the ethanol and leave to dry at room temperature.
5. Add one drop of the first reagent (a fluorochrome-labeled monoclonal antibody) to each spot so that the spot is just covered and leave in a humidity chamber for 30 min (the spots must not dry out). Hereon treat the control slide supplied with the kit in like manner.
6. Very gently, wash off the reagent with the PBS or gently dip into a PBS washing jar.
7. Tilt the slide and leave to dry. Excess liquid may be absorbed with a tissue, but do not touch the cell area.
8. Add one drop of the supplied mounting fluid and cover with a cover slip.
9. View as soon as possible after staining to get maximum resolution.

12.1.2. Hoechst Stain

12.1.2.1. Materials

1. Sterile Petrie dishes 6-cm in diameter; one for each culture to be investigated.
2. Pasteur pipets.
3. Forceps.
4. Hoechst 33258 stain (Sigma-Aldrich Company, Fancy Road, Poole. UK).

5. Sterile glass cover slips.
6. 20 mL Carnoy's fixative (3:1 methanol/glacial acetic acid).
7. Mounting fluid.
8. Glass microscope slides.
9. Wear disposable gloves.

12.1.2.2. Method

1. Place two or three sterile cover slips in the Petri dishes so that they do not overlap.
2. Add 10 mL cell culture to be investigated to each Petri dish so that the cover slips are covered and do not float.
3. Incubate for 3–4 h or overnight for the cells to settle on the cover slips.
4. Very carefully remove the Petri dishes to the bench so that the cells are not disturbed. Leave for a short while if necessary for the cells to settle.
5. Add 2 mL of fixative to the perimeter of the Petrie dish with a Pasteur pipette so that there is minimal disturbance to the cells. Leave for 3 min.
6. Pipet off the fixative/medium from the Petri dishes. Flush the fixative down the sink with plenty of water.
7. Add another 3 mL of fixative to the Petri dishes and leave for 3 min.
8. Pipet off the fixative and leave the cover slips to dry in the air by using forceps to prop up the cover slips, cell side uppermost, against the rim of the lid of the respective Petri dish.
9. Place the cover slips flat in the Petri dish lid with the cells uppermost and add 2 mL of Hoechst stain. Cover to exclude light. Leave for 5 min.
10. Pipet off the stain and dispose of down the sink with plenty of water.
11. Place one drop (one for each cover slip) of mountant on a glass slide. Place a cover slip cell side down onto each drop of mountant and leave at 37°C for 30 min to set.
12. Using a microscope with ultraviolet light, look for extra-nuclear fluorescence.

12.2. Elimination

There are a number of reagents which are commercially available. They are usually one or a mix of antibiotics which are marketed as "mycoplasma removal agents." Follow the manufacturer's instructions. It is possible to remove Mycoplasma by the in vivo methodology of "cell rescue"(*see* **Subheading 10.2.**).

ICN (Thame Park Business Center, Thame, Oxon. UK. market an oxo-carboxylic acid derivative as a Mycoplasma removal agent which is effective within 7 d. The organism is killed rather than its growth inhibited and it is nontoxic against the usual hybridoma fusion partner cells.

References

1. Kohler, G. and Milstein, C. (1975) Continuous cultures of fused cells secreting antibody of predefined specificity. *Nature* **256,** 495–497.
2. Kohler, G. and Milstein, C. (1976) Derivation of specific antibody - producing tissue culture and tumor lines by cell fusion. *Eur. J. Immunol.* **6,** 511–519.

3. Galfre, G., Howe, S. C., Milstein, C., Butcher, G. W., and Howard, J. C. (1977) Antibodies to major histocompatibility antigens produced by hybrid cell lines. *Nature* **266,** 550–552.
4. Galfre, G. and Milstein, C. (1981) Preparation of monoclonal antibodies: strategies and procedures, in: *Methods of Enzymology 73. Immunochemical Techniques (part B)* (Langone, J. J. and van Vunakis, H., eds.), Academic, London, pp. 1–46.
5. Goding, J. W. (1996) Production and application of monoclonal antibodies in cell biology, biochemistry and immunology, in *Monoclonal antibodies: Principles and Practice.* 3rd ed. Academic, London.
6. Stewart-Tull, D. E. S. (ed.) (1995) Theory and Practical Application of Adjuvants. Wiley, New York.
7. Stewart-Tull, D. E. S. (1996) The use of adjuvants in experimental vaccines, II. Water-in-oil emulsions: Freund's complete and incomplete adjuvants, in *Methods in Molecular Medicine: Vaccine Protocols.* (Robinson, A., Farrar, G. H., and Wiblin, C. N., eds.), Humana, Totowa, NJ, pp. 141–145.
8. Stewart-Tull, D. E. S. (1997) The use of adjuvants in the production of polyclonal antisera to protein antigens, in A*nimal Alternatives, Welfare and Ethics: Procedings of the Second World Congress on Alternatives and Animal Use in the Life Sciences.* October 1996. Utrecht. (Van Zutphen, L. F. M. and Balls, M., eds.), Elsevier, Amsterdam.
9. Luben, R. A. and Mohler, M. A. (1980) In vitro immunisation as an adjunct to the production of hybridomas producing antibodies against the lymphokine osteoclast activating factor. *Mol. Immunol.* **17,** 635–639.
10. Heeneman, S., Deutz, N. E. P., and Buurman, W. A. (1993) The concentrations of glutamine and ammonia in commercially available cell culture media. *J. Immunol. Meth.* **166,** 85–91.
11. Catty, D. and Raykundalia, C. (1989) ELISA and related enzyme immunoassays in *Antibodies Vol. II: A Practical Approach* (Catty, D., ed.), IRL Press, Oxford, pp. 97–154.
12. Ling, N. R., Bishop, S., and Jefferis, R. (1977) Use of antibody-coated red cells for the sensitive detection of antigen and in rosette tests for cells bearing surface immunoglobulins. *J. Immunol. Meth.* **15,** 279–289.
13. Goodall, M. (1998) A simple hollow fibre bioreactor for the "in house" production of monoclonal antibodies, in *Methods in Molecular Biology Vol. 10. Immunochemical Protocols 2nd ed.* (Pound, J. D., ed.), Humana Press, Totowa, NJ.

8

Cell Bank Preparation and Characterization

Stephen J. Froud

1. Introduction

One or more properties of a cell line are likely to change during an extended period of continuous passage. In order to ensure consistency of the properties of a cell line or of a product derived from it, it is, therefore, necessary to cryopreserve and characterize a reference stock of cells. Thereafter, throughout a research, development, or manufacturing program cell cultures are discarded after a defined passage interval and the cryopreserved stock used to reinitiate fresh cultures. Although a small number of ampoules in a single bank may be sufficient for a small research program, to ensure continuity of supply of a cell line from a culture collection or for the manufacture of a biopharmaceutical, it is usual to prepare a two-tiered cell bank system. This consists of a master cell bank (MCB) of 50–400 ampoules and manufacturer's working cell banks (WCB) each derived in an identical manner from one ampoule of the MCB (**Table 1**). In this chapter the methods of preparation, storage, and characterization of such cell banks are described.

2. Cryopreservation

2.1. Methodology

Detailed procedures are presented in **Subheading 2.2.** Key points are discussed below. It is important to cryopreserve only highly viable cultures that are in the exponential growth phase. Cells with low viability or not in the exponential growth phase can result in poor viability or altered properties upon recovery from cryopreservation. In order to ensure that the cells recovered from any ampoule of a cell bank will give the same result, the contents of each ampoule in a cell bank must be identical. Thus, it is critical to ensure that the

From: *Methods in Biotechnology, Vol. 8: Animal Cell Biotechnology*
Edited by: N. Jenkins © Humana Press Inc., Totowa, NJ

Table 1
Two-Tier Cell Bank System Prepared for a Batch Manufacturing Process

Lot	Initial number of ampoules	Number after testing & retains	Maximum output
Master cell bank (MCB)	200	180	180 WCB
Manufacturer's working cell banks (WCB)	200	180	90 production lots from each WCB[a]
Product lots			

Note: The two-tiered cell bank system ensures the long term availability of sufficient ampoules to produce product. Although a master cell bank may consist of 200 ampoules, 20 of these may be used for tests and for retention samples. Thus up to 180 WCB can be produced from a single MCB. Allowing for testing and retention samples, and assuming a second back-up ampoule is used for each production run, then each WCB could produce up to 90 lots of product. Thus this two tiered system could produce 180 × 90 = 16,200 production lots. The estimate of 16,200 lots represents 324 yr of production at 50 lots per year, well beyond the conceivable lifetime of any product. This period would be 22 mo if the MCB only was used for manufacture (single tier system), or 9 yr if both MCB and WCB were of 50 ampoules.

[a]Assumes the use of a backup WCB ampoule per lot.

cells from multiple cultures are pooled, mixed, and remain well mixed throughout dispensing into individual ampoules.

During cryopreservation, there is a danger of ice crystal formation within the cells and, furthermore, the cells are subjected to osmotic shock as the intracellular water content is decreased. Therefore, to minimize cell damage during cryopreservation, the rate of freezing must be controlled carefully and cryoprotectants are used. Dimethylsulfoxide (DMSO) or glycerol is used routinely, usually in conjunction with fetal calf serum, serum albumin, or polysaccharides *(1)*. If the cell line is to be used for the manufacture of a biopharmaceutical, animal-derived raw materials should be obtained from a relatively short list of preferred countries (e.g., New Zealand), from a reputable supplier using methods that minimize the risk of contamination with undesirable biological agents. In many cases, it is possible (and thus preferable) to cryopreserve cells in the absence of animal-derived raw materials *(2)*. As the cells are not cultured in the cryopreservation mixture, it may not be necessary to include selective agents (e.g., methotrexate) or medium components required for growth (e.g., interleukin-6 [IL-6]).

The rate of freezing can be controlled using a programmable controlled-rate cell freezer *(3)*. Although such freezers are expensive, they produce a record of the temperature profile that can be used to confirm correct performance of the cryopreservation procedure. Inexpensive alternatives include the Nalgene Cryo 1°C freezing containers (Nalge Company, Rochester, New York) or polystyrene boxes placed inside a freezer at –70°C. These latter methods are rarely used for large cell banks. If a –70°C freezer is used, it is critical to remove the cells to the liquid-nitrogen storage dewar as soon as possible after freezing. The viability of cells decreases very rapidly at –70°C (e.g., recombinant mouse myeloma cells die after 14 days of storage at this temperature).

Most liquid-nitrogen contains extraneous particles. Sealed glass ampoules, although more difficult to prepare and recover, eliminate the risk of cross-contamination of cells with such particles when the ampoules are stored immersed in the liquid. Recovery, however, requires some care and the presence of liquid nitrogen within an incompletely sealed glass ampoule presents a hazard many institutions will not accept. Plastic ampoules with screw caps allow liquid nitrogen to enter far more frequently than do glass ampoules, but this poses less risk to the scientist. The risk of contamination of the ampoule contents by such seepage into plastic ampoules can be eliminated by storing the plastic ampoules in the vapor over liquid nitrogen.

2.2. Cryopreservation and Recovery Procedures

1. Grow sufficient culture for the number of ampoules and the cell concentration required. It is usual to prepare 1-mL ampoules with between 5 and 10 × 10^6 viable cells per ampoule. Allow for up to a 20% loss during manipulations. The culture should be in the exponential growth phase (typically 50% of the maximum viable cell concentration achieved in that medium for suspension cultures, or well before a confluent monolayer is formed for an attached culture). The viability should be at the upper end of the normal range, e.g., >80%.
2. Prepare the sterile cryopreservation mixture: 80% (v/v) basal growth medium, 10% (v/v) fetal calf serum, 10% (v/v) dimethylsufoxide.
3. Aseptically remove the spent growth medium from the cells by centrifugation and resuspend the cells in the cryopreservation mixture at room temperature.
4. Mix the resuspended cells to produce a homogenous pool and aliquot into previously labeled ampoules. Continue mixing the pool throughout this step.
5. Freeze cells slowly (approx –1°C/min) to –70°C or below.
6. Transfer to liquid-nitrogen storage dewar immediately. Stand the frozen ampoules in 2 cm of liquid nitrogen if the transfer process takes more than a few minutes. A face visor and thermal gloves should be worn whenever liquid nitrogen is used.
7. Cells can be recovered using the following procedure. Warm rapidly to 37°C (e.g., by standing the ampoule in warm sterile water). Transfer the cells to a plastic container (growth flask or sterile centrifuge tube). Add the prewarmed growth medium slowly (e.g., by running the contents of a 10-mL pipet down the side of the tube).

3. Control of Protential Contaminating Agents

Cell banks used in the manufacture of biopharmaceuticals or for gene therapy are the starting point for the manufacturing process. As such, they need to be prepared and stored according to current good manufacturing procedures (cGMP) as specified in guidelines prepared by a number of regulatory authorities ***(4–7)***. Such documents focus on the control of, and subsequent testing for, potential contaminating agents.

Cells received from outside the cell bank laboratory should be quarantined and tested for bacteria, fungi, mycoplasma, and, possibly, specific viruses of concern. Results from one US testing laboratory found mycoplasma in 10% of the cultures sent to it for analysis ***(1)***. Although we have observed a lower level in cell lines sent from clients to my laboratory for biopharmaceutical production (three contaminated cell lines out of many hundreds received over a 10-year period), this still represents a significant risk. Traditional sterility tests using Tryptone Soya and Sabouraud broths and agar are often supplemented with other media (e.g., sheep's blood agar). These are used to detect fastidious bacterial contaminants that are difficult to observe by microscopic examination but may be found propagating at a low level in animal cell cultures.

The systems used for the control of potential contaminating agents at LONZA Biologics multiproduct facility are summarized in **Table 2**. For additional security, cell banks are placed in a segregated dewar immediately after preparation. They remain segregated from other cell banks until a further set of tests for bacteria, fungi, mycoplasma, and viruses of concern have been completed. Only then will they be moved to the vapor-phase liquid-nitrogen dewars used for the storage of similarly tested cell banks. Prior to large scale operations, we also perform a number a general and specific virus tests to ensure the safety of personnel and other cell lines (*see* **Subheading 5.**).

4. Storage

The objective of a cell bank storage system is to ensure the long term storage of cell lines (for decades) while minimizing risk of the following:

- Compromising operator safety
- Incorrect ampoule release (e.g., from the wrong cell bank)
- Changes occurring in the stored cells
- Cross-contamination
- Catastrophic loss of the bank

In order to allow an appropriate level of control, every ampoule of a cell bank should be uniquely numbered and coded with details specific to that bank. Subsequent release of ampoules should be documented such that the fate of each individually numbered ampoule is known. Audits can then be performed at intervals to ensure that erroneous release has not occurred. In addition, such annotation and documentation allows data collation and analysis should an investigation into the cell bank's performance over time become necessary.

Table 2
Precautions to Prevent Contamination of Cell Banks

Discrete Quarantine laboratory separate from cell bank preparation area for cell line receipt.
Discrete cell stock dewars for ampoules after passing quarantine requirements.
Cell banking laboratory within cGMP facility.
Access to cell bank laboratories restricted to trained, authorized individuals.
Scientists in cell bank laboratory wear dedicated suit, footwear, and disposable headcover.
During cell culture activities, scientists also wear a single-use, nonshedding laboratory coat, disposable gloves, and a face mask.
Cell culture rooms at positive air pressure to maintain airflow out of the room.
Strict aseptic technique is observed within a vertical laminar flow (VLF) cabinet.
Cabinet is cleaned with disinfectant before and after use.
All materials are obtained from accredited suppliers and further tested as required.
All materials are sterilized and all equipment is wiped thoroughly with disinfectant before entry to VLF cabinet.
No other cell lines handled simultaneously in the laboratory.
The following methods are used to monitor the environment within the VLF cabinet:
Airborne contaminants using settle plates (during manipulation of cultures)
Finger tips by touch plates (before and after culture manipulations)
Viable particles (during simulated manipulation of cultures)
Nonviable particles (resting)
Discrete locked cell bank dewars in locked rooms separate from other cell stocks.
Cell banks stored in the vapor phase above liquid nitrogen.
Sterility tests over 14 days on several ampoules using a variety of broths and agars.
Identity, mycoplasma, and virus tests on each cell bank.

In case of failure of the storage device, the ampoules of a cell bank should be divided between two separate vessels or freezers. An alarm system to detect failure is advisable. To protect against the adverse financial and ethical consequences of interrupted manufacture due to loss of the cell bank, cell lines used for biopharmaceutical manufacture should be stored in geographically separate sites.

In order to achieve biochemical stasis it is necessary to store cells at or below the glass temperature of water, –130°C to –133°C. Furthermore, cells may be damaged when the temperature is cycled above –130°C, for example, during the release of another ampoule from within the same dewar. The three most common storage systems are discussed below. The key differences concern the storage temperature and the potential of liquid nitrogen to enter the ampoules.

4.1. Vapor-Phase Liquid-Nitrogen Dewars

These systems offer two major advantages. There is a minimal risk of cross-contamination in the gaseous phase, and liquid nitrogen cannot leak into the ampoules. Because of the latter, there is no risk of an explosion that may occur if ampoules are removed from immersion within liquid nitrogen into warm air. The disadvantages, however, include the following. There is a very short reaction period for the cell bank curator to respond to high-temperature alarms. In addition, temperature-mapping experiments for the validation of cell bank storage systems in my laboratory have shown that several (but not all) commercially available dewars are unable to maintain temperatures below –130°C throughout the volume used to store cells (**Fig. 1**). The temperature within ampoules stored in such dewars rises rapidly when any part of the inventory system is removed and replaced during the release of an ampoule. Furthermore, it can then require up to 6 h for the temperature to return to the required low level. Such temperature variation may have implications for long-term storage. Trend analyses of both the viability and number of viable cells within ampoules stored over a 7-year period within my laboratory, however, have demonstrated that this concern may be theoretical. Analysis of several different cell lines has shown no statistically significant decline in either viability or number of viable cells within the ampoule throughout this 7-year storage period.

4.2. Electrical Freezers

The discussion on temperatures within vapor systems also applies to electrical freezers except that the active compressor may reduce the temperature within the ampoules faster than in passive nitrogen-vapor dewars. When opened, the temperature within a –130°C electrical freezer rises rapidly above the glass transition temperature of water. Thus, theoretically, stasis cannot be assured. Nevertheless, a number of companies do use such systems. Freezers at –140°C are relatively new to the market but may have an advantage in this area. The absence of liquid nitrogen minimizes the risk to operators. The emergency backup systems (usually a liquid-nitrogen dump into the cabinet) are such that a catastrophic failure is unlikely. With such large storage units, however, it is not normally possible to segregate cell lines based on their source and testing status.

4.3. Liquid-Phase Liquid-Nitrogen Dewar

For long term storage, stasis is maintained during immersion in liquid nitrogen at –196°C. Furthermore, temperature fluctuations are minimized after opening the dewar for removal of an ampoule, as reimmersion in the liquid nitrogen leads to rapid temperature equilibration of the remaining ampoules. This system, however, presents two potential hazards. Liquid nitrogen may

	Static mode	After ampoule release	Time to return to static temperature (hours)
	Top	Top	Top
Top tray	-113	-79	5.2
	-133	-94	4.8
	-141	-103	3.4
	-157	-123	1.7
	-169	R	0.7
Bottom Tray	-177	-156	0.0
	Level of liquid nitrogen		Level of liquid nitrogen

Fig. 1. Static and in-use temperature profile of a commercially available vapor-phase liquid-nitrogen dewar. The temperatures were measured continuously inside ampoules stored in each of six trays. These six trays were stored in the vapor phase at different heights above the liquid-nitrogen surface. After equilibration in the "static mode," an ampoule was recovered from the fifth tray from the top. The temperature in the other trays was recorded immediately after this release and the times taken for the temperature to return to that of the "static mode" were measured. R = ampoule removed from this tray. Temperature in remainder of tray rose to –95°C during removal procedure.

enter ampoules upon immersion and this can expand explosively upon warming of the ampoule. In addition, there is a potential risk of cross-contamination by adventitious agents through the liquid phase. That such contamination can occur has been demonstrated by hepatitis B transmission within a contaminated cryopreservation tank ***(8)***.

5. Characterization

The objective of cell bank characterization is to ensure that the cells in the cell bank meet the following criteria:

are viable and capable of further propagation.
are free from contaminating agents.
still have the desired characteristic(s) for which they were selected.
are suitable for their intended purpose.
provide a reference against which cells derived from the cell bank can be compared (e.g., after use in a production process for the manufacture of a biopharmaceutical).

Obtaining as much baseline data as can be practically gathered within a few weeks of preparing a cell bank is recommended. Such a reference database is essential because the cell bank curator of a successful cell line will have the unenviable task of ensuring that their charges have been appropriately stored for decades. The details of the characterization scheme will depend on the cell type, the previous history of the cell line, and the purpose for which it is to be employed. Before embarking on characterization reference should be made to the current guidance prepared by the relevant regulatory authorities *(5,6,9,10)*.

The number and type of samples to be analyzed will, again, depend on the intended use of the cell bank. A researcher preparing a small bank of cells to ensure consistency among collaborators may only test a few ampoules from this single bank for viability, identity and freedom from mycoplasma, bacteria, molds, and yeasts. At the opposite end of the scale will be a curator for a pharmaceutical company that is preparing to market a product. They will utilize a wide variety of complementary analytical methods in an extensive and well-documented characterization of the cell banks and of cells derived from these cell banks sampled at different points in the production process (**Tables 3** and **4**). The following discussion will focus on the requirements common to these two positions, and only briefly include the more specialized techniques. The reader must assess which tests should be applied to meet their requirements.

5.1. Samples to be Tested

As a minimum, every cell bank should be tested for viability, identity, and freedom from mycoplasma, bacteria, and fungi. These tests should be performed on the cells from one or more ampoules, as discussed below. In a two-tiered cell bank system, it is usual to perform extensive characterization of the MCB. The testing of the WCB can then be limited to confirmation of the identity of the cell line and to an abbreviated test program to ensure that adventitious agents have not been introduced during preparation of the WCB.

For pharmaceutical use, it is necessary to test cells at the beginning (i.e., the cell bank) and at the end of production (end of production cells [EPC]). Thus it can be assured that the cell line has an appropriate level of stability and that adventitious agents are not introduced during, or induced by, the production process. If the cell viability remains high throughout the production process, then the end of production cells can be taken when the last harvest of product has been collected (e.g., after 185 days) *(11)*. High cell viability is not retained throughout some processes, however, and a small number of tests require highly viable cells (e.g., isoenzyme analysis, electron microscopy, cocultivation of the cell line with a detector cell line). In these cases, the sampling regime must be designed to obtain cells of high viability as late as possible from the production process *(12)*. This may require the performance of a specific production run for characterization purposes. To ensure compliance with the "end of production" concept, this characterization run should be completed at the end of, or even beyond, the interval allowed for subsequent production. This extensive testing of end of production cells is usually only performed once, although it may be necessary to repeat it after any major process change. Having performed this extensive characterization, an abbreviated test regime can then be used routinely to test every production lot.

5.2. Viability and Propagation

The viability of a cell bank should be established immediately after preparation. Methods to determine cell viability are discussed in Chapter 10. In addition to confirming that the cells recovered from an ampoule are viable, it is advisable to demonstrate that they perform satisfactorily. In this way, the cell bank curator can be confident that the cells will be fit for their intended purpose. For example, cells from a working cell bank (WCB) should be tested in the intended production process, or in a scaled-down simulation of this process. Should the cell line then subsequently fail to perform as expected, this must be due to a change in the process or to a change that has occurred during storage. Furthermore, the cells from several different ampoules should be tested, ideally from ampoules at the beginning, end, and at several time points during the process of dispensing the cells into the ampoules. By using several ampoules, one can establish ranges for viability, growth, and marker or product characteristics against which the performance of subsequent recovered cells can be compared. In addition, for larger cell banks (over 50 ampoules), the curator can thereby be assured that no adverse events have occurred during cell bank preparation that affected only a proportion of the cell bank.

In order to ensure that the cell bank will be viable for its intended period of use, the viability of the cells in a cell bank should be determined over time. In my laboratory, critical cell banks are given a re-evaluation date 5 years after preparation. At the end of this period the viability and number of viable cells recovered from each ampoule during the intervening period is assessed. The trend of these parameters with time is then analyzed for statistically significant changes, and in the rare cases when a change has been seen, the useful lifetime is estimated. As long as this exceeds 5 years, the cell bank is then given a further re-evaluation date 5 years from the last data point. To prevent unnecessarily removing ampoules specifically to collect these data, ampoules are released from the cell bank solely for analysis only if there is insufficient recent viability data or if the results warrant further investigation.

5.3. Identity

The identity (or authenticity) of this baseline stock of cells must be assured. Unfortunately results from a testing laboratory in the United States showed that 35% of the 275 cell lines tested contained, or were exclusively, cells of a different cell type or even a different species than that expected by the donor laboratory *(1)*. The test regime will depend on the potential risk of cross-contamination with other cells. For example, isoenzyme analysis is commonly performed to confirm the species of a cell line. This, however, cannot be considered as confirmation of identity from a laboratory handling many cell lines of the same species each with different characteristics (e.g., mouse/mouse

Table 3
Summary of Tests for Adventitious Agents

Test	Purpose	MCB	WCB	EPC	Bulk harvest
Sterility	Bacteria and fungi	nd[a]	nd	nd	nd
Mycoplasma cultivation and stain	Mycoplasma	nd	nd	nd	nd
In vitro cell based virus assays	Adventitious viruses (broad-range assay)	nd	nd	nd	nd
In vivo virus assay	Adventitious viruses (broad-range assay)	nd	nd	nd	—[b]
Electron microscopy	Broad adventitious agent screen	Report[c]	—	Report	—
Quantification of viral-like particles by electron microscopy	Quantification of virus-like particles (rodent cell lines)	—	—	—	Report
Cocultivation with detector cell line	Depends on detector cell line; usually for rodent and human retroviruses and human viruses	nd	—	nd	—
Reverse transcriptase	Retroviruses	Report	—	—	Report
S^+L^- focus forming assay	Xenotropic murine retrovirus	Report	—	Report	Report
XC plaque assay	Ecotropic murine retrovirus	Report	—	Report	Report
Mus Dunni and detection	Amphoteric murine retrovirus (xenotropic and most ecotropic murine retrovirus)	Report	—	Report	Report
XC plaque assay	Ecotropic murine retrovirus	Report	—	Report	Report
MAP, RAP or HAP	Panel of up to 16 mouse, rat, or hamster viruses	nd	—	—	nd
Human and primate retroviruses	Includes PCR or other assays for human retroviruses such as HIV, HTLV, and SIV	nd	—	—	—

Specific human virus assays	Includes PCR or other assays for human viruses such as Epstein-Barr virus, cytomegalovirus, hepatitis A, B, and C and human herpes 6	nd	—	—	—
Specific primate virus assays	Includes PCR or other assays for simian viruses: herpes, cytomegalovirus, encephalomyocarditis, haemorrhagic fever, varicella, adenovirus, SV40, monkeypox, rubella and ebola	nd	—	—	—
Specific bovine virus assays	Bovine viruses from cell culture media	nd	—	—	—
Specific porcine virus assays	Porcine viruses from enzymes	nd	—	—	—
Specific ovine virus assays	Ovine viruses from cultivation method	nd	—	—	—
Identity	Cross-contamination with another cell line	Report	Report	Report	—

[a]nd: Usually tested. The expected result is no contaminating agent detected.

[b]—: Not usually tested.

[c]Report: Often tested. The result may be quantitative, species-specific, or vary with the cell type under test; for example, rodent cell lines are often positive in retrovirus assays due to the presence of endogenous rodent retroviral particles.

Table 4
Summary of Typical Tests for Genetic Characterization and Assessment of Stability

Test	Purpose	Initial construct	MCB	WCB	EPC or cells after extended passage
DNA fingerprint	Demonstrate absence of change	—[a]	Reference[b]	Same as MCB[b]	Same as MCB[b]
Phenotypic marker	Demonstrate absence of change	—	Reference[b]	Same as MCB[b]	Same as MCB[b]
Gene copy number	Demonstrate absence of change	—	Reference	Same as MCB	Same as MCB
Size of restriction endonuclease digests	Confirmation that deletions or insertions have not occurred	—	Reference	Same as MCB	Same as MCB
Junction analysis	For single-copy inserts: confirmation that the insertion site has not changed during passage; for low-copy-number inserts: confirmation that the number of integration sites has not changed	—	Reference	—	Same as MCB
mRNA number and length(s)	Demonstrate absence of change	—	Reference	—	Same as MCB
Protein coding sequence	Demonstrate absence of change	Reference	No change	—	Same as MCB[c]
Peptide mapping	Detection of low-level changes in coding sequence	—	Reference[b]	—	Same as MCB[b]
Amino acid sequence	Detection of low-level changes in coding sequence	—	Reference[c]	—	Same as MCB[b,c]
Rate of production or quantity of product	Demonstrate absence of change	—	Reference	Same as MCB	Same as MCB
Identity of product (e.g., activity, carbohydrate analyses)	Demonstrate absence of change	—	Reference	Same as MCB	Same as MCB

Reference: A sample of cells analyzed and used as a reference standard against which later samples are compared.
[a]—: Testing may not be required.
[b]Tests that may not be necessary but may be considered to provide additional data for a specific cell line.
[c]It may be possible to test either the nucleic acid or analysis of the protein product.

hybridomas expressing different antibodies). For such laboratories, confirmation of species and expression of the desired product may be appropriate. DNA fingerprinting provides a useful baseline against which future cultures can be compared. For the cell bank curator preparing this baseline, however, it must be supplemented with other tests because it cannot confirm that the correct cell line has been cryopreserved. Other unusual phenotypic characteristics can also be considered (e.g., the ability to grow in glutamine-free medium for a mouse myeloma transfected with the glutamine synthetase gene) ***(13)***. Morphological analysis may be useful for cell lines that grow attached to a substratum but not for suspension cultures. The data from genetic analyses of recombinant cell lines can also be considered as a test of identity (*see* **Subheading 5.6.**). Identity testing is usually performed on the cells from one ampoule of the cell bank.

5.4. Mycoplasma, Bacteria, and Fungi

Results from a testing service in the United States showed that 10% of the 34,697 cell cultures tested contained mycoplasma ***(1)***. Such data have been instrumental in ensuring that most cell culture laboratories now routinely test for mycoplasma. Kits are available and several companies offer relatively inexpensive testing services. Regulatory authorities provide details of the extensive protocols required for cell lines used in pharmaceutical production ***(5)***. These include inoculation of the cell culture into liquid broths and onto solid media designed to cultivate mycoplasma. Because these media do not support the growth of all known mycoplasma, these tests are supplemented with assays using indicator cells and the fluorescent dye Hoechst 33258. Mycoplasma testing is usually performed on the cells from one or two ampoules of the cell bank.

Similarly, there are standard procedures for testing cell cultures for bacteria and fungi ***(14)*** (often referred to as sterility tests). Examples of media used routinely include Tryptone Soya broth (or soybean-casein digest) and Tryptone Soya agar for the detection of a wide variety of micro-organisms, and Fluid Thioglycollate medium for the detection of aerobic and anaerobic micro-organisms ***(14)***. These may be supplemented with Sabouraud-dextrose media specifically to detect yeasts. Although these media allow the detection of a broad range of bacteria and fungi, it is possible to find fastidious organisms at a low level in some cell cultures. Such organisms can be difficult to detect, as they can propagate slowly over many subcultures and thereby remain at the limit of detection by visual observation or using a good quality light microscope. To detect such organisms, some laboratories supplement the above broad-range media with media that better simulate the conditions in cell cultures. Examples include blood agar and peptone yeast glucose. Sterility testing is usually performed on the cells from at least two ampoules of the cell bank.

5.5. Viruses

The potential for contamination with viruses is a major consideration in the use of animal cell technology. Methods for testing cell lines for the presence of viral contaminants are described in Chapter 3. A cell bank should be tested to an appropriate level to ensure the safety of persons using the cell line. For example, to safeguard the scientist who may handle large-scale cultures, a human cell line may be tested using a reverse transcriptase assay (to detect retroviruses), a broad-range virus test (e.g., in vitro assays using a variety of detector cell lines) and specific tests for viruses of concern (e.g. hepatitis B). To ensure the safety of patients, considerably more extensive testing of cell lines used in the production of pharmaceuticals is required. Common virus assays are described in **Table 3** ***(5,6,9,10)***. It should be stressed that only a selection of the tests in **Table 3** is relevant to a specific cell line. Note that the history of the cell line needs to be examined in order to define an appropriate scheme. For example, assays may need to be included for viruses that may have been introduced from the culture medium (e.g., bovine viruses in fetal calf serum), from enzymes used in cultivation (e.g., porcine viruses from trypsin) or from the laboratory environment (e.g., amphotrophic mouse retroviruses used as gene therapy vectors), as well as viruses capable of infecting the cell line (e.g., parainfluenza types 1, 2, and 3 for CHO cells) ***(15)*** or endogenous to the cell line itself [e.g., the infectious mouse retroviruses found in all cell lines derived from mouse myelomas ***(12)***].

To account for such diversity, it is usual to perform several in vitro virus assays utilising detector cell lines of different species. Examples of such detector cell lines include the human normal MRC-5, human tumor HeLa, porcine PT-1, bovine turbinate, mouse NIH 3T3, rabbit RK13, and hamster CHO-K1 cell lines. To detect viruses that may not be amenable to in vitro cultivation, a small number of in vivo assays may be performed using chicken eggs, mice, and, sometimes, guinea pigs or rabbits. Rodent cell lines are usually tested using a panel of species-specific assays (MAP, RAP or HAP) as well as a variety of sensitive tests to detect murine retroviruses (reverse transcriptase, S^+L^- focus forming assay, XC plaque assay, *Mus Dunni* cultivation and subsequent retrovirus detection, and cocultivation with one or more detector cell lines). Human and primate cell lines are extensively tested for retroviruses and adventitious viruses (*see* **Table 3**).

The action required by a positive result will depend on the cell type under test. For example, although a positive result would be expected for the reverse transcriptase assay from a rodent cell line, the same result from a human cell line would be a matter for concern. Electron microscopy is a relatively insensitive method, but it is capable of detecting potential adventitious agents missed by the other methods. Thus, it is used to examine both cell bank samples and to estimate the number of retroviral particles released from all rodent cell lines.

5.6. Genetic Characterization and Cell Line Stability

An acceptable level of stability of the desired characteristics must be established for all cell lines. A maximum passage level must be defined so that the researcher can be sure that he or she is investigating a culture of consistent properties. For a cell line used in a production process, it is necessary to demonstrate that the cell line will consistently produce an economically viable quantity of product with the correct characteristics. Commonly used analytical approaches are described in **Table 4**. As noted earlier for virus assays, it should be stressed that only a selection of the tests in **Table 4** is relevant to a specific cell line. Conversely, no single method can detect all of the possible modifications to a cell line or protein product. Thus, an extensive analysis will usually include a combination of methods involving analysis of genotypic and/or phenotypic markers, the stability of product formation, nucleic acid data and the characteristics of any protein product.

For nonrecombinant cell lines, the identity tests described above can be used to demonstrate stability of phenotypic and genotypic markers. Having established a baseline by analysis of the initial construct or of the cells in the MCB, the stability of a cell line can be established by comparison of these data with those obtained after extensive passage of the cell line (e.g., the EPC).

The stability of product formation can be measured by a variety of methods. Common examples include the following:

Establishing that cells express product at a similar rate before and after extensive passage.

Demonstrating that the amount of product produced does not vary significantly when cultures are analyzed at intervals during passage from the MCB up to, and sometimes beyond, the limit of in vitro cell age.

Analyzing between 100 and 300 clones derived from extensively passaged cultures to demonstrate that all of the cells in the population produce product.

For recombinant DNA cell lines, some genetic characterization will be required. This is to ensure that the correct coding sequence has been incorporated into the cell line and is maintained for the period of use. Restriction endonuclease mapping will provide information on copy number, insertions, deletions, and the number of integration sites. For cell lines with multiple copies additional analysis may be required to determine the copy number (e.g., Southern blotting) and the number of insertion sites (e.g., fluorescently labeled *in situ* hibridization [FISH; *see* **Chapter 5**]). The presence of a single transcript of the expected length can be demonstrated by Northern blotting. Verification of the protein coding sequence is usually performed by determining the sequence of mRNA, as sequencing genomic DNA is only practical for a protein with a single gene copy. Although most comparisons use the MCB as the baseline, the coding sequence can be compared to that obtained or expected in the original construct, as these data should be readily available. Although it is possible to analyze multiple clones of the cell population such an undertaking would require considerable resources. It is usual, therefore, to analyze nucleic acid or protein from the bulk population, knowing that a low-level of genetic variants will not be detected.

Lower-level variants can sometimes be detected by peptide mapping or amino acid sequencing of a protein product. Such characterization is essential, as none of the other techniques provide information on potential errors during posttranslational modification. At this point, the characterization of a cryopreserved cell line begins to merge with detailed characterization of the protein product (e.g., carbohydrate analyses, as described in Chapter 22) and of the production process (e.g., consistency studies on multiple production lots).

Acknowledgment

I would like to thank Robert Kallmeier for translating the genetic characterization guidelines into someting that can be understood by mere scientists.

References

1. Hay, R. J. (1996) Animal cells in culture, in *Maintaining Cultures for Biotechnology and Industry* (Hunter-Cevera, J. C. and Belt, A., eds.), Academic Press, London, pp. 161–178.
2. Merten, O.W., Petres, S., and Couvé, E. (1995) A simple serum-free freezing medium for serum-free cultured cells. *Biologicals* **23:2,** 185–189.
3. Doyle, A. and Morris, C. B. (1994) Cryopreservation, in *Cell & Tissue Culture: Laboratory Procedures* (Doyle, A., Griffiths, J. B., and Newell, D. G., eds.), Wiley, Chichester, U.K., pp. 4C:1.1–4C:1.7.
4. Quality of biotechnological products: derivation and characterization of cell substrates for the production of biotechnological/biological products (1997) ICH, London.
5. Points to consider in the characterization of cell lines used to produce biologicals (1993) CBER, FDA, Washington, DC.
6. Production and quality control of monoclonal antibodies (1994) European Commission, Brussels.
7. Requirements for use of animal cells as in vitro substrates for the production of biologicals (1997) WHO, Geneva.
8. Tedder, R. S., Zuckerman, M. A., Goldstone, A. H., Hawkins, A. E., Fielding, A., Briggs, E. M., et al. (1995) Hepatitis B transmission from contaminated cryopreservation tank. *Lancet* **346,** 137–140.
9. Viral safety evaluation of biotechnology products derived from cell lines of human or animal origin (1997) ICH, London.

10. Points to consider in the manufacture and testing of monoclonal antibody products for human use (1997) CBER, FDA, Washington.
11. Böedeker, B. G. D., Newcomb, R., Yuan, P., Braufman, A., and Kelsey, W. (1994) Production of recombinant factor VIII from perfusion cultures: I. Large-scale fermentation, in *Animal Cell Technology: Products of Today, Prospects for Tomorrow* (Spier, R. E., Griffiths, J. B., and Berthold, W., eds.), Butterworth-Heinemann, Oxford, pp. 580–583.
12. Froud, S. J., Birch, J., Mclean, C., Shepherd, A. J., and Smith, K. T. (1997) Viral contaminants found in mammalian cell lines used in the production of biological products, in *Animal Cell Technology: From Vaccines to Genetic Medicine* (Carrondo, M. J. T., Griffiths, B., and Moreira, J. L. P, eds.), Kluwer, Dordrecht, pp. 681–686.
13. Bebbington, C. R., Renner, G., Thomson, S., King, D., Abrams, D., and Yarranton, G. T. (1992) High level expression of a recombinant antibody from myeloma cells using a glutamine synthetase gene as an amplifiable selectable marker. *Bio/technology* **10,** 169–175.
14. Code of Federal Regulations 21, 610.12, FDA, Washington, DC.
15. Wiebe, M. E., Becker, F., Lazar, R., May, L., Casto, B., Semense, M., et al. (1989) A multifaceted approach to assure that recombinant tPA is free of adventitious virus, *in Advances in Animal Cell Biology and Technology for Processes* (Spier, R. E., Griffiths, J. B., Stephenne, J., and Crooy, P. J., eds.), Butterworths, Oxford, pp. 68–71.
16. Quality of biotechnological products: analysis of the expression construct in cells used for production of r-DNA derived protein products (1995) ICH, London.

9

DNA Fingerprinting and Characterization of Animal Cells

Glyn N. Stacey

1. Introduction

The subject of cell characterization is obviously of great importance in the utilisation of cell lines from a scientific perspective in order to demonstrate the validity and limitations of a particular cell line as a model of in vivo tissue or as a cell substrate in a biological assay or production process. However, the history of in vitro animal cell culture is littered with cases where more diligent scientists have identified "cross-contamination" of cultures which, in many cases, probably represented simple mislabeling or switching of cultures. Nelson-Rees et al. published the seminal review of this problem which focused scientific attention on the hazards of "cross-contaminated" cell lines *(1)*. Their results were based on painstaking studies involving karyology and isoenzyme analysis which revealed a large number of cell lines, reported to be of diverse types and origins, which, in fact, turned out to be subclones of the HeLa cell line. Thus, it is important that all workers using cell lines be aware of this problem and be able to take steps to avoid the use of bogus cell lines This chapter is intended to give the reader an overview of some of the more popular DNA identification tests available, to put them in context with other more traditional tests, and to give some guidance on the most appropriate selection of an identity test which may vary depending on the type of work and facilities available.

1.1. The Development of Cell Identification Techniques

Early descriptions of cell lines were heavily dependent on the microscopic appearance and morphology of individual cells. The development of a reliable karyotyping technique for cells by Hsu *(2)* and Tjio and Levan *(3)* enabled the visualization of the entire genome in the form of premitotic condensed chro-

From: *Methods in Biotechnology, Vol. 8: Animal Cell Biotechnology*
Edited by: N. Jenkins © Humana Press Inc., Totowa, NJ

mosomes. This technique enabled rapid identification of the species of origin and also revealed unique genetic markers for an individual cell line. Subsequently, a wide variety of characterization techniques were developed for the identification of cells, including species-specific antigen immunofluorescence *(4)* and isoenzyme analysis *(5)*. Of these, isoenzyme analysis has achieved widest use and this has been promoted by the availability of a standardized kit (Authentikit™, Innovative Chemistry). This technique is based on the visualization of certain enzymes that vary between species in their electrophoretic mobility and can thus be discriminated by their differential migration rates. Different isoenzyme profiles represent the products of different gene alleles that may also be influenced by posttranslational modification or hybridisation in hybrid cells.

From the 1970s the rapid development of molecular biology delivered a host of new methods for cell identification based on restriction fragment length polymorphisms (RFLPs). A particularly useful technique for discriminating human cells was human leucocyte antigen (HLA) typing *(6)*. In 1985, Jeffreys et al. first described the technique of DNA fingerprinting, which was unique in its capability to differentiate between human individuals *(7)*. This technique was based on RFLPs derived from hypervariable minisatellite sequences called variable number tandem repeats (VNTRs) and revealed patterns of hybridization on Southern blots of genomic DNA that are inherited in a Mendelian fashion. Furthermore, the observed patterns were only the same in the case of identical twins. Since the publication of this method, a range of techniques has been developed for DNA typing based on hypervariable DNA sequences and other RFLP approaches. The following protocols will describe the basis of the DNA fingerprinting and profiling techniques and will provide validated protocols for cell identification. Furthermore, approaches to selecting an appropriate identification system to meet the reader's own purposes will be discussed.

1.2. Utilization of Different Types of Variable DNA for Cell Typing

1.2.1. Variable DNA

Repetitive DNA comprises the majority of the genome of many higher organisms. Although the function of many types of repetitive DNA has yet to be elucidated, some have proven highly useful for the identification of individuals and the cell lines derived from them.

Two groups of repetitive DNA called minisatellite and microsatellite DNA have proven particularly useful for cell identification. Both occur as VNTRs throughout the genome of most higher eukaryotes, but differ in their structure (i.e., core repeat units are 1–5 base pairs for microsatellites and 10–15 base pairs for minisatellites). A variety of single locus probes and polymerase chain reaction (PCR) primers are available for specific VNTR loci. These approaches suffer from the need to use different probes or primers for each species analyzed. Nevertheless, the use of probe cocktails for Southern blot hybridization and multiplex PCR providing data on multiple human VNTR loci can give very specific identification.

A number of other types of DNA RFLP sequences have been used to good effect for organism identification. These include ribosomal RNA, interspersed sequences (i.e., nonVNTR repeat sequences related to transposons) ***(8)***, analysis of variation of a single gene using cDNA probes ***(9)***, and PCR for conserved introns ***(10)***. A further group of methods that has been very popular in recent years is based on the use of random sequence primers for random amplified polymorphic DNA (RAPID) detection ***(11)*** and simple tandem repeat (STR) analysis. Despite their popularity, these techniques require very careful scientific validation for reliability and reproducibility before they can be applied in a routine setting.

1.2.2. Multilocus DNA Fingerprinting

Certain VNTR sequences such as the Jeffreys probes for the human myoglobin locus, 33.15 and 33.6 ***(12)***, and the microsatellite sequence from the M13 phage protein III gene ***(13)***, will cross-hybridize (under the appropriate experimental conditions) with a wide range of families of VNTR sequences. These probes have proven especially successful in delivering specific identification methods for a wide range of species, including plants and animals ***(14)***. Multilocus DNA fingerprinting benefits from the presence of related satellite sequences present in the genomes of a wide spectrum of eukaryotic organisms. These sequences are revealed in Southern blot hybridization in which the stringency of hybridization is set to permit visualisation of DNA–DNA hybrids that have homology but are not completely complementary (i.e., derived from related satellite families) and reveal RFLPs from a range of genetic loci widely distributed in the genome (**Fig. 1**).

Multilocus DNA fingerprinting has been applied to cell culture in research ***(15,16)***, culture collections ***(17,18)***, and in the manufacture of biological products from animal cells ***(19,20)***. This technique is now providing a valuable tool for assuring the quality of cell culture processes by excluding cross-contamination of cell lines between master and working banks and screening for common contaminants such as HeLa ***(21,22)***. The following protocols give examples of a validated multilocus DNA fingerprinting method (*see* **refs.** *21* and *22*) and a PCR-based method that has been applied to the identification of cell lines from a wide range of species ***(8)***.

2. Methods

2.1. Multilocus Fingerprinting Protocol

2.1.1. Preparation for Southern Blots of Genomic DNA

For DNA fingerprinting, it is important to obtain undegraded high-molecular-weight genomic DNA. This may be obtained using one of the many com-

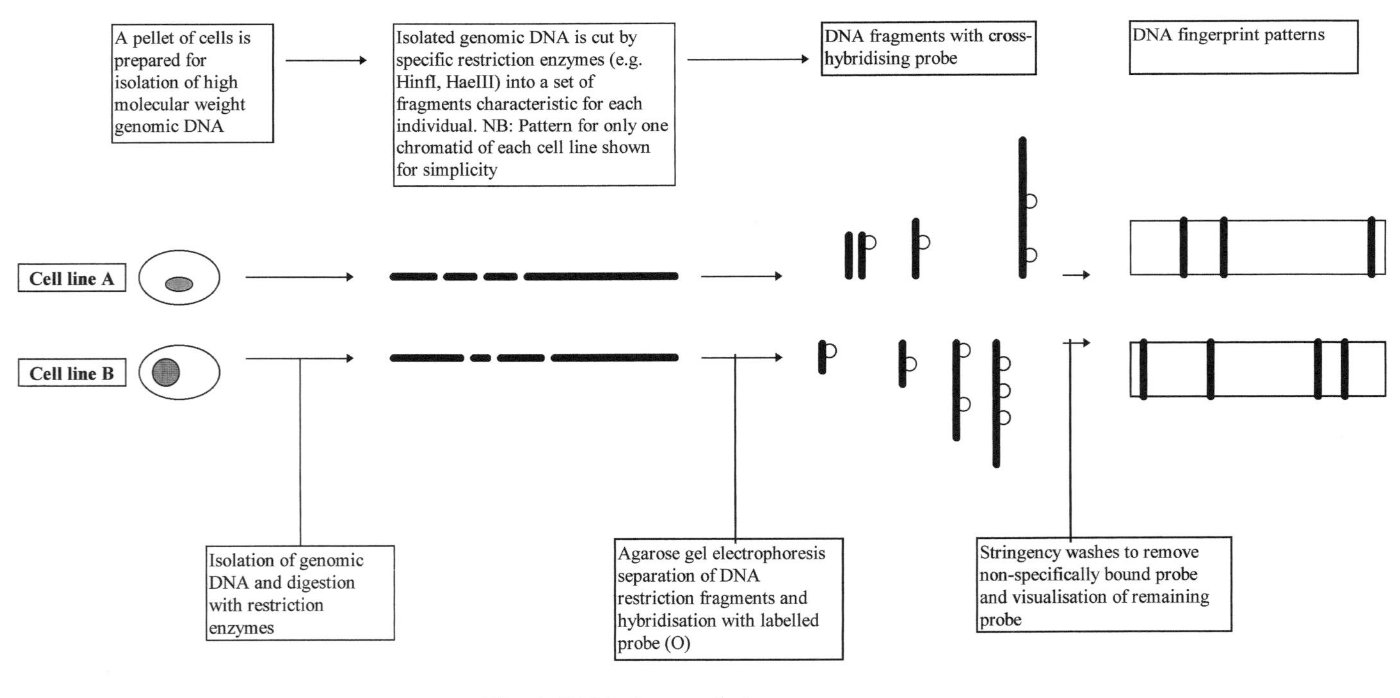

Fig. 1. DNA fingerprinting process.

mercially available kits for DNA extraction, but the quality should be checked by confirming the absence of significant amounts of degradation. An inexpensive and safe alternative minipreparation method ***(8)*** that avoids the use of phenol is as follows:

1. Resuspend a pellet of at least $10^6 \times 5$ cells in 1ml phosphate buffered saline (PBS) pH 7.6 in a sterile 1.5 mL microtube, microfuge (9,000*g* for 1 min) to pellet the cells and aspirate the PBS.
2. Resuspend the pellet in 1 mL ice-cold sterile distilled water, microfuge as before for 2 min and aspirate the supernatant.
3. Resuspend the pellet in ice-cold sucrose/detergent buffer (0.32 *M* sucrose, 10 m*M* Trizma-base pH 7.5, 5 m*M* $MgCl_2{\cdot}6H_2O$, 1% (v/v) TritonX-100), microfuge at 12,000*g* for 5 min and again aspirate the supernatant.
4. Resuspend the pellet in 1mL Lysis Buffer (75 m*M* NaCl, 24 m*M* Na_2EDTA), and after mixing by inversion add 20 μL Proteinase K (20 mg/mL, Life Technologies), 20 mL RNAse A (10 mg/mL, Sigma), and 10 mL 10%SDS.
5. Remix by inversion and incubate at 60°C for 1 h or 37°C overnight.
6. After incubation, cool the tube on ice and add 400 μL of 5 *M* NaCl and repeatedly invert vigorously to yield a white precipitate.
7. Microfuge at 12,000*g* for 5–10 min and transfer the clear supernatant to a fresh sterile microtube.
8. Add two volumes of cold ethanol and mix by inversion at least 10× to precipitate the DNA before microfuging at 12,000*g* for 5 min and aspirating the ethanol.
9. Wash the pellet in 80% ethanol and microfuge at 12,000*g* for 2 min before partially drying the pellet at 37°C.
10. Suspend the pellet in 30 μL TE buffer (10 m*M* Trizma base pH 7.5, 1 m*M* Na_2 EDTA) and dissolve the DNA at 37–40°C with occasional mixing.
11. The DNA is then quantified by spectrophotometer (260 nm) and 10 μL digested with *HinfI* (or *HaeIII*) restriction enzymes (Life Technologies) according to the enzyme manufacturer's instructions. NB: for each set of digests standard DNA samples (e.g., HeLa, K562) should be run in parallel so that the DNA fingerprints on each blot can be checked for consistency.
12. Run an agarose Minigel to check the high molecular weight of uncut genomic DNA (1 μL) in parallel with restriction digest (1 μL) to confirm satisfactory digestion: only low molecular weight DNA below 5 Kb should be visible.

2.1.2. Southern Blots of Genomic DNA

1. Quantify the DNA in each digest (*see* **Subheading 2.1.1., step 11** above), then run an analytical agarose gel (electrophoresis buffer, TBE, is 20 m*M* Tris-HCl, 20 m*M* Na borate, 2 m*M* EDTA, 0.25% SDS, 0.5 mg/L ethidium bromide) of at least 20 cm length. Electrophorese 5 μg of digested DNA with loading buffer (15% Ficoll 400 and 0.25% bromophenol blue, in TBE, 15% v/v). Allow the 2.3 Kb fragment of a *Lambda/HinDIII* molecular weight marker to run the full length of the gel.

2. Gently agitate the gel with intermediate distilled water washes in the following buffers: 0.25 *M* HCl (15 min); 1.5 *M* NaCl, 0.5 *M* NaOH (30 min); 3 *M* NaCl, 0.5 *M* Tris-HCl pH 7.5 (30 min).
3. Lay the gel onto support paper (two sheets of Whatman 3MM, Whatman-Labsales, UK, or equivalent), drawing from a 20x SSC buffer reservoir, 20x SSC is 175 g/L NaCl, 88.2 g/L Tri-sodium citrate, pH 7.4.
4. Place a Nylon membrane (20 cm × 20 cm, Hybond N, Genetic Research Instruments Ltd, or equivalent) over the top surface of the gel and then cover with four similar sized sheets of Whatman 3MM paper soaked in the 20x SSC transfer buffer. Place two piles of absorbent paper hand towels (e.g., Kimberly-Clark, U.K.) over the soaked sheets and apply a 1 kg weight evenly on top of the towels.
5. Transfer of DNA fragments should be complete after blotting overnight, when the nylon membrane can be removed, dried, and fixed over a uv-transilluminator (315 nm for 5 min).

2.1.3. Visualization of DNA Fingerprints by Chemiluminescence with the Multilocus Probe 33.15

1. Wet Southern membranes (up to 10 per hybridisation) in 1x SSC (20xSSC in distilled water, 5% v/v) and place into 500 mL sterile prehybridisation solution (0.5 *M* Na_2HPO_4, pH 7.2, containing 0.1% SDS, w/v) and agitate at 50°C for 20 min.
2. Transfer each membrane individually to 160 mL hybridization buffer (900 mL/L prehybridization buffer, 100 mL/L of 100 g/L casein Hammarsten [Sigma]) in stringency wash Solution 2, (13.8 g/L maleic acid and 8.7 g/L NaCl in distilled water at pH 7.5) add the contents of one NICE™ probe (Cellmark Diagnostics-Zeneca) vial and gently agitate at 50°C for 20 min.
3. Transfer the membranes individually to 500 mL prewarmed Stringency wash solution 1 (160 mL/L 0.5 *M* Na_2HPO_4, pH 7.2 with 10 mL/L of 10%SDS) and gently agitate at 50°C for 10 min.
4. Repeat **step 3**.
5. Transfer each membrane into 500 mL Stringency wash solution 2 and gently agitate at room temperature for 10 min.
6. Repeat **step 5**.
7. Drain each membrane, and place DNA side up on a transparent polyester sheet, apply 3 mL Lumiphos 530 luminescent reagent (Cellmark Diagnostics) (in a fume cabinet), place another polyester sheet on top of the gel, squeezing out excess Lumiphos. Seal the membrane between the sheets with tape. The sealed gels are then put into an X-ray development cassette.
8. Fix two sheets of X-ray film (e.g., Fuji-RX) over the sealed membranes and incubate at 37°C for 3–5 h.
9. Develop the top X-ray film according to the manufacturer's instructions, and check that all the expected bands in a standard HeLa DNA fingerprint (e.g., European Collection of Cell Cultures, Salisbury, UK) are present and clear.
10. If increased exposure is required, reincubate the second X-ray film as before (additional films can be exposed for up to 3 d after the Lumiphos has been added).

Membranes can be stripped with 0.1% SDS at 80°C and rinsed in 1x SSC for reprobing.

2.2. Cell Typing by PCR of Intron G of the Aldolase Gene

1. Set up PCR reactions as described by Lessa and Applebaum ***(10)*** using the primers *Ald1* (TGTGCCCAGTATAAGAAGGATGG) and *Ald2* (CCCATCAGG GAGAATTTCAGGCTCCACAA) with negative controls (no primers, no template, unrelated plasmid template) and a standard genomic DNA of known profile (e.g., HeLa, *see* **Subheading 2.1.3.**).
2. Perform 35 cycles of PCR as follows: 94°C for 1 min (denaturation), 57°C for 45 s (annealing), 72°C for 2 min (extension). An initial denaturation (at 94°C for 3 min) and a final extension (at 72°C for 5 min) are also carried out.
3. PCR products are visualised in an ethidium bromide stained minigel (1.4% agarose) following electrophoresis at 75 v for 2 h in parallel with Molecular Weight Marker VI (Boehringer-Mannheim).
4. Typical PCR products are as follows:

Human and primate	0.50Kb, 0.38Kb, 0.30Kb, 0.19Kb
Mouse and rat	0.49Kb, 0.38Kb, 0.31Kb, 0.19Kb
Rabbit	0.48Kb, 0.30Kb, 0.19Kb
Dog	0.55Kb, 0.31Kb, 0.19Kb
Cat	0.56Kb, 0.32Kb
Frog	0.29Kb
Mosquito	0.99Kb, 0.85Kb, 0.60Kb

Further results showing the value of this technique for testing cell lines from a wider range of species are given in **ref. *8***.

3. Notes

There are several important questions which will be critical in the selection of cell identification system, they include the following:

What are the species of origin of the cell lines in use in the laboratory?
Is there a need for information on characteristics other than cell identification, such as genomic stability?
What resources, technology, and reagents are available to perform cell identity tests?

The ideal cell identification system would be capable of identifying the species or strain of origin of each cell line while also providing a unique marker, code, or profile for each individual cell line. Systems approaching this ideal would be expected to be developed in reference facilities such as culture collections (ATCC in the United States, ECACC in the United Kingdom, DSMZ in Germany, JCRB in Japan). The specific requirements for identification methods will vary between different laboratories, depending on the type of work performed and the available resources. For example, a laboratory dedicated to working with a large number of cell lines from human tumors will have a specific set of demands very different to those in a laboratory where cell lines from many different species are in use. While meeting the special needs for identification within a particular laboratory, it is vital that the identification system used will also immediately expose a culture that has been cross-contaminated or mistaken for another of completely different origin.

Analysis of minisatellite or microsatellite DNA loci using multiplexed PCR or multiple probes in Southern blot hybridization has provided a highly specific identification of a range of organisms and cell lines. Some commercially available simple tandem repeat (STR) kits are available which analyze nine different human loci simultaneously (e.g., AmpFLSTR Profiler Plus Amplification Kit, from Perkin Elmer, Foster City, CA). It is important to be aware that the specificity of identification by such methods will depend on matching the origin of the cells tested with the normal cell populations of origin for which the methods were validated Furthermore, while cross-hybridization will occur for some species, the probe sets used for human analysis are unlikely to be useful for cell lines from other species. Even when using a test fully validated for the species and population of origin, unusual results may occur, because cell lines may have genetic defects which will confuse these methods of identification. Nevertheless, the use of multiple probes for specific VNTR loci provides clearly-interpretable data, given that there will be a defined set of expected allele sizes for each VNTR locus.

Multilocus DNA fingerprinting of minisatellite or microsatellite DNA offers a major advantage in that a single method, such as that given above, will give specific banding patterns for a diverse range of species ***(23)***. However, microsatellite probes [e.g., $(GATA)_4$, $(GACA)_4$, $(GTG)_3$] often show a restricted species range for which they will yield useful intra-species differentiation. The "Jeffreys" probes 33.15 and 33.6 and the M13 phage protein III gene microsatellite probe ***(24)*** have proven valuable in identifying individuals for a remarkably wide range of species in ecological studies. Thus, these are excellent candidates for a general DNA fingerprinting method, where cell lines from many species are handled. However, as for the other VNTR-based methods, although the multilocus DNA fingerprint of a cell line is extremely valuable for identity testing and, to some extent for determining genetic stability, ***(18,20,21)*** it cannot be used to identify cross-contaminated cell lines without appropriate samples from the individual of original or standard profiles from contaminating cells such as HeLa.

Use of RAPD methods should be very carefully validated before they can be accepted to give reproducible results. However, some of the methods for PCR of conserved intron sequences, such as the method given here, have shown some potential in reproducible identification of cell lines for a wide range of species, although genes may not be sufficiently polymorphic to differentiate cells from different individuals ***(8)***. An additional disadvantage with PCR methods that is often overlooked is that competing PCR primers or template DNA competing for the same primers may result in a poor level of sensitivity ***(8)***. Thus, cross-contamination of cell lines may go undetected when using PCR that might be identified by Southern blot hybridization methods or the more traditional techniques of isoenzyme analysis and karyology.

When setting up an identification system, the methods of isoenzyme analysis and karyology should not be forgotten, as they both provide rapid and reliable identification of species of origin. These methods are identified in the guidelines from regulatory bodies, which is of particular interest for cells used in manufacturing processes ***(25,26)***. Karyology may also reveal unique marker chromosomes for identification and has been used as an indicator of genetic stability. The main restriction on the use of karyology is the need for experience and expertise if dealing with cell lines from a range of species. Even so, the rapidly developing area of molecular cytogenetics including techniques such as comparative genomic hybridization (CGH) and fluorescent *in situ* hybridization (FISH, *see* Chapter 5) may provide valuable new techniques for the future ***(27)***. Isoenzyme analysis is an extremely reliable and straightforward technique which can also provide confirmation of the species composition of hybrid cells. The major disadvantages of this technique, however, are the improbability of achieving specific identification of an individual cell line and that in the absence of a chromosome carrying a key isoenzyme, identification, even to the level of species, may not be possible.

4. Conclusion

Cell identification techniques generally fall into one of two categories:

1. Those that will give specific identification (e.g., DNA fingerprinting).
2. Those that identify a particular group or species (e.g., isoenzyme analysis).

Very often, the molecular techniques that give highly specific identification will be limited to a restricted group of species (as with single-locus satellite probes). Multilocus DNA fingerprinting with the Jeffreys probes or the M13 probe is useful over a wide range of species; however, it does not determine the species of origin. It is often necessary to decide whether to set up a new system for each species, such as STRs, or run two techniques in parallel: one giving a specific profile (such as multilocus fingerprinting) and the other identifying the species of origin (e.g., species-specific PCR, isoenzyme analysis). The latter course is the one often selected in culture collections where there is the extreme combination of many species and many cell lines from each species.

References

1. Nelson-Rees, W. A., Daniels, D. W., and Flandermeyer, R. R. (1989) *Science* **212,** 446–452.

2. Hsu, T. C. (1952) *J Hered.* **43,** 167–172.
3. Tjio, J. H. and Leven, A. (1956) *Hereditas* **42,** 1–6.
4. Simpson, W. F., Stulberg, C. S., and Petersen, W. D. (1978) *Tissue Culture Association Manual* Vol. 4, American Type Culture Collection. Rockville, MD, pp. 771–774.
5. O'Brien, S. J., Kleiner, G., Olson, R., and Shannon, J. E. (1977) *Science* **195,** 1345–1348.
6. Ferrone, S., Pellegrino, M. A., and Reisfeld, R. A. A. (1971) *J. Immunol.* **107,** 613–615.
7. Jeffreys, A. J., Wilson, V., and Thein, S-L. (1985) *Nature* **316,** 76–79.
8. Stacey, G. N., Hoelzl, H., Stephenson, J. R., and Doyle, A. (1997) *Biologicals* **25.**
9. Christensen, B., Hansen, C., Debiek-Rychter, M., Kieler, J., Ottensen, S., and Schmidt, J. (1993) *Br. J. Cancer.* **68,** 879–884.
10. Lessa, E. P. and Applebaum, G. (1993) *Mol. Ecol.* **2,** 119–129.
11. Williams, J. G. K., Kubelik, A. R., Livak, K. L., Rafalski, J. A., and Tingey, S. V. (1990) *Nucl. Acids Res.* **18,** 6531–6535.
12. Jeffreys, A. J., Wilson, V., and Thein, S-L. (1985) *Nature* **314,** 67–73.
13. Vassart, G., Georges, M., Monsieur, R., Brocas, H., Lequarre, A. S., and Christophe, D. (1987) *Science* **235,** 683–684.
14. Stacey, G. N., Bolton, B. J., and Doyle, A. (1991) DNA fingerprinting: Approaches and Applications. (Burke, T., Dolf, G., Jeffreys, A. J., and Wolf, R., eds.), Birkhauser, Basel, pp. 361–370.
15. Thacker, J., Webb, M. B. T., and Debenham, P. G. (1988) *Som. Cell Mol. Genet.* **14,** 519–525.
16. Van Helden, P. D., Wiid, I. J. F., Albrecht, C. F., Theron, E., Thornley, A. L., and Hoal-van Helden, E. G. (1988) *Cancer Res.* **48,** 5660–5662.
17. Stacey, G. N., Bolton, B. J., and Doyle, A. (1992) *Nature* **391,** 261–262.
18. Gilbert, D. A., Reid, Y. A., Gail, M. H., Pee, D., White, C., Hay, R. J., and O'Brien, S. J. (1990) *Am. J. Hum. Gen.* **47,** 499–517.
19. Racher, A. J., Stacey, G. N., Bolton, B. J., Doyle, A., and Griffiths, J. B. (1994) Genetic and biochemical analysis of a murine hybridoma in long term continuous culture, in *Animal Cell Technology: Products for Today Prospects for Tomorrow.* (Spier R. J., Griffiths J. B., and Berthold W., eds.), Butterworth-Heinemann Ltd., Oxford, pp. 69–75.
20. Doherty, I., Smith, K. T., and Lees, G. M. (1994) DNA fingerprinting as a quality control marker for the genetic stability of production cells, in *Animal Cell Technology: Products for Today Prospects for Tomorrow.* (Spier R. J., Griffiths J. B., and Berthold W., eds.), Butterworth-Heinemann Ltd., Oxford, pp. 76–79.
21. Stacey, G. N., Bolton, B. J., Morgan, D., Clark, S. A., and Doyle, A. (1992) *Cytotechnology* **8,** 13–20.
22. Stacey, G. N., Bolton, B. J., Doyle, A., and Griffiths, J. B. (1992) *Cytotechnology* **9,** 211–216.
23. Burke, T., Dolf, G., Jeffreys, A. J., Wolf, R. (Eds.) (1990) *DNA fingerprinting: Approaches and Applications,* Birkhauser, Basel.

24. Stacey, G. N. (1994) Contributed protocol in: *Culture of Animal Cells* (3rd ed.). (Freshney, I. R., ed.), Wiley-Liss, New York, pp. 210.
25. CBER (1993) Points to consider in the characterisation of cell lines used to produce biologicals. Centre for Biologics Evaluation and Research, NIH, Bethesda, MD.
26. World Health Organisation expert committee on biological standardisation and executive board (1998) Requirements for the use of animal cells as *in vitro* substrates for the production of biologicals. World Health Organisation, Geneva.
27. Kallioniemi, A., Kallioniemi, O-P., Sudar, D., Rutovitz, D., Gray, J. W., Waldman, P., and Pinkel, D. (1992) *Science* **258,** 818–821.

III

Cell Evaluation Protocols

10

Cell Counting and Viability Measurements

Michael Butler

1. Introduction

The growth of mammalian cells in culture can be monitored by a number of parameters related to the increase of cellular biomass over time. The simplest method is by cell counting at regular intervals. In routine cultures this would be performed once a day, which corresponds to the approximate doubling time of mammalian cells during the exponential growth phase. This would establish an overall growth profile of a culture. More frequent counts would be required to follow more subtle changes that may, for example, be associated with the cell growth cycle.

The two direct cell counting methods routinely used are performed visually through a microscope or electronically by a particle counter. Both methods depend on obtaining a sample of an even distribution of cells in suspension. Therefore, it is extremely important to ensure that the culture is well mixed by stirring or shaking before taking a sample.

Indirect methods of estimating cell growth rely on the measurement of an intracellular cell component such as DNA or protein or, alternatively, an extracellular change such as nutrient depletion or an enzyme activity released by the cells. Indirect methods of growth estimation depend on a relationship between the measured parameter and cell concentration. However, it is important to realize that these relationships are rarely linear over the course of a culture. It is well documented that the total protein content and specific enzyme activity levels measured on a per cell basis vary substantially over the course of a culture owing to changes in the growth rate and composition of the culture medium.

In some situations (such as those that may occur in immobilized cell bioreactor systems), an indirect measurement of cell growth may be the only

From: *Methods in Biotechnology, Vol. 8: Animal Cell Biotechnology*
Edited by: N. Jenkins © Humana Press Inc., Totowa, NJ

available option. This can be used to monitor the progress of a culture. However, care must be taken if such data are used in comparative analysis between cultures, as differences may be a reflection of changes in metabolic or functional activity rather than of cell concentration.

Viability is a measure of the metabolic state of a cell population, which is indicative of the potential of the cells for growth. One of the simplest assay types is dye exclusion, which is an indication of the ability of the cell membrane to exclude a dye. This may be included in the protocol for microscopic cell counting. More sophisticated measures involve the ability of cells to synthesize DNA or protein. A further metabolic assay measures the intracellular adenylate nucleotide concentrations. This allows determination of the energy charge, which is an index of the metabolic state of the cells. These viability assays are described in **Subheading 4.**

2. Direct Methods of Cell Counting

2.1. Cell Counting by Hemocytometer

The improved Neubauer hemocytometer consists of a thick glass plate that fits onto the adjustable stage of a microscope. A grooved calibrated grid is observed through the microscope on the hemocytometer surface (**Fig. 1**). A cell suspension is put onto the grid by touching the end of a capillary tube containing the cell suspension at the edge of a cover slip placed on the upper surface of the hemocytometer. The cells are then counted in a standard volume (usually 5 × 0.1 µL) as defined by the area of the grid. A hand-held tally counter helps in counting.

Trypan blue is often added to the cell suspension before counting ***(1)***. The dye penetrates the membrane of nonviable cells, which are stained blue and which can therefore be distinguished from viable cells.

2.1.2. Materials

1. Phosphate-buffered saline (PBS; 0.1 *M* NaCl, 8.5 m*M* KCl, 0.13 *M* Na_2HPO_4, 1.7 m*M* KH_2PO_4, pH 7.4).
2. Trypan blue reagent: 0.2% w/v trypan blue in PBS.

2.1.3. Method

1. Add an equal volume of trypan blue reagent to a cell suspension and leave for 2 min at room temperature.
2. Introduce a sample into the hemocytometer chamber by a Pasteur pipet.
3. Count cells on each of five grid blocks defined by triple lines in the hemocytometer chamber (*see* **Note 1**).
4. Determine the cell concentration (cells/mL) in the original sample = (2 × total count/5) × 10^4. (The calculation is based on the volume of each grid block = 0.1 µL.)

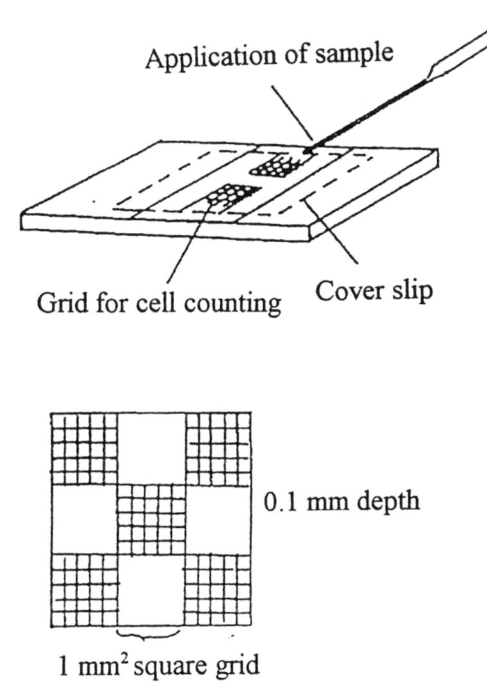

Fig. 1. The hemocytometer. A sample of cells is loaded between the plate and the cover slip. The cells are then counted over five large grids, each of which corresponds to a sample volume of 0.1 µL.

The percentage of cells that are not stained with trypan blue is a measure of the viability.

2.2. Nuclei Counting by Hemocytometer

A modification of the hemocytometer method involves counting nuclei. Incubation of cell samples in a mixture of citric acid and crystal violet causes cells to lyse and the released nuclei to stain purple *(2)*. Nuclei counting is well suited to the determination of anchorage-dependent cells, for example, when they are attached to microcarriers.

2.2.1. Materials

1. Crystal violet reagent: 0.1% w/v crystal violet in 0.1 *M* citric acid.

2.2.2. Method

1. Allow microcarriers from a culture sample (1 mL) to settle to the bottom of a centrifuge tube.
2. Remove clear supernatant by aspiration.

3. Add 1 mL of crystal violet reagent.
4. Incubate at 37°C for at least 1 h.
5. Introduce a sample into the hemocytometer chamber and count the purple-stained nuclei as for whole cells (*see* **Note 2**).

2.3. Coulter Counter

The principle of an electronic cell counter (or Coulter counter) is that a predetermined volume (usually 0.5 mL) of a cell suspension diluted in buffered saline is forced through a small hole (diameter 70 μm) in a tube by suction. The cells interrupt the current flow between two electrodes, one inside and one outside the glass tube. This produces a series of pulses recorded as a signal on the counter. Particles smaller than cells can be eliminated from the count by setting a lower threshold of detection. The largest particle size is determined by the size of the hole in the tube.

2.3.1. Materials

1. Saline solution: 0.7% NaCl, 1.05% citric acid, 0.1% mercuric chloride in distilled water.

2.3.2. Method

1. Add 0.5 mL of a cell suspension (10^5–10^6 cells/mL) to 19.5 mL of the saline solution.
2. Introduce the suspension into a Coulter counter (Courter Electronics, Luton, UK).
3. From standard settings of the counter, 0.5 mL of the suspension is counted. Multiply this count by 40 to give the original cell concentration (*see* **Note 3**).

3. Indirect Methods of Cell Determination

A number of colorimetric methods are based on the measurement of cell components. These are relatively simple and are suitable for multiple samples. However, the contents of cells can vary dramatically during culture. For example, the protein and enzyme content per cell will be high during exponential growth but lower in the lag or stationary phases.

3.1. Protein Determination

Total cell protein can be used as a measure of biomass (total cellular material). The protein content of a mammalian cell is typically 100–500 pg/cell. These measurements are also useful in the determination of specific enzyme activities, which are commonly expressed as the maximum measured reaction velocity of an enzyme per total cell protein.

The most common colorimetric assays are the Lowry and Bradford methods. Of these the Bradford assay is favored because of its speed and sensitivity as well as the negligible interference from other cell components *(3)*. By this method lysed cells are added to the reagent, Coomassie blue. A blue color, which develops within 10 min, can be measured by a colorimeter or spectrophotometer and compared with standard proteins.

3.1.1. Materials

1. Bradford's reagent: dissolve 100 mg of Coomassie brilliant blue G (Sigma, Poole, UK) in 95% ethanol (50 mL) and 85% phosphoric acid (100 mL). After the dye dissolves make the solution up to 1 L with distilled water. Alternatively, a dye (Coomassie) reagent liquid concentrate can be purchased from Bio-Rad (Hercules, CA).

3.1.2. Method

1. Homogenize or sonicate a cell suspension (10^6 cells/mL).
2. Add 5 mL Bradford's reagent to 100 µL of the lysed cell sample (0–0.5 mg/mL protein).
3. Incubate for 10 min at room temperature.
4. Measure the absorbance at 595 nm.
5. Determine the sample concentrations from a standard curve, which is established from standard solutions of bovine serum albumin (BSA) at 0–0.5 mg/mL protein.

3.2. DNA Determination

A commonly used protocol involves treatment of the solubilized cells with fluorescent reagents that bind to DNA. Fluorescence detection offers high sensitivity with reagents such as Hoechst 33258 (4) or 4',6-diamidino-2-phenylindole (DAPI; **ref. 5**), both from Sigma.

3.2.1. Hoechst Method

3.2.1.1. Materials

1. Buffer: 0.05 *M* $NaPO_4$, 2.0 *M* NaCl, 2 m*M* EDTA, pH 7.4.
2. Hoechst reagent: 0.1 µg/mL Hoechst 33258 in buffer.
3. Standard DNA solution: 8 mg/mL of calf thymus DNA (Sigma) in distilled water.

3.2.1.2. Method

1. Homogenize or sonicate to lyse a cell suspension (10^5 cells/mL) in buffer.
2. Dilute lysate or standard DNA solution 1:10 in Hoechst reagent.
3. Measure fluorescence with an excitation λ of 356 nm and emission λ of 492 nm.
4. Determine DNA concentration by reference to standard DNA.

3.2.2. DAPI Method

3.2.2.1. Materials

1. Buffer: 5 m*M* HEPES, 10 mM NaCl, pH 7.0.
2. DAPI reagent: a stock solution (100×) contains 300 mg DAPI in buffer.
3. Standard DNA solution: 8 mg/mL of calf thymus DNA (Sigma) in distilled water.

3.2.2.2. METHOD

1. Homogenize or sonicate to lyse a cell suspension (10^5 cells/mL).
2. Dilute 150 μL lysed cell suspension with 850 μL buffer.
3. Prepare a DAPI solution (10×) by diluting 100 μL of DAPI stock solution with 900 μL of buffer and mix well. Prepare a DAPI working solution by adding 0.5 mL of DAPI (10×) to 4.5 mL of buffer.
4. Add 50 μL of DAPI working solution to each cell suspension or standard DNA (up to 0.8 μg) in a tube kept dark by a foil cover.
5. Vortex the tubes and let stand for 30 min.
6. Measure fluorescence with an excitation λ of 372 nm and emission λ of 454 nm.
7. Determine DNA concentration by reference to the standard DNA (*see* **Note 4**).

3.3. Glucose Determination

Cell growth can be monitored by changes in the concentration of key components of the culture medium. The rate of change in the glucose content of the medium may be suitable for such an assay as an indirect measure of cell concentration. Alternatives include measurement of lactic acid production or oxygen consumption.

Correlations have been shown between cell concentration and rates of consumption or production of these components. This relationship may be constant for a particular cell line under a given set of conditions. However, if the cell line or any of the culture conditions are altered, the relationship between substrate consumption or product formation and cell number will change.

3.3.1. Glucose Oxidase Assay

Glucose can be determined by a colorimetric assay utilizing the two enzymes glucose oxidase and peroxidase ***(6)***:

$$\text{D-glucose} + H_2O + O_2 \rightleftharpoons \text{D-gluconic acid} + H_2O_2 \quad (1)$$

$$H_2O_2 + \text{reduced } o\text{-dianisidine} \rightleftharpoons 2H_2O + \text{oxidized o-dianisidine (brown)} \quad (2)$$

$$\text{oxidized } o\text{-dianisidine (brown)} + H_2SO_4 \rightleftharpoons \text{oxidized o-dianisidine (pink)} \quad (3)$$

Reaction 1 is catalyzed by glucose oxidase (GOD) and reaction 2 by peroxidase (POD). The dye *o*-dianisidine hydrochloride is reduced by hydrogen peroxide to a product that has a pink color in the presence of sulfuric acid (reaction 3) and is measured colorimetrically. The glucose oxidase kit from Sigma contains glucose oxidase/peroxidase reagent and *o*-dianisidine reagent.

3.3.1.1. MATERIALS

1. Glucose oxidase/peroxidase reagent: dissolve the contents of a reagent capsule (Sigma) in 39.2 mL of distilled water. Each capsule contains 500 U of glucose oxidase and 100 U of peroxidase.
2. *o*-Dianisidine reagent: dissolve the contents of a vial of *o*-dianisidine (Sigma) in 1 mL of dissolved water. Each vial contains 5 mg of *o*-dianisidine dihydrochloride.

3. Assay reagent: mix 0.8 mL of *o*-dianisidine reagent with 39.2 mL of glucose oxidase/peroxidase reagent.
4. Glucose standard solution: 1 mg/mL of D-glucose.
5. Sulfuric acid, 12 *M*.

3.3.1.2. Method

1. Start the reaction by adding 2 mL of assay reagent to glucose standard or culture media supernatant (0.01–0.1 mL). Make the assay volume up to 3 mL with distilled water.
2. Allow the reaction to proceed for 30 min at 37°C.
3. Stop the reaction by adding 2 mL of 12 *M* H_2SO_4.
4. Measure the absorbance at 540 nm.
5. Determine the glucose concentration of the media samples against a standard value obtained with the glucose solution.

3.3.2. Hexokinase Assay

Glucose can also be measured enzymatically in the following two reactions catalyzed by hexokinase (HK) and glucose 6-phosphate dehydrogenase (G6PDH; **ref.** *7*).

$$\text{D-glucose} + \text{ATP} \rightleftharpoons \text{glucose-6P} + \text{ADP} \quad (4)$$

$$\text{glucose-6P} + \text{NAD} \rightleftharpoons \text{6-phosphogluconate} + \text{NADH} + \text{H}^+ \quad (5)$$

HK converts glucose into glucose 6-phosphate (G-6P) in the presence of adenosine triphosphate (ATP; **Reaction 4**). The G-6P is immediately converted into 6-phosphogluconate by G-6P dehydrogenase (**Reaction 5**). The associated formation of reduced nicotinamide adenine dinucleotide (NADH) is monitored by the change in absorbance at 340 nm, and this is proportional to the concentration of glucose originally present. The hexokinase kit from Sigma contains a hexokinase/G6PDH reagent. The kit includes a glucose standard solution (1 mg/mL).

3.3.2.1. Materials

1. Glucose (HK) assay reagent: dissolve the contents of a reagent vial (Sigma) into 20 mL of distilled water. The dissolved reagent contains 1.5 m*M* NAD, 1.0 m*M* ATP, 1 U/mL HK, and 1 U/mL G6PDH.
2. Glucose standard solution: 1 mg/mL of D-glucose.

3.3.2.2. Method

1. Mix 10–200 µL of standard glucose solution or sample of culture media with 1 mL of assay reagent. Make the total assay volume up to 2 mL with distilled water.
2. Incubate at room temperature for 15 min.
3. Measure the absorbance at 340 nm.
4. Determine the glucose concentration of the media samples against a standard value obtained with the glucose solution (*see* **Note 5**).

3.3.3. The Glucose Analyzer

A modification of the glucose oxidase assay system is used in an analyzer such as the YSI model 27 Industrial Analyzer (Yellow Springs Instrument, Yellow Springs, OH). The instrument is provided with various membranes containing immobilized enzymes appropriate for measuring a particular analyte such as glucose or lactic acid. The sample is injected into a membrane that converts the glucose to hydrogen peroxide, which can be determined by a sensor system based on a Clark electrode. The latter consists of a platinum electrode that measures the hydrogen peroxide amperometrically:

$$H_2O_2 \rightleftharpoons 2H^+ + O_2 + 2e^- \quad (6)$$

$$AgCl + e^- \rightleftharpoons Ag + Cl^- \quad (7)$$

Current flow in the platinum anode is linearly proportional to the local concentration of hydrogen peroxide. This electrode is maintained at an electrical potential of 0.7 V with respect to a silver/silver chloride reference electrode, the potential of which is determined by **Reaction 7** above. The signal current, which is proportional to the quantity of injected glucose, is converted to a voltage by the instrument circuitry.

3.3.3.1. Method

1. Fit the appropriate membrane into the analyzer for glucose analysis.
2. Calibrate the instrument with standard glucose solutions (2–5 g/L).
3. Inject 25 μL of a cell-free sample of culture supernatant into the instrument and compare with standard readings (*see* **Notes 6** and **7**).

4. Viability Measurements

Viability is a measure of the proportion of live, metabolically active cells in a culture, as indicated by the ability of cells to divide or to perform normal metabolism. The viability is measured by an indicator of the metabolic state of the cells (such as energy charge) or by a functional assay based on the capacity of cells to perform a specific metabolic function.

The viability may be determined from simple assays such as dye exclusion, whereby cells are designated as either viable or nonviable. The viability is usually expressed as a percentage of viable cells in a population (viability = [number of viable cells/total number of cells] × 100).

4.1. Dye Exclusion

Cell counting by hemocytometer as described earlier in this chapter can be adapted to measure viability. The most common is the dye exclusion method, in which loss of viability is recognized by membrane damage resulting in penetration of the dye, trypan blue. Other dyes that can be used include erythrosin B, nigrosin, and fluorescein diacetate.

4.2. Tetrazolium Assay

The tetrazolium assay is a measure of cellular oxidative metabolism. The tetrazolium dye 3-(4,5-dimethylthiazol-2-yl)-2,5-diphenyltetrazolium bromide (MTT) is cleaved to a colored product by the activity of NAD(P)H-dependent dehydrogenase enzymes, and this indicates the level of energy metabolism in cells ***(8)***. The color development (yellow to blue) is proportional to the number of metabolically active cells. The assay response may vary considerably between cell types. Cells of biotechnological importance such as CHO cells and hybridomas can be monitored with the MTT assay.

4.2.1. Materials

1. MTT reagent: 5 mg/mL of the tetrazolium dye (Sigma) MTT in PBS, pH 7.4.
2. Sodium dodecyl sulfate (SDS) reagent: 45% w/v *N,N*-dimethyl formamide in water adjusted to pH 4.5 with glacial acetic acid.

4.2.2. Method

1. Remove the media from adherent cells in a multiwell plate and add 0.1 mL MTT reagent. Alternatively, add 0.1 mL MTT reagent to a 1-mL cell suspension in PBS.
2. Incubate for 2 h at 37°C.
3. Add 600 µL of SDS reagent and mix to solubilize the formazan crystals.
4. Measure the absorbance at 570 nm (*see* **Notes 8–10**).

4.3. Colony-Forming Assay

The most precise of all the methods of viability measurement is the colony-forming assay. Here the ability of cells to grow is measured directly ***(9)***. A known number of cells at low density is allowed to attach and grow on the surface of a Petri dish. If the cell density is kept low, each viable cell will divide and give rise to a colony or cluster of cells. From this the plating efficiency is determined as the number of colonies scored/100 cells plated × 100. Although the colony-forming assay is time consuming, it has been widely used in cytotoxicity studies.

A less precise method of determining the viability by the cellular reproductive potential is from the lag phase of a growth curve. **Figure 2** shows that by extrapolation from the linear portion of a growth curve to time zero, the derived cell number can be compared with the original cell count. This method can be easily adapted to determine how a particular treatment (such as addition of a toxic compound) affects cell viability.

4.4. Lactate Dehydrogenase Determination

A decrease in viability of cells is usually associated with a damaged cell membrane, which causes the release of large molecules such as enzymes from the cell into the medium. Thus the loss of cell viability may be followed by an increase in enzyme activity in the culture medium ***(10)***. Lactate dehydrogenase

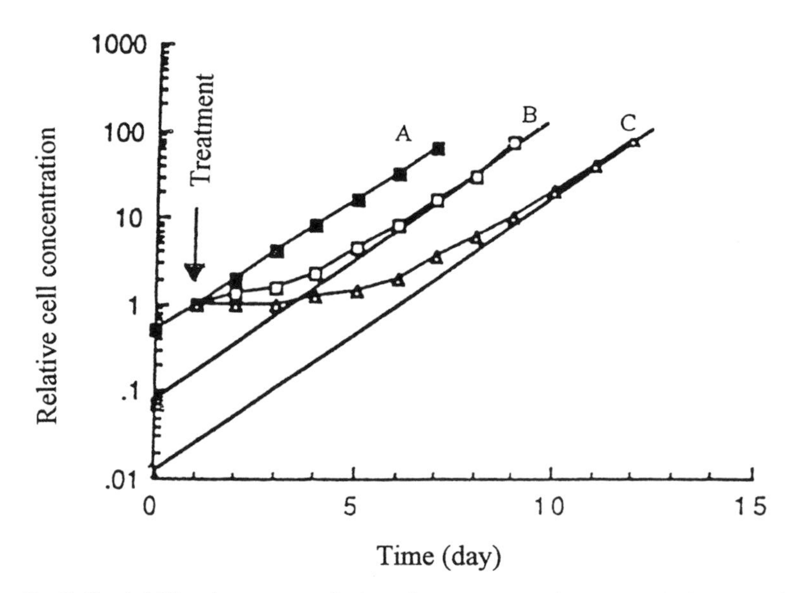

Fig. 2. Cell viability by extrapolation from a growth curve. Culture A is of an untreated control. Cultures B and C have different treatments at d 1, such as addition of a toxic compound. The relative surviving fraction of the treated cultures can be used as a measure of the effect of the treatment on cell viability. B = 0.1/0.8 = 0.125; C = 0.01/0.8 = 0.0125 (from **ref. 9**).

(LDH) activity is the enzyme most commonly measured in this technique. The enzyme activity can be easily measured by a simple spectrophotometric assay involving the oxidation of NADH in the presence of pyruvate. The reaction is monitored by a decrease in ultraviolet (UV) absorbance at 340 nm.

$$\text{pyruvate} + \text{NADH} + \text{H}^+ \xrightleftharpoons{\text{LDH}} \text{lactate} + \text{NAD}^+$$

NADH absorbs at $\lambda = 340$ nm.

4.4.1. Method

1. Mix 2.8 mL Tris-HCl (0.2 *M*), pH 7.3, 0.1 mL NADH (6.6 m*M*), and 0.1 mL sodium pyruvate (30 m*M*) in a cuvette.
2. Preincubate for 5 min at the desired reaction temperature (25°C or 37°C).
3. Start reaction by adding 50 µL of sample or standard LDH enzyme (Sigma).
4. Record enzyme activity as an absorbance decrease at 340 nm (*see* **Note 11**).

4.5. Intracellular Energy Charge

The energy charge is an index based on the measurement of the intracellular levels of the nucleotides adenosine monophosphate (AMP), adenosine diphosphate (ADP), and ATP. The energy charge = ([ATP] + 0.5 × [ADP])/([ATP] + [ADP] + [AMP]) and is based on the interconversion of the three adenylate nucleotides in the cell: AMP ⇄ ADP ⇄ ATP. This index varies between the theoretical limits of 0 and 1. For normal cells values of 0.7–0.9 would be expected, but a gradual decrease in the value gives an early indication of loss of viability by a cell population.

These nucleotide concentrations can be measured by chromatography (high-performance liquid chromatography) or by luminescence using the luciferin-luciferase enzyme system ***(11,12)***. The luminescence assay is dependent on the emission of light resulting from the enzymic oxidation of luciferin, a reaction requiring ATP.

$$ATP + LH_2 + O_2 \rightarrow AMP + PPi + CO_2 + L + light$$

ADP and AMP can also be measured by the luciferase assay after conversion to ATP by coupled enzymic reactions:

1. Pyruvate kinase: ADP + PEP → ATP + pyruvate.
2. Myokinase: AMP + CTP → ADP + CDP.

Where PEP is phosphoenol pyruvate, CTP is cytosine triphosphate, and CDP is cytosine diphosphate.

4.5.1. Materials

1. ATP monitoring reagent (ATP-MR; LKB/BioOrbit, Turku, Finland) contains a lyophilized mixture of firefly luciferase, D-luciferin, bovine serum albumin, magnesium acetate, and inorganic pyrophosphate. Reconstitute each vial with 4 mL buffer plus 1 mL potassium acetate (1 *M*).
2. ATP standards (LKB): ATP (0.1 μmol) and magnesium sulfate (2 μmol).
3. Buffer: 0.1 *M* Tris-acetate, pH 7.75.
4. Pyruvate kinase (PK)-PEP reagent: 55 μL tricyclohexylammonium salt of PEP (0.2 *M*) + 50 μL PK (500 U/mg) in Tris buffer.
5. Myokinase (MK)-CTP reagent: 95 μL myokinase (2500 U/mg) + 10 μL CTP (110 m*M*) in Tris buffer.

4.5.2. Method

1. Extract soluble nucleotides by addition of 0.1 mL perchloric acid (20% v/v) to 1 mL of a cell culture sample (10^6 cells/mL).
2. Place on ice for 15 min and centrifuge for 5 min at 10,000*g*.
3. Remove supernatant and neutralize with 5 *M* KOH.
4. For ATP determination: mix 860 μL buffer, 100 μL ATP-MR, and 10 μL sample.
5. For ADP determination: add a further 10 μL PK-PEP.
6. For AMP determination: add a further 10 μL MK-CTP.
7. For standardization: add a further 10 mL ATP standard.
8. Measure the light emission in a luminometer (for example, LKB 1250) after 1 min of each stage of addition (*see* **Note 12**).

4.6. Rate of Protein Synthesis

The rate of protein synthesis of intact cells can be measured by incubation in standard culture medium to which is added a radioactively labeled amino acid. Any radioactive amino acid is suitable, but those most commonly used are [^{3}H]leucine or [^{35}S]methionine.

4.6.1. Method

1. Add [^{3}H]leucine or [^{35}S]methionine (Amersham, Little Chalfont, UK) at a final specific activity of 20–40 μCi/mL to cell suspension at 5–10 × 10^6 cells/mL.
2. Remove 5–10 × 10^5 cells at each time point up to 4–6 h.
3. Isolate cell pellet by centrifugation in a microcentrifuge tube and wash in PBS.
4. Precipitate protein by addition of 500 μL trichloroacetic acid (TCA; 5%) containing unlabeled amino acids.
5. Wash the protein precipitate three times in the TCA solution.
6. Add 30 μL NCS™ tissue solubilizer (Amersham) to the pellet and leave for 60 min.
7. Cut tip of tube and place in scintillation fluid for radioactive counting (*see* **Note 13**).

5. Notes

1. The hemocytometer counting method is the most commonly used assay for cell viability. The method is simple and effective but can be laborious for multiple samples. At least 100 cells should be counted for statistical validity of the final value.
2. Care must be taken in interpreting nuclei counts as cells can become binucleated, particularly when growth is arrested. As a result the nuclei concentration may be higher than the cell concentration *(13)*.
3. The major advantage of the Coulter counter method is its speed of analysis, making it suitable for counting a large number of samples. The method is based on the number of particles contained in suspension, and consequently the proportion of viable cells in the sample cannot be determined. It must be ensured that cell aggregates are not present in the sample; otherwise, the cell count will be underestimated. The Coulter counter can also be used to determine the size distribution of a cell population by careful control of the threshold settings of the instrument.
4. The DNA content of diploid cells is usually constant, although variations can occur as a result of the distribution of cells through the cell cycle. Cells in the G_1 phase have the normal diploid content of DNA, which is typically 6 pg/cell. DNA measurement is probably one of the best indicators of cell concentration in solid tissue *(14)*.
5. The sensitivity of the HK assay for glucose can be increased by measuring the rate of increase of absorbance at 340 nm. This can be achieved with a recording spectrophotometer or using the kinetic mode of a multiwell plate reader.
6. The glucose analyzer is particularly suitable for the analysis of glucose in multiple samples of culture medium.
7. By selection of the appropriate membrane in the analyzer, various analytes can be determined, such as glucose, sucrose, starch, lactose, galactose, glycerin, lactate, or ethanol.

8. The tetrazolium method is particularly convenient for the rapid assay of replicate cell cultures in multiwell plates. Plate readers are capable of measuring the absorbance of each well of a standard 96-well plate.
9. It is important to ensure that the colored formazan salt formed from MTT is completely dissolved in the SDS reagent.
10. Alternative tetrazolium salts, such as XTT and WST-1 (available from Boehringer-Mannheim, Lewes, UK), can be used in this assay. These salts form soluble colored products.
11. The LDH assay is well suited for the determination of multiple samples, particularly if a multiwell plate reader is available. Care must be taken when interpreting the results by this method because the LDH content per cell can change considerably during the course of batch culture. The loss of cell viability can be expressed as the activity of LDH in the medium as a proportion of total LDH in the culture.
12. The measurement of energy charge is more time consuming than the routine counting procedures discussed earlier but can allow a means of monitoring the decline in the energy metabolism of a cell culture that occurs during the loss of viability.
13. The cells should be incubated in the medium for sufficient time to measure radioactivity in the extracted cell pellet. Normally 4–6 h is sufficient, but this may be longer ***(15)***. The rate of DNA synthesis of a cell population can be determined in an assay similar to that described for protein synthesis but using a radioactively labeled nucleotide precursor such as tritiated thymidine ([^{3}H]TdR or deoxycytidine ([^{3}H]CdR; Amersham, Buckinghamshire, UK). The exposure period may be short (30–60 min) for DNA synthesis rate determinations, and a specific activity of 1 μCi/mL of culture is sufficient. Higher specific activities may be required if culture media contain the corresponding nonradioactive components such as methionine or thymidine.

References

1. Patterson, M. K. (1979) Measurement of growth and viability of cells in culture. *Methods Enzymol.* **58,** 141–152.
2. Sanford, K. K., Earle, W. R., Evans, V. J., Waltz, H. K., and Shannon, J. E. (1951) The measurement of proliferation in tissue cultures by enumeration of cell nuclei. *J. Natl. Cancer Inst.* **11,** 773–795.
3. Bradford, M. (1976) A rapid and sensitive method for the quantitation of microgram quantities of protein using the principle of protein-dye binding. *Anal. Biochem.* **72,** 248–254.
4. Labarca, C. and Paigen, K. (1980) A simple, rapid and sensitive DNA assay procedure. *Anal. Biochem.* **102,** 344–352.
5. Brunk, C. F., Jones, K. C., and James, T. W. (1979) Assay for nanogram quantities of DNA in cellular homogenates. *Anal. Biochem.* **92,** 497–500.
6. Bergmeyer, H. U. and Bernt, E. (1974) *Methods of Enzymatic Analysis*, 2nd ed., vol. 3 (Bermeyer, H. U., ed.), VCH, Weinheim, pp. 1205–1212.

7. Kunst, A., Draeger, B., and Ziegenhorn, J. (1984) *Methods of Enzymatic Analysis*, 3rd ed., vol. 8 (Bermeyer, H. U., ed.), VCH, Weinheim, pp. 163–172.
8. Mosmann, T. (1983) Rapid colorimetric assay for cellular growth and survival: application to proliferation and cytotoxicity assays. *J. Immunol. Methods* **65,** 55–63.
9. Cook, J. A. and Mitchell, J. B. (1989) Viability measurements in mammalian cell systems. *Anal. Biochem.* **179,** 1–7.
10. Wagner, A., Marc, A., and Engasser, J. M. (1992) The use of lactate dehydrogenase (LDH) release kinetics for the evaluation of death and growth of mammalian cells in perfusion reactors. *Biotechnol. Bioeng.* **39,** 320–326.
11. Holm-Hansen, O. and Karl, D. M. (1978) Biomass and adenylate energy charge determination in microbial cell extracts and environmental samples. *Methods Enzymol.* **57,** 73–85.
12. Lundin, A., Hasenson, M., Persson, J., and Pousette, A. (1986) Estimation of biomass in growing cell lines by adenosine triphosphate assay. *Methods Enzymol.* **133,** 27–42.
13. Berry, J. M., Heubner, E., and Butler, M. (1996) The crystal violet nuclei staining technique leads to anomalous results in monitoring mammalian cell cultures. *Cytotechnology* **21,** 73–80.
14. Kurtz, J. W. and Wells, W. W. (1979) Automated fluorometric analysis of DNA, protein and enzyme activities: application of methods in cell culture. *Anal. Biochem.* **94,** 166–175.
15. Dickson, A. J. (1991) Protein expression and processing, in *Mammalian Cell Biotechnology: A Practical Approach* (Butler, M., ed.), Oxford University Press, Oxford, pp. 85–108.

11

Monitoring Animal Cell Growth and Productivity by Flow Cytometry

Mohamed Al-Rubeai

1. Introduction

Flow cytometry provides a rapid and accurate means of analyzing individual cells in suspension. Several parameters can be detected simultaneously and analyzed to provide information on the relationship between them. Fluorescently labeled cells can give quantitative data on specific target molecules and their distribution in the cell population. Moreover, subpopulations of cells with specific attributes can be discriminated and physically sorted for further development. These advantages have given flow cytometry the potential to be applied at all stages of process development, from cell line selection to the development of monitoring and control strategies for animal cell bioreactor operation.

The principal function of monitoring techniques is to provide accurate and timely measurements. In cell culture, this is generally achieved by examination of cells removed from a culture flask or a fermenter using microscopic or biochemical (metabolic or enzymatic) approaches. These approaches impose severe limitation, as the former is slow and only provides static qualitative information, and the latter does not provide any information on the heterogeneity of the cell population, nor can it detect rare events. The development of flow cytometric techniques has greatly advanced our ability to identify and characterize the morphological and biochemical heterogeneity of cell populations, permitting a rapid and sensitive cell analysis.

In this chapter, flow cytometric applications in cell culture are described for Chinese hamster ovary (CHO) and mouse hybridoma cell lines grown on a laboratory scale, but the principles can also be employed for the monitoring of other cell types in suspension up to the large industrial scale.

From: *Methods in Biotechnology, Vol. 8: Animal Cell Biotechnology*
Edited by: N. Jenkins © Humana Press Inc., Totowa, NJ

2. Materials

1. 70% ethanol (ice cold).
2. DNase solution: dissolve 100 mg DNase (Sigma, Poole, UK) in 200 mL SMT (2.43 g Tris, 85.6 g sucrose, and 1.01 g $MgCl_2$ in 1 L distilled water, pH 6.5). Store at –20°C in 1 mL aliquates.
3. Phosphate-buffered saline (PBS), pH 7.2.
4. RNase solution: dissolve 5 mg RNase (Sigma) in 50 mL distilled water. Store at –20°C in 0.5-mL aliquots. Incubate in a boiling water bath for 10 min before use to eliminate contaminated DNase. Otherwise, use chromatographically purified RNase.
5. Propidium iodide solution (PI; Sigma): 10 mL of 1 mg/mL PBS.
6. Stock staining solution for determination of cell number: fluorospheres (1.1×10^5/mL Immuno-Check [Coulter, Luton, UK] beads, 10 µm, and PI (250 µg/mL) in PBS.
7. 4% paraformaldehyde stock solution (Sigma). Dilute in PBS to make a final concentration of 1%.
8. Saponin solution (Sigma): 10% w/v in PBS.
9. Fluorescein isothiocyanate (FITC)-conjugated goat anti-mouse (H+L) IgG (Sigma): 1:10 dilution in PBS.

3. Methods

3.1. Cell Cycle Analysis as a Performance Predictor

Analysis of the cell cycle can provide a reliable index for the prediction of growth potential and changes in cell number with time during the cultivation period ***(1)***. In a continuously proliferating culture, cells will progress through the cell cycle and divide, a process that leads to an increase in cell number and biomass. Consequently, the percentage of S-phase cells could be used as an indicator for the change in the apparent specific growth rate. S-phase cell measurement is normally accomplished by analysis of the cell cycle using one of a variety of DNA fluorochromes, such as PI, ethidium bromide, Hoechst 33258, mithramycin, and at least another 15 other dyes. PI is a widely available fluorochrome that excites at 488 nm. It intercalates into DNA to produce fluorescence with a maximum at 620 nm. The quality of the DNA histogram for many cell types is extremely good and can be achieved with a relatively inexpensive, low-power laser flow cytometer. Particular caution is recommended when handling PI and other DNA fluorochromes as they are potential carcinogens. A typical pattern of cell cycle distribution during batch cultivation of animal cells shows a substantial decrease in the percent of cells in the S and G_2 fractions during the stationary and decline phases of the culture. The same pattern also occurs during low feeding rate in perfusion culture and in stressed cultures.

The accuracy of cell cycle analysis depends largely on the sharpness of the profiles, as measured by the coefficient of variation (CV) across each peak ***(2)***. It is also important that doublets, aggregates, and debris be excluded from the analysis, and that the G_1 peak have a CV of 3 or lower ***(3)***. Most flow cytometry

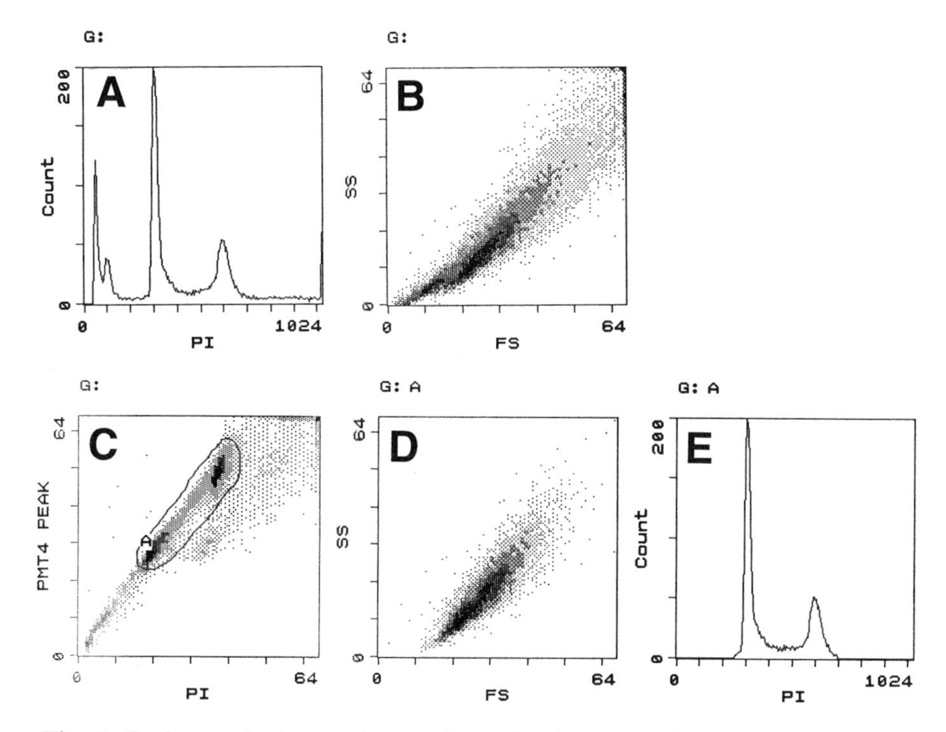

Fig. 1. Gating analysis to reduce noise and enhance resolution of DNA distributions. CHO cells were stained with propidium iodide **(A)** and analyzed by means of dual-parameter analysis of peak height vs integral fluorescence **(C)**. Selective gating of cells (gate A) to exclude debris and aggregates produces a typical DNA distribution **(E)** more suitable for cell cycle analysis. **(B)** and **(D)** are side scatter vs forward scatter of ungated and gated populations, respectively.

systems are now able to distinguish between cells and "noise" by comparing the signal height with its area. **Figure 1** shows a cell cycle distribution of CHO cells that is analyzed by constructing a gate in the dot plot of peak height against integral fluorescence to elimate the cellular debris and aggregates. Another requirement for the use of flow cytometry to effectively monitor growth and productivity is to elimiate nonviable cells from the analysis. This can be achieved by treating cells before fixation with DNase to eliminate DNA only from necrotic cells, leaving DNA intact in live cells ***(4)***. The quality of the cell cycle histogram is also very much linked to the fixative employed and the length of storage time; we have found that fixation with ice-cold ethanol for at least 15 min but not more than a week in –20°C is particularly effective in producing a better CV and therefore better discrimination. Moreover, RNase is used to eliminate double-stranded RNA, which is stained by intercalating dyes such as PI, with a consequent distortion of cell cycle distribution.

DNA distribution in the selected gate can then be mathematically analyzed to yield the percentages of cells in the G_1, S and G_2/M compartments of the cell cycle *(5)*. Many computer programs are now available for fast and consistent analysis of normal and perturbed distributions including the Multicycle (Phoenix Flow Systems, San Diego, CA) software program, which is based on a polynomial S-phase algorithm with an iterative, nonlinear least-square fit.

3.1.1. Analysis of DNA Distributions of Suspension Cells in Batch Culture

1. Remove cell samples at frequent intervals from the batch culture, centrifuge at 100*g* for 5 min, and wash in 5 mL PBS.
2. Centrifuge and resuspend cells in 1 mL stock DNase solution. Incubate at 37°C for 15 min and then place on ice for 5–10 min.
3. Centrifuge and resuspend cells in cold 70% ethanol with vigorous shaking. At this stage cells may be kept in the freezer (–20°C) prior to analysis, if required.
4. Centrifuge and wash in 5 mL PBS.
5. Centrifuge and resuspend in 4.5 mL PBS. Add 0.5 mL stock RNase to make a final concentration of 10 μg/mL. Incubate at 37°C for 20 min.
6. Centrifuge and wash in 5 mL PBS.
7. Centrifuge and resuspend in 4.75 mL PBS. Add 250 μL stock PI solution and incubate for 15 min at room temperature.
8. Centrifuge and resuspend in 1 mL PBS.
9. Analyze cells using flow cytometry. Argon ion laser with excitation at 488 nm is used to excite the dye, and a 620-nm bandpass interference filter is used to collect the emission of PI fluorescence.

3.1.2. Prediction of Growth Rate in CHO Batch Culture

The percentge of S cells measured at any time during a CHO batch culture can be used to predict the future specific growth rate. This interesting feature of cell cycle kinetics provides an early indicator of growth long before it can be assessed by cell counting and is based on data published recently *(1)*. The specific growth rate (per hour) at time t + 10.5 h (t is the time at which S phase cells are measured) can be obtained from the equation: $\mu = (9.9 \times 10^{-4})S - 0.018$, where μ is the specific growth rate at t + 10.5 h, and S is percentage of S-phase cells at time t.

3.2. Determination of Cell Number and Viability

A number of methods for the determination of cytotoxicity, viability, and survival in animal cell culture are available including dye exclusion, 3-(4,5-dimethylthiazol)-2,5-diphenyltetrazolium bromide (MTT) assay, ^{51}Cr release after cell lysis, radioisotope incorporation ([^{3}H]thymidine or [^{125}I]iododeoxyuridine) during cell proliferation, and protein content deter-

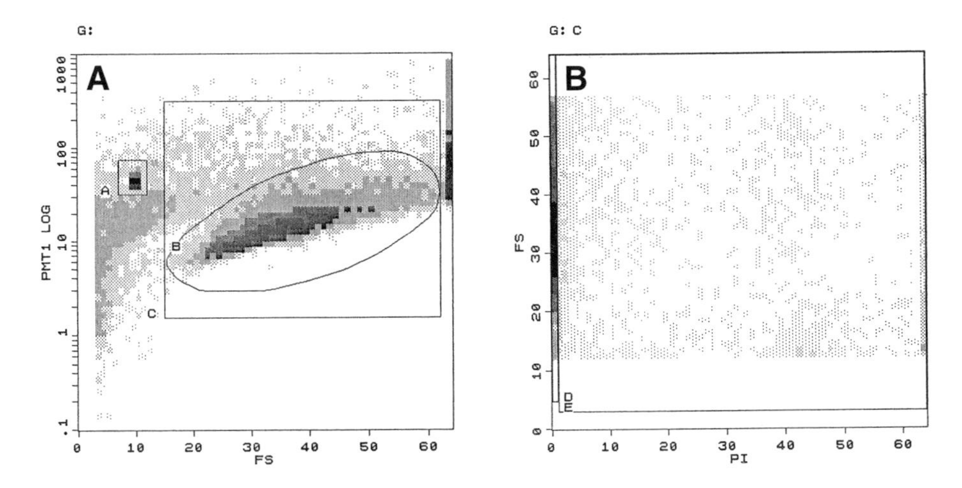

Fig. 2. Evaluation of cell number and viability by flow cytometry. Cultured hybridoma cells were stained with propidium iodide. The data were taken directly from the display screen of an EPICS Elite flow cytometer. **(A)** Dot plot of side scatter (PMT1 LOG) vs forward scatter (FS). **(B)** Dot plot of FS vs propidium iodide fluorescence (PI). (A) shows the gates for beads (A), living cells (B), and living and dead cells (C). (B) Gates for viable cells (D) and dead cells (E).

mination ***(6)***. *See* also Chapter 10. All of these methods are labor intensive and time consuming, and, with the exception, of dye exclusion all are based on the assessment of viable cell function and therefore cannot give an estimation of the nonviable fraction. A rapid and reliable approach is proposed for assessing the number and viability of cells, as well as cell size, in suspension culture by the use of flow cytometry ***(7)***. The main advantage of this method is that it permits the processing of a large number of samples and the detection of viable cells quantitatively in a heterogeneous culture of living and dead cells and debris.

3.2.1. Flow Cytometry of Cell Number and Viability

1. Add 500 µL of cell suspension to 100 µL stock staining solution (*see* **Subheading 2., item 6**) and mix by gentle shaking.
2. Analyze by loading an appropriate software protocol (**Fig. 2**) for the acquired parameters: forward scatter (FS), log side scatter (SS), and PI fluorescence integral. Most flow cytometers permit the display of any combination as dual-parameter dot plots. The SS vs FS plot presents the analyzed sample as subpopulations of cells, debris, and beads, and the FS vs PI plot presents viable and dead cells. Analytic software is available from the manufacturers.
3. Gate areas in the FS vs log SS dot plot in which beads, living cells, and total cells appear (gates A, B, and C in **Fig. 2A**, respectively).

4. Calculate cell number from:

$$\text{cell density } (n/\text{mL}) = \frac{\%\text{ cells in gate B} \times \text{beads in staining solution } (n/\text{mL})}{\alpha \times \text{area gate A}}$$

where n = number and α = dilution factor of staining solution.
5. To determine cell viability, gate areas in the PI vs FS dot plot in which viable cells (PI unstained) and dead cells (PI stained) appear (gates D and E in **Fig. 2B**, respectively). Calculate the ratio of viable to dead cells to obtain percent viability.

3.3. Monitoring of Cell Line Stability

Many cell lines lose secreted protein productivity during long cultivation periods. In hybridoma cells, this loss is often attributed to the loss of a functional immunoglobulin structural gene or its expression in at least a portion of the cell population. Such instability in productivity will have a significant effect if the nonproducer cells have a competitive advantage over the producer cells, resulting in the overgrowth of cells secreting little or no product *(8)*. The tendency of cell lines to lose productivity in long-term culture (**Fig. 3**) emphasizes the need for a rapid and reliable method to monitor the changes in the fraction of nonproducing cells and also to sort the high producing cells, which can be performed during cell line development and process optimization. It is also of considerable importance that the assay developed to monitor stability be able to pick up subtle changes with relative ease and to operate as a monitor during the production processes to identify any changes in the cell population homogeneity. In many ways, flow cytometry can be used to monitor productivity and to isolate cells with specific characteristics in a rapid and efficient manner.

3.3.1. Monitoring of IgG Productivity in Hybridoma Culture

1. Remove cell samples at intervals from the culture, centrifuge at 100*g* for 5 min, and resuspend in 1% paraformaldhyde. Cells can be stored at 4°C until analysis for up to 3 weeks.
2. Centrifuge and wash in 5 mL PBS.
3. Resuspend in 0.1 mL PBS.
4. Add 50 µL of FITC-conjugated goat antimouse IgG diluted 1:10 in PBS. Incubate for 30 min at room temperature. Alternatively, add to the mixture 1.5 µL saponin (a detergent treatment that improves accessibility of epitopes and therefore enhances IgG intensity).
5. Wash with 0.5% Tween-20 in PBS. Resuspend in 1 mL PBS.
6. Analyze samples by collecting the fluorescence readings on the logarithmic scale. Use FS and SS signals to gate intact single cells.

3.3.2. Dual Analysis of Cellular IgG and DNA

Formaldhyde is a good fixative for protein, but when used with DNA stains such as PI it yields poor quality staining, as indicated by broad G_1 and G_2 peaks

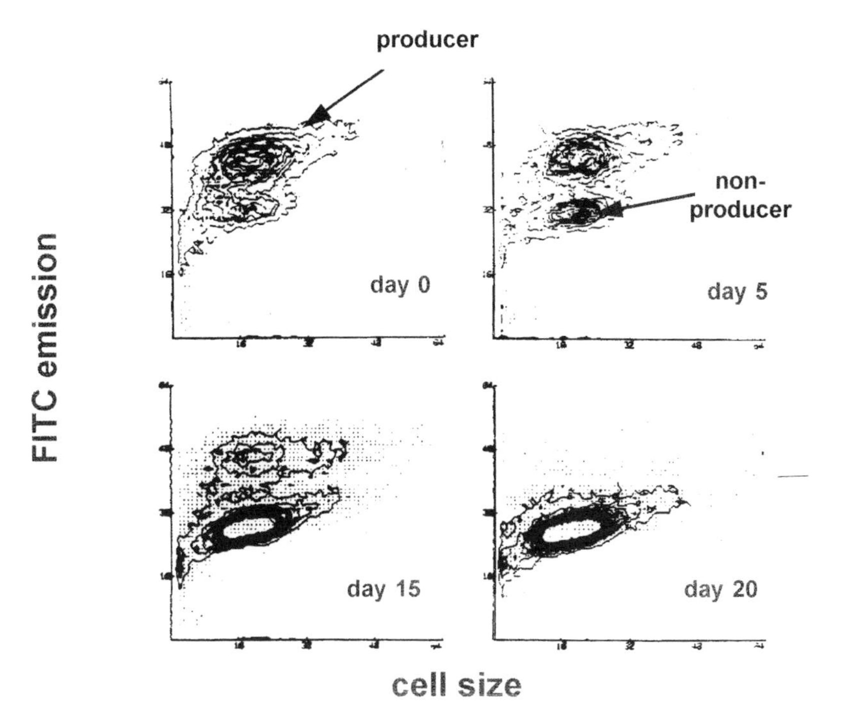

Fig. 3. Unstable hybridoma cells producing monoclonal antibody during long-term cultivation in continuous culture. The dot plots represent intracellulare IgG vs cell size measured for cell samples removed at days 0, 5, 15, and 20 days. FITC emission indicates the log fluorescence emission of FITC-conjugated goat antimouse IgG.

(9). If DNA staining is required, the cells can be removed from the formaldehyde fixative after 15 min by centrifugation and resuspended in cold 70% ethanol. The cells can then be treated as follows:

1. Repeat **steps 2–5** in the previous protocol (**Subheading 3.3.1.**).
2. Add 3.75 mL PBS and 250 µL stock PI solution and incubate for 15 min at room temperature.
3. Centrifuge and resuspend in 1 mL PBS.
4. Analyze samples by gating out debris and aggregates using peak vs integral PI fluorescence, and then display log FITC vs integral PI fluorescence.

An example of this type of analysis is shown in **Fig. 4**.

4. Conclusions

Automated flow cytometry provides an efficient and reliable monitoring means of in vitro animal cell bioprocesses. The changes in the cell population mean value and heterogeneity of DNA and intracellular product content of

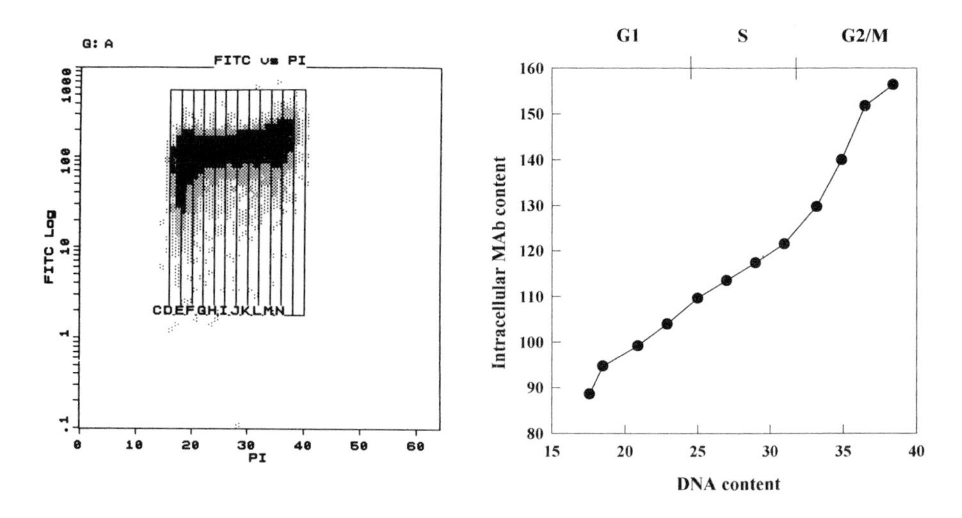

Fig. 4. Relationship between cell cycle and intracellular monoclonal antibody (MAb) in hybridoma cells. Exponentially grown hybridoma cells were fixed and stained with FITC-conjugated goat antimouse IgG (FITC Log) and propidium iodide (PI). The relationship between cell cycle and the amount of MAb can be studied by measuring the mean values of cell populations present in rectangular compartments constructed on the dot plot of FITC Log vs PI.

individual cells can be analyzed to obtain useful information on the dynamic of animal cell population. The technique can also be extended with relative ease to monitor other cellular activities such as protein and RNA content, intracellular pH, mitotic index, and respiratory activity.

References

1. Leelavatcharamas, V., Emery, A. N., and Al-Rubeai, M. (1996) Monitoring the proliferative capacity of cultured animal cells by cell cycle analysis, in *Flow Cytometry Applications in Cell Culture* (Al-Rubeai, M. and Emery, A. N., eds.), Marcel Dekker, New York, pp. 1–15.
2. Ormerod, M. G. and Imrie, P. R. (1990) Flow cytometry, in *Methods in Molecular Biology,* vol. 5 (Pollard, J. W. and Walker, J. M., eds.), Humana Press, Totowa, NJ, pp. 543–558.
3. Grogan, W. M. and Collins, J. M. (1990) *Guide to Flow Cytometry Methods,* Marcel Dekker, New York.
4. Frankfort, O. S. (1990) Flow cytometric measurement of cell viability using DNase exclusion, in *Methods in Cell Biology*, vol. 33 (Darzynkiewicz, Z. and Crissman, H. A., eds.), Academic, San Diego, pp. 13–18.
5. Rabinovitch, P. S. (1995) Analysis of flow cytometric DNA histograms, in *Cell Growth and Apoptosis* (Studzinski, G. P., ed.), IRL, Oxford, pp. 45–58.

6. Wilson, A. P. (1986) Cytotoxicity and viability assays, in *Animal Cell Culture: a practical approach* (Freshney, R. I., ed.), IRL, Oxford, pp. 183–216.
7. Al-Rubeai, M., Welzenbach, K., Lloyd, D. R., and Emery, A. N. (1997) A rapid method for evaluation of cell number and viability by flow cytometry. *Cytotechnology,* **24,** 161–168.
8. Frame, K. K. and Hu, W.-S. (1991) Comparison of growth kinetics of producing and nonproducing hybridoma cells in batch culture. *Enzyme Microbiol. Technol.* **13,** 690–696.
9. Bauer, K. D. (1990) Analysis of proliferation-associated antigens, in *Methods in Cell Biology*, vol. 33 (Darzynkiewicz, Z. and Crissman, H. A., eds.), Academic, San Diego, pp. 235–247.

12

Measurement of Cell Death in Culture

Afshin Samali and Thomas G. Cotter

1. Introduction

Cells can die via two distinct pathways, apoptosis and necrosis. Cells that are exposed to severe traumatizing conditions usually undergo an uncontrolled swelling and lyse via necrosis *(1)*. However, a cell dying under less traumatizing conditions has limited control over its own demise and utilize a mode of programmed cell death (apoptosis) first characterized by Kerr et al. *(2)*. It has been demonstrated that cells undergoing apoptosis usually share a number of common features, such as cell shrinkage, externalization of phosphatidylserine, nuclear condensation, nuclear fragmentation, and degradation of the DNA into nucleosomal fragments *(3–5)*.

Cells may die as a result of changes in culture conditions, such as accumulation of metabolic wastes, serum depletion, changes in the pH or temperature, and so forth. However, for experimental purposes one may want to induce cell death by other cytotoxic agents. It has been demonstrated that a wide variety of stimuli such as removal of growth factors from the culture environment *(6)*, heat shock *(7)*, radiation *(8)*, various genotoxic drugs *(9)*, and anti-Fas antibody *(10)* can all induce cell death. All these agents, with the exception of anti-Fas antibody which kills cells via apoptosis, are capable of inducing both apoptosis and necrosis in a dose- and time-dependent manner *(7)*.

A variety of techniques have been developed to measure cell viability. Most of these assays are based on loss of membrane integrity. Membrane disruption can be detected by the uptake of vital dyes or release of cell components into the media. However, during apoptosis the cell membrane remains intact and therefore some of these assays will not detect this type of cell death. Alternatively assays that detect all types of cell death do not dis-

From: *Methods in Biotechnology, Vol. 8: Animal Cell Biotechnology*
Edited by: N. Jenkins © Humana Press Inc., Totowa, NJ

tinguish between apoptosis and necrosis. Therefore more specific techniques have been developed to determine cell death and distinguish between the two separate types. These techniques rely on the morphology and biochemical/ molecular changes associated with the two processes. In this chapter some of the most commonly used techniques to detect cell death in culture systems are described, including the trypan blue exclusion assay, morphological assessment, flow cytometric analysis of cell size and cell cycle, *in situ* DNA nick end-labeling (TUNEL), and finally the annexin V binding assay.

2. Materials

1. An improved Neubauer (Brand, Germany) hemocytometer.
2. Cover slip.
3. Phosphate-buffered saline (PBS) tablets (Sigma, Poole, UK); make a 1× and 2× solution, pH 7.2.
4. Trypan blue (Sigma).
5. Microscope slide.
6. Cytospin and cytospin cup (Shandon, Runcorn, UK).
7. RAPI-DIFF stain kit (Diagnostic Developments, Burscough, UK).
8. DPX mountant (BDH).
9. 70% ethanol.
10. Propidium iodide (Sigma); make a 50 μg/mL solution in PBS.
11. RNase A (Sigma); make a 25 mg/mL stock solution.
12. Centrifuge.
13. Flow cytometer (e.g., Becton Dickinson) and tubes.
14. 1% formaldehyde in PBS (pH 7.2).
15. TUNEL reaction mixture: 0.1 *M* Na-cacodylate, pH 7.0; 0.1 m*M* dithiothreitol; 0.05 mg/mL bovine serum albumin (BSA); 2.5 m*M* $CaCl_2$ (all from Sigma); 0.4 m*M* bio-16-d uridine triphosphate (UTP) and 0.1 U/mL terminal deoxynucleotidyl transferase (TdT) enzyme (both from Boehringer-Mannheim, Germany).
16. TUNEL staining buffer: 2.5 mg/mL fluoresceinated avidin; 4× concentrated saline-sodium citrate buffer; 0.1% Triton X-100; and 5% (w/v) low fat dry milk (all from Sigma).
17. Annexin V-FITC (Bender, Vienna, Austria).
18. HEPES buffer: 10 m*M* HEPES-NaOH, pH 7.4; 150 m*M* NaCl; 5 m*M* KCl; 1 m*M* $MgCl_2$; 1.8 m*M* $CaCl_2$ (all from Sigma).

3. Methods

3.1. Trypan Blue Exclusion Assay

Trypan blue exclusion is a cell viability assay based on the ability of live cells to exclude the vital dye trypan blue ***(11)***. Therefore this assay allows one to distinguish between cells with intact and those with disrupted membrane (*see* **Note 1**).

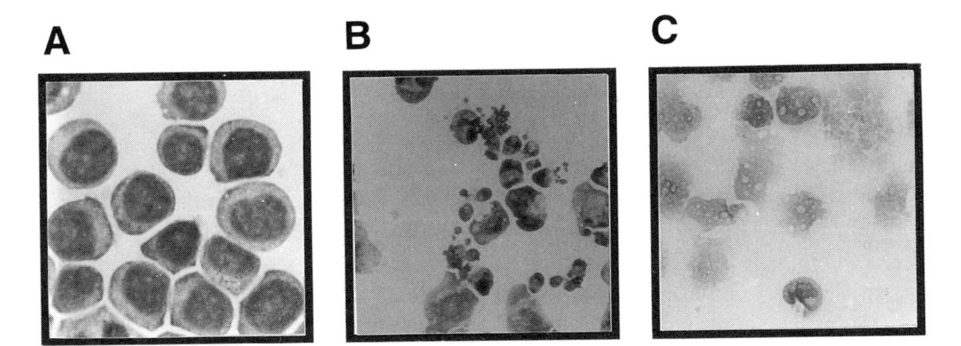

Fig. 1. Morphological features of **(A)** untreated HL-60 cells, and UV-irradiated HL-60 cells for **(B)** 15 min or **(C)** 1 h. Cells exposed to UV for 15 min were incubated for 4 h, to allow development of apoptotic features, and then cytospun onto glass slides and stained. Apoptotic cells display a condensed or fragmented nucleus. Necrosis was induced after 1 h of irradiation with UV. The cells displayed swelling and disrupted membranes immediately after irradiation.

1. Take a sample from cell suspension.
2. Count cells by mixing an equal volume of cells with an equal volume of a 0.2% solution of trypan blue prepared in 2× PBS.
3. Load the sample onto the hemocytometer.
4. Calculate the number of cells/mL of culture medium is calculated by multiplying the average number of cells per field by the dilution factor of 2 and hemocytometer index (10^4). The percentage viability is determined as follows:

$$\% \text{ viability} = \frac{\text{number of nonstained cells}}{\text{total number of cells}} \times 100$$

3.2. Morphological Assessment of Cell Death

Apoptotic and necrotic cells are identified using morphological criteria described previously *(3,4,12)*. These include cell shrinkage, nuclear condensation and nuclear fragmentation for apoptotic cells, nuclear swelling, chromatin flocculation, and loss of nuclear basophilia for necrotic cells. The levels of both apoptosis and necrosis in a particular cell culture can be estimated by RAPI-DIFF staining of cytocentrifuge cell preparations (**Fig. 1**).

3.2.1. Cell Preparation

1. Take a sample from the cell suspension.
2. Pellet the cells by centrifugation (200*g* for 5 min).
3. Resuspend in PBS to 5 × 10^5 cells/mL.
4. Add 100 µL of sample to the prepared cytospin cup and slide setup.
5. Cytocentrifuge at 100*g* for 2 min.

3.2.2. Staining

1. Fix cells in solution A (100% methanol) and immerse the slide 10× in the solution.
2. Stain nucleus by dipping the slide in solution B (acid dye containing 0.1% w/v Eosin Y, 0.1% w/v formaldehyde, 0.4% w/v sodium phosphate dibasic, 0.5% w/v potassium phosphate monobasic) and immerse the slide 10× in the solution.
3. Stain cytoplasm by dipping the slide in solution C (basic dye containing 0.4% w/v methylene blue-polychromed, 0.4% w/v azure, 0.4% w/v sodium phosphate dibasic, 0.5% w/v potassium phosphate monobasic) of the kit and immerse the slide 10× in the solution.
4. Wash the slides with water, air-dry, and mount the slides using DPX mountant (*see* **Note 2**).

3.3. Simultaneous Assessment of Changes in Cell Size/ Granularity and Cell Cycle by Flow Cytometry

Many reports have stated that a reduction in cell size and buoyant density accompany apoptotic cell death *(13)*. Thus early apoptotic cells will appear smaller and denser than their normal counterparts. These changes are accurately detected in most cells by their light-scattering properties, as measured by a flow cytometer (**Fig. 2A**). The characteristics of forward scatter, which indicates cell size, and side scatter, which reveals the degree of granularity of the cell, are employed to monitor cells undergoing apoptosis *(13–15)*.

The cell cycle profile of a cell population determines the DNA content and therefore position in the cell cycle, i.e., G_1/G_0, S, or G_2/M. Dead cells contain a higher amount of subdiploid DNA, which accumulates in the pre-G1 area of the cell cycle histogram (**Fig. 2B**). The protocol described here will permit simultaneous assessment of changes in cell size/granularity and cell cycle by flow cytometry on fixed cells.

3.3.1. Cell Preparation and Fixation

1. Take a sample from cell suspension.
2. Pellet 10×10^5 cells by centrifugation (1000 rpm for 5 min), fix in 1 mL of ice-cold 70% (v/v) ethanol (add ethanol dropwise while vortexing to avoid clumping), and incubate on ice for 30 min.
3. Pellet cells and decant ethanol.

3.3.2. Staining and Analysis

1. Incubate in 1 mL of propidium iodide, containing 20 μL of RNase A stock, for 15 min.
2. Analyze forward and side scatter for cell size/granularity and FL-2 (560–640) for DNA content during the cell cycle on a flow cytometer (*see* **Note 3**).

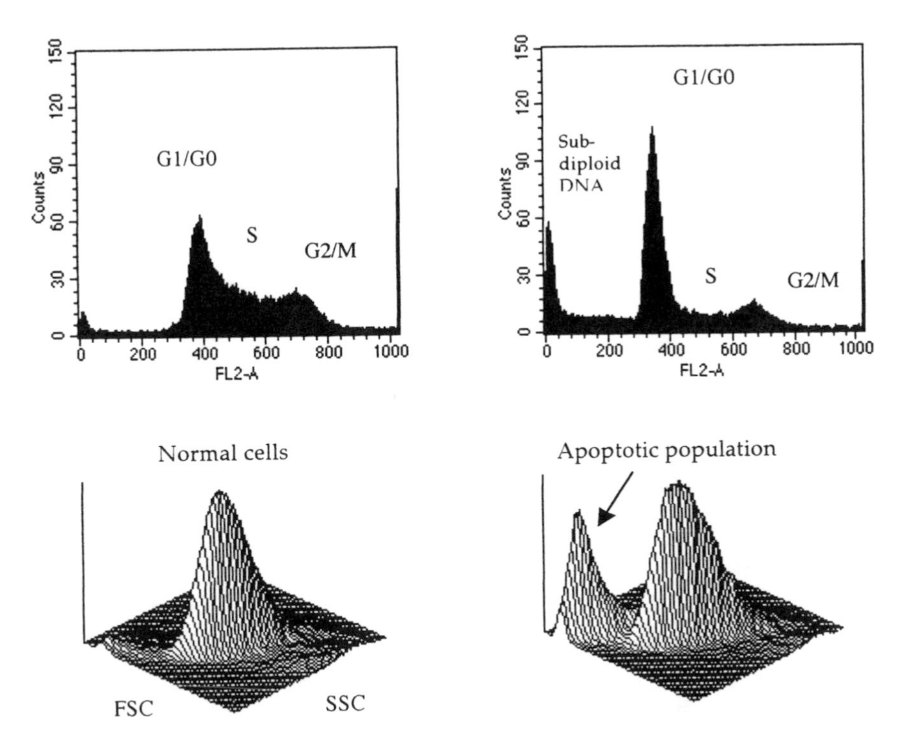

Fig 2. Cell cycle profile and three-dimensional plot representation of light scattering propfiles of **(A)** normal and **(B)** UV-treated (5 min) HL-60 cells. The increase in subdiploid DNA of UV treated cells on the cell cycle histogram is an indicitive of cell death. On the three-dimensional plot, forward scatter is plotted against side scatter. As the level of apoptosis increases, the population representing normal cells disappears, and there is a concomitant increase in the population representing apoptotic cells and apoptotic bodies.

3.4. In Situ *DNA Nick End-labeling*

DNA fragmentation during cell death occurs in three stages, which include the nicking of single strands of DNA, followed by DNA fragmentation into large (50–200 kbp) and finally nucleosome size fragments of 180–200 bp size ***(16–19)***. On the basis of these observations, the in situ labeling of nicked DNA serves as an assay for detection of cell death. A modified version of the TUNEL was described by Gorczyca et al. ***(17)*** and is routinely used in the study of cell death ***(15,17)***. This protocol involves three stages, fixation, elongation, and staining.

3.4.1. Fixation

1. Take a sample from the cell suspension.
2. Pellet 5×10^5 cells by centrifugation (200*g* for 5 min) and wash in PBS.

3. Fix in 1 mL of formaldehyde for 15 min on ice.
4. Wash cells in PBS before resuspension in 70% ice-cold ethanol.

3.4.2. Elongation

1. Wash cells in PBS and incubate in 50 μL reaction mixture for 30 min at 37°C .

3.4.3. Staining

1. Wash cells in PBS and incubate in 100 μL staining buffer for 30 min at room temperature in the dark.
2. Wash stained cells in PBS before analysis for fluorescence intensity by a flow cytometer, FL-1 (515–545 nm) (*see* **Note 4**).

3.5. Annexin V Binding to Externalized Phosphatidylserine Assay

A typical feature of apoptosis is rearrangement and loss of the plasma membrane asymmetry, as a result of externalization of phosphatidylserine (PS) from the inner leaflet of the plasma membrane, where it is mostly found, to the outer leaflet ***(5)***. It has been demonstrated that annexin V binding to PS may serve as an assay for detection of apoptosis at the early stages of the process based on the above observation ***(20)***.

1. Take a sample from cell suspension, pellet 2–4 × 10^5 cells by centrifugation (200*g* for 5 min), and remove supernatant.
2. Incubate cells in 200 μL of annexin V-FITC (1–3 mg/mL), diluted in HEPES buffer, for 5 min at room temperature.
3. Analyze samples by a flow cytometer (FL-1, 515–545 nm) (*see* **Note 5**).

4. Concluding Remarks

The methods used to assess cell death are many and varied and typically are based on assays that measure membrane integrity. Techniques that measure apoptotic cell death are based on the ability to measure one or more of a series of specific degenerative changes that result in cell death. These cellular modifications include a number of striking morphological changes such as cell shrinkage, nuclear condensation, and nuclear fragmentation.

The precise kinetics of the events that occur within a cell undergoing apoptosis may vary between cell types. Here we have described a limited number of quick assays to measure cell death, some of which can determine the type of cell death. However, in most cases the cell death assays do not distinguish between the two modes of cell death and therefore one or more of these assays should be combined.

5. Notes

1. A characteristic feature of apoptosis is that membrane integrity of the dying cell is maintained long after the process has been engaged. However, a reduction in membrane integrity usually occurs during the late stages of apoptosis in a process commonly called secondary necrosis. Therefore assays such as the trypan blue exclusion assay, which rely on the uptake of vital dyes, will underestimate the levels of apoptosis in culture.
2. If the cells are understained after a cycle of fixation and staining, then the procedure can be repeated until the desired staining is achieved. If the cells are overstained, wash the cells in water, immerse in the first solution for 5 s, and rinse in water. Cells are viewed under the light microscope (×40 magnification), and 3× 100× cells are counted from separate fields of view and scored for viable, apoptotic, or necrotic cells based on the following morphological characteristics: *apoptotic cells*—cell membrane blebbing, chromatin condensation and nuclei shrinkage, cytoplasmic constriction and loss of cell volume, and formation of apoptotic bodies; *necrotic cells*—nuclear swelling, chromatin flocculation, cell membrane disruption, and cell lysis resulting in "ghost cells." This is the quickest and easiest way to identify dead cells, if the samples are stained properly.
3. If ethanol is added too quickly, the cells may clump together. On the cell cycle histograms the pre-G_1 area or unstained DNA represents the DNA from dead cells. On the three-dimensional diagrams the peak to the left represents a population of condensed and shrunken apoptotic cells. Although fixation involves dehydration and shrinkage of cells, since both normal and dead cells shrink to the same extent, this should not interfere with a comparative analysis of cell size. An increase in the levels of pre-G_1 DNA indicates cell death in general; however, simultaneous analysis of cell size/granularity allows determination of mode of cell death.
4. Cells can be stored overnight at –20°C after addition of ethanol. The TdT primer-dependent DNA polymerase catalyzes the repetitive addition of deoxyribonucleotide from deoxynucleotide triphosphate to the terminal 3'-hydroxyl of a DNA or RNA strand with the release of inorganic pyrophosphate. Cells with extensive DNA nicking incorporate Bio-16-dUTP to a greater extent than their unnicked counterparts (**Fig. 3**).
5. Binding of annexin V to PS is Ca^{2+}-dependent and therefore the correct amount on Ca^{2+} ions is required for optimal binding. This assay is very quick, as no fixation or permeabilization is required to allow entry of reagents (as in the TUNEL assay). Another advantage of this assay is the detection of apoptosis during the early stages of the process (**Fig. 4**). However, care should be taken when analyzing samples in the late stages of apoptosis, when the apoptotic cells are undergoing secondary necrosis. During this stage the membrane is damaged, and annexin V can enter the cell and bind to PS in the inner leaflet of the membrane. Thus it is not possible to distinguish between cells undergoing primary and secondary necrosis. Therefore this assay should

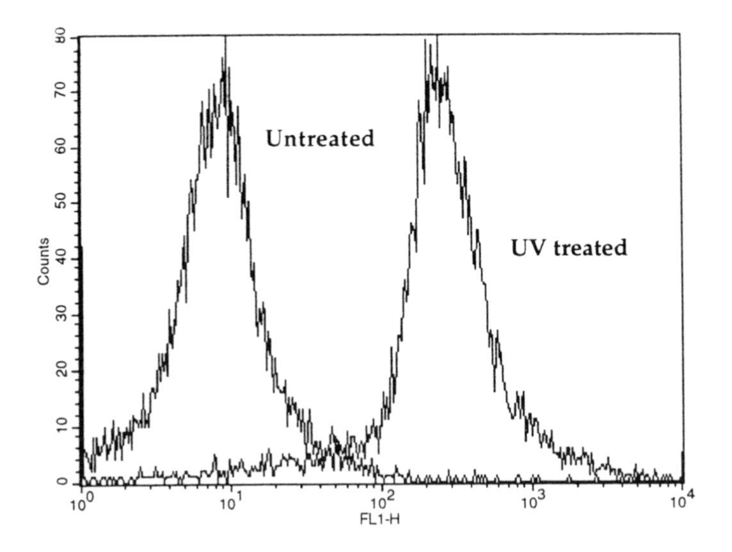

Fig 3. Fluorescence profile of untreated and UV-treated HL-60 cells in an *in situ* terminal deoxynucleotidyl transferase assay of nicked DNA. Cells were irradiated for 5 min and incubated under normal conditions for 4 h. Cells (5×10^5/sample) were fixed and analyzed for DNA nicks by measuring fluorescence intensity using CellQuest software on a flow cytometer.

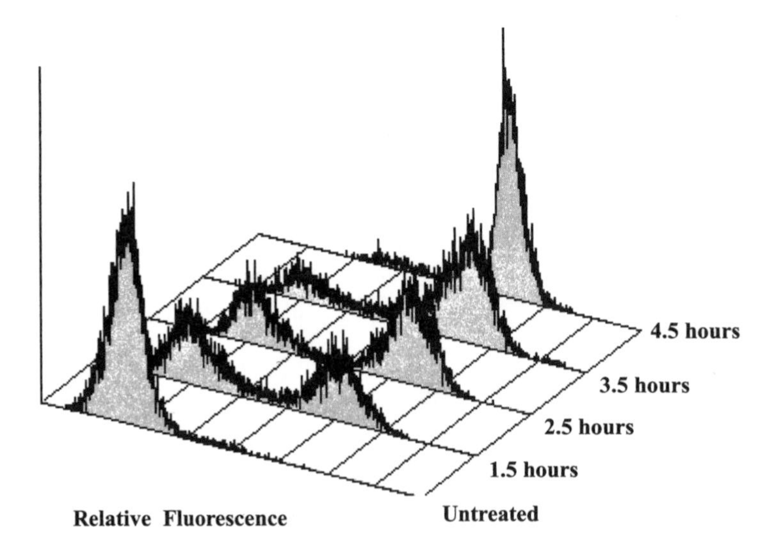

Fig. 4. Time-course of changes in plasma membrane phosphotidylserine distribution during Fas-induced apoptosis. Jurkat cells were treated with anti-Fas antibody, and samples were tested for annexin V-FITC binding by measuring fluorescence intensity using CellQuest software on a flow cytometer.

always be accompanied by the vital die exclusion assay or morphological assessment of the cells.

References

1. Trump, B. F. and Ginn F. L., (1969) The pathogenesis of subcellular reaction to lethal injury, in *Methods and Achievements in Experimental Pathology*, vol. 4 (Bajusz, E. and Jasmin, G., eds.), Karger, Basel, pp. 1–10.
2. Kerr, J. F. R., Wyllie, A. H., and Currie, A. R. (1972) Apoptosis: a basic biological phenomenon with wide ranging implications in tissue kinetics. *Br. J. Cancer* **26,** 239–257.
3. Wyllie, A. H. (1980) Glutocorticoid-induced thymocyte apoptosis is associated with endogenous endonuclease activation. *Nature* Basel: Karger 555,556.
4. Cohen, G. M., Sun, X.-M., Snowden, R. T., Dinsdale, D., and Skilleter, D. N. (1992) Key morphological features of apoptosis may occur in the absence of internucleosomal DNA fragmentation. *Biochem. J.* **286,** 331–334.
5. Fadok, V. A., Voelker, D. R., Campbell, P. A., Cohen, J. J., Bratton, D. L., and Henson, P. M. (1992) Exposure of phosphatidylserine on the surface of apoptotic lymphocytes triggers specific recognition and removal by macrophages. *J. Immunol.* **148,** 2207–2216.
6. Rodriguez-Tarduchy, G., Collins, M., and Lopez-Rivas, A. (1990) Regulation of apoptosis in interleukin-3 dependent hemopoietic cells by interleukin-3 and calcium ionophores. *EMBO J.* **9,** 2997–3002.
7. Lennon, S. V., Martin, S. J., and Cotter, T. G. (1991) Dose-dependent induction of apoptosis in tumour cell lines by widely divergent stimuli. *Cell Prolif.* **24,** 203–214.
8. Martin, S. J. and Cotter, T. G. (1991) Ultraviolet B irradiation of human leukaemia HL-60 cells *in vitro* induces apoptosis. *Int. J. Radiat. Biol.* **59,** 1001–1016.
9. Hickman, J. A. (1992) Apoptosis induced by anticancer drugs. *Cancer Metastasis Rev.* **11,** 121–139.
10. Trauth, B. C., Klas, C., Peters, A. M. J., Matzku, S., Moller, P., Falk, W., Debatin, K.-M., and Krammer, P. H. (1989) Monoclonal antibody-mediated tumor regression by induction of apoptosis. *Science* **245,** 301–305.
11. McGahon, A. J., Martin, S. V., Bissonnette, R. P., Mahboubi, A., Shi, Y., Mogil, R. J., Nishioka, W. K., and Green, D. R. (1995) The end of the (cell) line: methods for the study of apoptosis *in vitro*, in *Cell Death* (Schwartz, I. M. and Osborne B. A., eds.), Academic, San Diego, pp. 153–185.
12. Martin, S. J., Lennon, S. V., Bonham, A. M., and Cotter, T. G. (1990) Induction of apoptosis (programmed cell death) in human leukaemic HL-60 cells by inhibition of RNA or protein synthesis. *J. Immunol.* **145,** 1859–1867.
13. Cotter, T. G., Lennon, S. V., Glynn, J. M., and Green, D. R., (1992) Microfilament-disrupting agents prevent formation of apoptotic bodies in tumour cells undergoing apoptosis. *Cancer Res.* **25,** 997–1005.
14. Swat, W., Ignatowicz, L., and Kieslow, P. (1991) Detection of apoptosis of immature $CD4^{+}8^{+}$ thymocytes by flow cytometry. *J. Immunol. Methods* **137,** 79–87.
15. Samali, A. and Cotter, T. G. (1996) Heat shock proteins increase resistance to apoptosis. *Exp. Cell Res.* **223,** 163–170.

16. Wyllie, A. H., Morris, R. G., Smith, A. L., and Dunlop, D. (1984) Chromatin cleavage in apoptosis: association with condensed chromatin morphology and dependence on macro molecular synthesis. *J. Pathol.* **142,** 67–77.
17. Gorczyca, W., Bruno, S., Darzynkiewicz, R. J., Gong, J., and Darzynkiewicz, Z. (1992) DNA strand breaks occurring during apoptosis: their early detection by terminal deoxynucleotide transferase and nick translation assays and prevention by serine protease inhibitors. *Int. J. Oncol.* **1,** 639–648.
18. Brown, D. G., Sun, X.-M., and Cohen, G. M. (1993) Dexamethasone-induced apoptosis involves cleavage of DNA to large fragments prior to internucleosomal fragmentation. *J. Biol. Chem.* **268,** 3037–3039.
19. Cohen, G. M., Sun, X.-M., Fearhead, H., MacFarlane, M., Brown, D. G., Snowden, R. T., and Dinsdale, D. (1994) Formation of large molecular fragments of DNA is a key committed step of apoptosis in thymocytes. *J. Immunol.* **153,** 507–516.
20. Koopman, G., Reutelingsperger, C. P. M., Kuijten, G. A. M., Keehnen, R. M. J., Pals, S. T., and Van Oers, M. H. J. (1994) Annexin V for flow cytometric detection of phosphatidylserine expression on B cells undergoing apoptosis. *Blood* **84,** 1415–1420.

13

Nuclear Magnetic Resonance Methods of Monitoring Cell Metabolism

Maria L. Anthony, Shane N. O. Williams, and Kevin M. Brindle

1. Introduction

Nuclear magnetic resonance (NMR), which was discovered in 1946, was used primarily by organic chemists for elucidation of the structure of relatively small organic molecules. The advent of Fourier transform NMR, coupled with the development of superconducting magnets with higher field strengths, opened the technique to a variety of biological and clinical applications. NMR is now a proven technique for monitoring metabolism in diverse systems—isolated cells and perfused organs, as well as the intact animal and humans *(1–4)*. Great scope exists, therefore, for the development of NMR applications in the biotechnology industry. A major analytical advantage of NMR spectroscopy is its unique ability to yield extensive information on a wide range of biologically important low-molecular-weight species simultaneously. Although many NMR-detectable nuclei exist, studies of cell metabolism have generally utilized the ^{31}P, ^{13}C, ^{1}H, and ^{15}N nuclei *(1,4)*. The basic principles of NMR spectroscopy, which are beyond the scope of this chapter, have been described extensively elsewhere (for example, *see* **ref. 5**). This chapter will focus on various NMR-based approaches for studying metabolism in cultured mammalian cells (including lines used by the biotechnology industry) and will highlight commonly used techniques for obtaining preparations of cells suitable for NMR studies.

1.1. Examination of Cellular Energetics using ^{31}P NMR Methods

^{31}P NMR is particularly well-suited to the study of cellular metabolism, as several phosphorylated metabolites are present in sufficiently high concentra-

From: *Methods in Biotechnology, Vol. 8: Animal Cell Biotechnology*
Edited by: N. Jenkins © Humana Press Inc., Totowa, NJ

tions to yield detectable NMR signals, within a reasonable time frame. ^{31}P NMR methods can be used to assess the energy status of a cell by monitoring the intensity of resonances from adenosine triphosphate (ATP), phosphocreatine, and inorganic phosphate. The chemical shift of the inorganic phosphate resonance has also been used ***(4)*** to measure intracellular pH. Information on a number of phosphorylated metabolites important in intermediary metabolism (e.g., glycolytic intermediates), phospholipid metabolism (phosphocholine, glycerophosphocholine, phosphoethanolamine, and glycerophosphoethanolamine) and protein glycosylation (uridine-diphosphate-sugars) can be obtained from ^{31}P NMR spectra ***(4,6)***.

1.2. Nutrient Metabolism and Metabolite Fluxes from Isotope Labeling Experiments

This NMR-based strategy is analogous to the use of labeled compounds in radiotracer studies. Cells are incubated with ^{13}C- and ^{15}N-labeled nutrients, and the redistribution of label among cellular metabolites is monitored using ^{13}C and ^{15}N NMR experiments. In contrast to radiotracer studies, in which further sample purification may be required, all metabolic intermediates can be simultaneously observed in the NMR spectrum. NMR can reveal which molecule is labeled and also, the position in the molecule that is labeled. It is therefore possible to focus specifically on the metabolic fate of the labeled precursor, allowing measurement of metabolite fluxes within cells ***(6)***. For example, flux in the glycolytic pathway and tricarboxylic cycle have been examined through the use of ^{13}C-labeled substrates ***(7,8)***. The sensitivity of label detection can be improved by observing the label indirectly *via* the ^{1}H nucleus using $^{1}H/^{13}C$ or $^{1}H/^{15}N$-NMR experiments. In these cases, quantitative information on fractional labeling is also obtained as both enriched and unlabeled species can be detected in the ^{1}H spectrum. Street et al. ***(9)*** applied this type of strategy to monitor the ^{15}N labeling of cellular metabolites in cultures of mammalian cells.

2. Materials

2.1. Cell Culture

1. Medium for adherent or suspension culture cells, typically supplemented with 2 m*M* L-glutamine, 100 U/mL penicillin, 100 μg/mL streptomycin and 10% fetal bovine serum.
2. Phosphate-buffered saline (PBS) and Trypsin (1x)-ethylenediamine tetra-acetic acid (EDTA) (0.02%) solution for routine sub-culturing of adherent cell lines.
3. T175 tissue culture flasks and 140 × 15 mm tissue culture dishes (Nunclon) for adherent cultures; siliconized spinner flasks and magnetic stirrer (Techne, Duxford, Cambridge, UK) for suspension cultures.

2.1.1. Perchloric Acid Extraction of Cells

1. 6% perchloric acid (PCA; Fisons).
2. 2 *M* K_2CO_3.
3. Ion exchange resin: Chelex™ 100 resin (200–400-mesh sodium form; Bio-Rad, Hercules, CA).

2.1.2. ^{31}P NMR Analysis of Cell Extracts

1. NMR tubes (typically 10 mm; Wilmad).
2. Extract buffer: 50 m*M* triethanolamine, pH 8.4, plus 15 m*M* EDTA.
3. Deuterium oxide (D_2O; Aldrich) (*see* **Note 1**).
4. NMR chemical shift and quantitation standard: 30 m*M* methylene diphosphonic acid (MDP) in 50 m*M* triethanolamine buffer, pH 8.4, plus 1 m*M* EDTA, contained in a sealed coaxial capillary tube. These tubes can be obtained from Wilmad (*see* **Note 2**).
5. High-field, high-resolution NMR spectrometer (*see* **Note 3**).

2.2. Culture of Intact Cell Systems

1. Culture medium, typically supplemented with 2 m*M* L-glutamine, 100 U/mL penicillin, 100 μg/mL streptomycin and 10% fetal bovine serum.
2. T175 tissue culture flasks (Nunclon), spinner flasks and magnetic stirrer (Techne).
3. Microcarrier beads, e.g., Cytodex-1™ (Pharmacia).

2.2.1. Hollow Fiber Bioreactor (HFBR) Perfusion System for NMR

1. HFBR (e.g., NMR bioreactor, Setec's Tricentric™, Livermore, CA).
2. NMR spectrometer (*see* **Subheading 2.1.2.**) (*see* **Note 4**).

3. Methods

3.1. Cell Culture and Preparation of Cell Extracts for NMR Analysis

The intrinsically low sensitivity of NMR means that cells must be grown to very high densities (of the order of 10^8 cells/mL) if spectra are to be obtained from intact cells. For this reason many studies are performed on protein-free cell extracts. These have the advantage that they are magnetically more homogeneous and therefore give better resolved spectra (compare **Fig. 1A** and **B**). The extracts can be concentrated prior to NMR examination and as there is, in principle, no limit on the NMR measurement time, the signal-to-noise ratio obtainable in the spectra of extracts is also much better. The most commonly used aqueous extraction procedure employs PCA, although TCA and acetone provide suitable alternative extraction methods. A ^{31}P NMR spectrum of a typical PCA cell extract is shown in **Fig. 1A** (*see* **Note 5**).

3.1.1. Perchloric Acid Extraction of Cells

1. For anchorage-dependent cells, remove cell culture medium from dishes and scrape cell monolayer (typically 0.5–1.0 × 10^9 cells, protein content approx

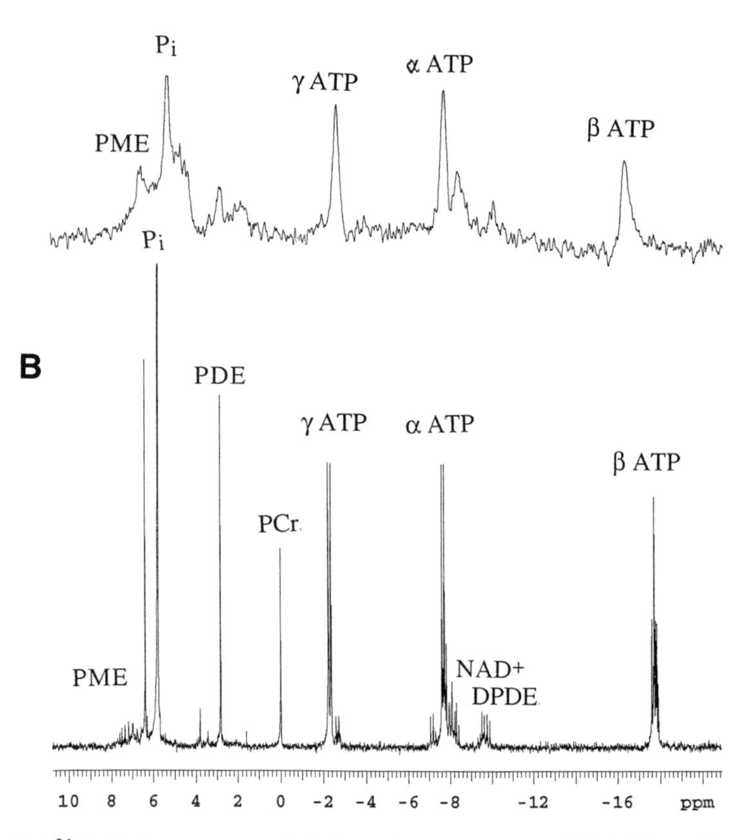

Fig. 1. **(A)** ^{31}P NMR spectrum of Chinese hamster ovary (CHO K1) cells growing in a hollow-fiber bioreactor. A ^{1}H NMR image, taken from a transverse slice through the reactor, is shown in **Fig. 2**. **(B)** ^{31}P NMR spectrum of a perchloric acid extract of CHO K1 cells. Note the improved resolution in the extract spectrum. Signal assignments are PME, phosphomonoesters; P_i, inorganic phosphate; PDE, phosphodiesters; PCr, phosphocreatine; ATP, adenosine triphosphate; NAD^+, nicotinamide adenine dinucleotide; DPDE, diphosphodiesters.

90 mg) into 30 mL of 6% PCA. Leave at 4°C for 20 min. For suspension cultures, centrifuge cell suspension (typically 0.5–1.0 × 10^9 cells in total) at 1800*g* for 5 min, discard culture medium and resuspend cell pellet in 30 mL of 6% PCA for 20 min at 4°C.

2. Centrifuge extract at 1800*g* for 5 min and neutralise supernatant with 2 *M* K_2CO_3.
3. Centrifuge and remove any paramagnetic ions present by treating supernatant with 2*g* of Chelex™ 100 resin. Leave at 4°C for 30 min.

4. Centrifuge to remove Chelex from the solution. Place supernatant on ice.
5. Wash the Chelex with an equal volume of water and leave at 4°C for a further 30 min. Centrifuge and pool the supernatant with the first.
6. Snap freeze sample in liquid nitrogen.
7. Lyophilize sample and store at –20°C prior to NMR analysis.

3.1.2. ^{31}P NMR Analysis of Cell Extracts

1. Dissolve cell extract in 5 mL of extract buffer and remove undissolved material by centrifugation.
2. Transfer 3.5 mL of the supernatant to a 10-mm-diameter NMR tube containing 0.5 mL D_2O. In the case of ^{31}P NMR, MDP contained in a coaxial tube can be used as a chemical shift and quantitation standard (*see* **Note 2**).
3. Acquire NMR spectrum, maintaining sample at a fixed temperature, typically 30°C. In the case of ^{31}P NMR, a 5 s interpulse delay and a 60° flip angle pulse is sufficient to ensure complete relaxation of the metabolite resonances (*see* **Note 6**).

3.2. Intact Cell Systems

Examination of intact cells by NMR has necessitated the development of cell perfusion systems that can be used to maintain viable cell populations at high densities, under well-defined and homogeneous conditions in the NMR spectrometer. Various cell perfusion methods are in general use (**Table 1**). Further information on these systems, covering how each has been developed for NMR applications and the type of metabolic data obtainable, can be found in a number of publications ***(10,11)***. There is considerable interest in the biotechnology industry in the development of systems for maintaining high-density mammalian cell cultures. In addition to their use as production systems for monoclonal antibodies, vaccines, and other proteins with diagnostic or therapeutic applications, they are also attracting attention as vehicles for growing human cells ex vivo for cellular therapies and artificial organs ***(12)***. The hollow-fiber bioreactor (HFBR), which is an important bioreactor among these intensive reactor systems, has been adapted by a number of groups for use with an NMR spectrometer (**Figs. 1** and **2**). This has allowed noninvasive NMR studies of cellular metabolism at various stages of cell growth ***(13–15)***. A primary objective of using HFBRs in these NMR studies was to allow access of the NMR technique to studies of isolated cell metabolism. However, it is clear that these studies may be equally important in understanding the factors that limit bioreactor performance in terms of the influence of reactor design and operation on cell metabolism. NMR spectroscopy ***(16,17)*** and also imaging (**Fig. 2**; **ref.** ***13)*** could thus provide important tools for assisting the future design and operation of these systems. This is likely to be of particular importance in the area of artificial organs, where the metabolic activity of the cells may be closely related to their function as a surrogate organ.

Table 1
Cell Perfusion Methods That Can Be Adapted for NMR Studies

Culture method	Cell culture type	Advantages	Disadvantages
Aeration of cells at normal or low temperatures	Suspension culture	Opportunity for on-line monitoring of metabolic events over a short time-course	Sedimentation of cells can occur with time. Close packing of cells not ideal physiologically
Surface oxygenation using solid or macroporous microbarrier beads as a cell support, followed by superfusion with nutrient medium in the spectrometer	Anchorage dependent	Direct contact between perfusate and each cell. Growth of cells at rates comparable to those in culture dishes	Relatively low cell densities attainable due to large volume occupied by beads. Limited rate of perfusion due to inability to with-stand harsh shear forces
Cell entrapment within a polymer matrix (e.g., agarose gel threads that are then perfused with medium in the spectrometer)	Anchorage dependent; suspension culture	Allows study at higher cell densities. Ease of immobilization/recovery. Cells are metobolically stable for a few days	Unclear if cells can divide, questioning whether their properties remain constant. Cells are perturbed by the immobilization process itself

Hollow fiber reactors. An array of semipermeable hollow fibers are used to perfuse a contained cell culture. Comparable to capillary perfusing a tissue	Anchorage dependent; suspension culture	Can obtain cultures near tissue density. Allows cell maintenance/growth in the NMR spectrometer over a long time scale. Permits study of the same population of cells over an extended time-course	Nutrient supply is via diffusion through fiber walls. Can lead to insufficient supply/concentration gradients of nutrients within the extracapillary space

(Adapted with permission from **ref.** ***1***, ***6***, and ***12***.)

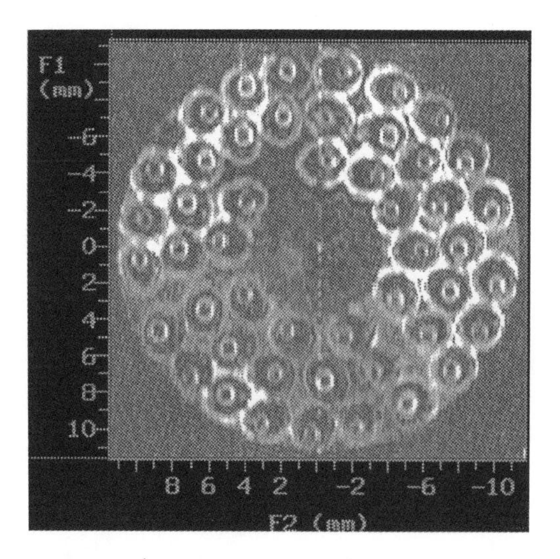

Fig. 2. Diffusion-weighted ^{1}H NMR image of a transverse section of a functional hollow-fibre bioreactor. The reactor contained approximately 1.4×10^9 CHO K1 cells. The image was obtained from a 2-mm-thick slice 35 d after the reactor was seeded with 5×10^8 cells. Water molecules with a relatively low apparent diffusion coefficient, corresponding to intracellular water and water within the fiber walls, give rise to higher signal intensities in these images ***(13)***.

3.2.1. Cell Culture in Preparation for the HFBR Perfusion System

1. In the case of adherent cells the cells can be grown in T-flasks before being transferred to spinner flasks containing Cytodex™ beads (1 g/L) (inoculate cells at 2×10^5 cells/mL).
2. Maintain cells in spinner flask until a cell density of approx 5×10^8 cells is reached.
3. Load cells on beads into the extracapillary cavity of a purpose-built HFBR.

3.2.2. HFBR Perfusion System and NMR Analyses

1. Pump medium from a batch fed stirred tank fermenter containing 2 L of medium through the HFBR. The flow rate required depends on the HFBR. For the Setec reactor we use a flow rate of 50 mL/min.
2. Gas medium with mixture of O_2, N_2, and CO_2 (*see* **Note 7**).
3. Monitor pO_2 and pH of medium in the fermenter using a polarographic oxygen electrode and a glass pH electrode, respectively.
4. Regulate dissolved oxygen tension, pH, and temperature of the perfusate.
5. Cells can be maintained in this type of system for several weeks and can achieve densities of $>10^8$/mL.
6. Multinuclear NMR spectra can be acquired throughout the duration of the culture ***(13,15,16)***.

7. Imaging techniques are as follows:
   a. Spin-echo NMR images can be used to monitor fiber distribution in the HFBR *(13)*.
   b. Diffusion-weighted images can be used to map cell distribution *(13)*.
   c. Flow imaging can be used to map flow rates in both the fibers and the extracapillary space *(18)*.
   d. Chemical shift imaging can be used to map the distribution of cellular metabolites *(19)*.
   e. ^{19}F NMR imaging of a perfluorocarbon probe molecule can be used to map dissolved oxygen concentration *(20)*.

4. Notes

1. The deuterium lock stabilizes the spectrometer against variation in its static magnetic field or radio frequency by maintaining the ratio of the two constant. This is achieved by monitoring the resonance frequency of the deuterium signal and varying the static magnetic field so that this frequency remains constant.
2. The inclusion of a standard allows signal intensities to be converted into concentrations and also provides a reference for chemical shift. Chemical shifts in the spectra shown in this chapter are expressed relative to the resonance of MDP, which was present as an external standard in a coaxial capillary. Phosphocreatine has also been used frequently as an internal chemical shift standard.
3. Relatively high magnetic fields (typically >7T) are required in order to obtain sensitivity and spectral resolution (i.e., machines that operate at >300 MHz) for protons.
4. Again, this should be a high-field instrument to obtain the best sensitivity. A wide-bore magnet is preferable as this can accommodate relatively large HFBRs (up to 25 mm in diameter). A wide-bore magnet can also be equipped with a magnetic field gradient set for imaging and localized spectroscopy experiments.
5. Although cell extraction is frequently adopted, information about molecules affected by the extraction procedure is necessarily lost. In the case of ^{31}P NMR spectroscopy, the level of inorganic phosphate in an extract can be significantly higher than that measured in the intact cell, due to hydrolysis of organic phosphates during the extraction procedure.
6. The inherent insensitivity of NMR imposes limitations on the concentrations of compounds that can be detected. Repeated data acquisitions are required to obtain adequate signal-to-noise ratios in the resulting spectra. However, to determine concentrations from resonance intensities either the relaxation times of the resonances must be known or acquisition conditions must be used that allow complete relaxation of the resonances between successive acquisitions, i.e., the delays between successive acquisitions must be relatively long. This frequently leads to lengthy data acquisition times.
7. The medium cannot be sparged directly with the gases as this causes foaming. Gas exchange is effected by passing the gases through silicone rubber tubing wound on a former in the mixing vessel.

Acknowledgments

The work from K. M. B.'s laboratory, which is described here, was supported by the Biotechnology and Biological Sciences Research Council, UK, and the European Community Framework IV Programme (Biotechnology-950207).

References

1. Avison, M. J., Hetherington, H. P., and Shulman, R. G. (1986) Applications of NMR to studies of tissue metabolism. *Ann. Rev. Biophys. Chem.* **15,** 377–402.
2. Nicholson, J. K. and Wilson, I. D. (1989) High resolution proton magnetic resonance spectroscopy of biological fluids. *Prog. NMR Spectrosc.* **21,** 449–501.
3. Cohen, J. S., Jaroszewski, J. W., Kaplan, O., Ruiz-cabello, J., and Collier, S. W. (1995) A history of biological applications of NMR spectroscopy. *Prog. NMR Spectrosc.* **28 (1),** 53–85.
4. Gadian, D. S. (1995) *NMR and Its Applications to Living Systems,* 2nd ed., Oxford Science Publications, Oxford.
5. Sanders, J. K. M., and Hunter, B. K. (1987) *Modern NMR Spectroscopy: A Guide for Chemists,* Oxford University Press, Oxford.
6. Callies, R. and Brindle, K. M. (1996) Nuclear magnetic resonance studies of cell metabolism in vivo, in *Principles of Cell Biology,* vol. 4, *Cell Chemistry and Physiology: Part II.* pp. 241–269.
7. Jans, A. W. H. and Liebfritz, D. (1989) A ^{13}C NMR study on the fluxes into the Krebs cycle of rabbit renal proximal tubular cells. *NMR Biomed.* **1,** 171–176.
8. Chauvin, M. F., Megnin-Chanet, F., Martin, G., Lhoste, J. M., and Baverel, G. (1994) The rabbit kidney tubule utilizes glucose for glutamine synthesis- A ^{13}C NMR study. *J. Biol. Chem.* **269 (42),** 26,025–26,033.
9. Street, J. C., Delort, A. M., Braddock, P. S. H., and Brindle, K. M. (1993) A ^{1}H/^{15}N NMR study of nitrogen metabolism in cultured mammalian cells. *Biochem. J.* **291,** 485–492.
10. Egan, W. M. (1987) The use of perfusion systems for nuclear magnetic resonance studies of cells, in *NMR Spectroscopy of Cells and Organisms* (Gupta, R. K., ed.) vol. 1, CRC, Boca Raton, FL, pp. 135–162.
11. Szwergold, B. S. (1992) NMR spectroscopy of cells. *Ann. Rev. Physiol.* **54,** 775–798.
12. McGovern, K. A. (1994) Bioreactors, in *NMR in Physiology and Biomedicine* (Gillies, R. J., ed.). Academic Press, London.
13. Callies, R., Jackson, M. E., and Brindle, K. M. (1994) Measurements of the growth and distribution of mammalian cells in a hollow-fibre bioreactor using nuclear magnetic resonance imaging. *Biotechnology* **12 (1),** 75–78.
14. Mancuso, A., Sharfstein, S. T., Tucker, S. N., Clark, D. S., and Blanch, H. W. (1994) Examination of primary metabolic pathways in a murine hybridoma with ^{13}C nuclear magnetic resonance spectroscopy. *Biotechnol. Bioeng.* **44 (5),** 563–585.
15. Galons, J. P., Job, C., and Gillies, R. J. (1995) Increase of GPC levels in cultured mammalian cells during acidosis. A ^{31}P NMR spectroscopy study using a continuous bioreactor system. *Mag. Res. Med.* **33 (3),** 422–426.

16. Gillies, R. J., Mackenzie, N. E., and Dale, B. E. (1989) Analysis of bioreactor performance by nuclear magnetic resonance spectroscopy. *Biotechnology* **7,** 50–54.
17. Gillies, R. J., Galons, J. P., McGovern, K. A., Scherer, P. G., Lien, Y. H., Job, C., Ratcliff, R., Chapa, F., Cerdan, S., and Dale, B. E. (1993) Design and application of NMR-compatible bioreactor circuits for extended perfusion of high density mammalian cell cultures. *NMR Biomed.* **6 (1),** 95–104.
18. Hammer, B. E., Heath, C. A., Mirer, S. D., and Belfort, G. (1990) Quantitative flow measurements in bioreactors by nuclear magnetic resonance imaging. *Bio/Technology* **8,** 327–330.
19. Constantinidis, I. and Sambanis, A. (1995) Towards the development of artificial endocrine tissues: ^{31}P NMR spectroscopic studies of immunoisolated, insulin-secreting AtT-20 cells. *Biotech. Bioeng.* **47,** 431–443.
20. Williams, S. N. O., Rainer, R. M., and Brindle, K. M. (1997) Mapping of oxygen tension and cell distribution in a hollow fiber bioreactor using magnetic resonance imaging. *Biotech. Bioeng.* **56,** 56–61.

IV

Specialist Techniques

14

Culturing Animal Cells in Fluidized Bed Reactors

Annette Waugh

1. Introduction

Biomass entrapment as a means of increasing cell density in bioreactors can take many forms. The growth of mammalian cells on microcarriers is a well-established technique and has been successfully used from laboratory to production scale in stirred tank reactors. However, as the loading of the microcarriers is raised to increase cell density further, adequate mixing of the culture environment by stirring becomes problematic. To achieve high loading of microcarriers, alternative means of obtaining mass transfer of oxygen, nutrients, and so forth are required.

One means of realizing this is to use weighted microcarriers and to fluidize the resulting bed of carriers with an upflow of medium. Such an arrangement forms the basis of a fluidized bed reactor (FBR). While the vertical recycle of the culture medium suspends the bed, a separate flow stream continually adds fresh medium and removes conditioned medium containing product. With such a configuration, the system can be used for the long-term perfusion culture of both anchorage-dependent and anchorage-independent cells.

In this chapter, the fluidized bed reactor described is a Cellex System 20 utilizing Cellex Microspheres™ as the carrier (Cellex Biosciences, Minneapolis, MN). However, the principles described can be applied to other makes and models. The complete 2.6-L system consists of a borosilicate glass reactor with a steel headplate and an external recycle loop, as shown in **Fig. 1**. The reactor houses the colonized bed, temperature probe, sampling septa, microsphere addition, and harvest overflow ports. The carrier bed is fluidized using a viscous drag, low-shear recycle pump in the loop outside the reactor. This loop also contains a hollow-fiber gas exchanger, dissolved oxygen and pH probes, a recycle flow meter, and medium and base addition lines. Temperature is

From: *Methods in Biotechnology, Vol. 8: Animal Cell Biotechnology*
Edited by: N. Jenkins © Humana Press Inc., Totowa, NJ

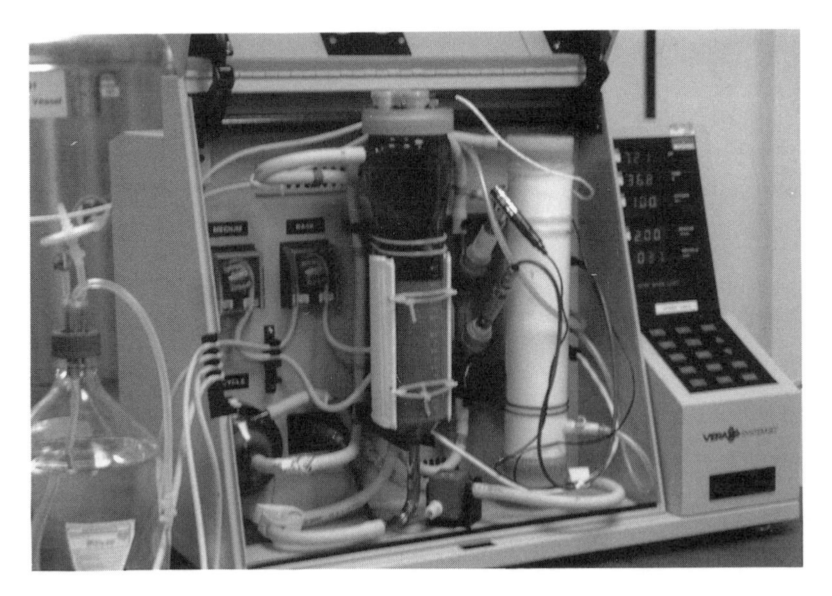

Fig. 1. The Cellex System 20.

controlled via a thermal jacket around the reactor and a forced hot air heater that warms the process compartment.

The culture medium is continually recycled through the loop, leaving the reactor above the culture bed and returning to the bottom of the reactor. The addition of oxygen, nitrogen, and carbon dioxide through the gas exchanger and base through a peristaltic pump conditions the medium to the correct oxygen tension and pH. **Figure 2** shows a schematic diagram of the system.

The microspheres, which have an average diameter of 500–700 µm and a pore size of 20–50 µm, are coated with collagen. To enhance cell attachment, the microspheres also contain fibronectin, a natural cell adhesion molecule. Metal weighting particles in the microspheres increase their density and so allow them to be fluidized without escaping through the top of the reactor. The cells grow in and on the beads, often reaching densities of >100 million cells/mL of microspheres. This is approaching tissue cell density. **Figure 3** shows a photomicrograph of a colonized bead, illustrating its gross structure.

2. Suppliers of Fluidized Bed Reactors

1. Cellex Fluidized Bed Reactors
   All reactors (System 1 [100 mL], System 3 [250 mL], System 20 [2.6 L], System 200 [4.5 L], and System 2000 [60 L]) are manufactured by Cellex Biosciences, 8500 Evergreen Boulevard, Minneapolis, MN 55433.

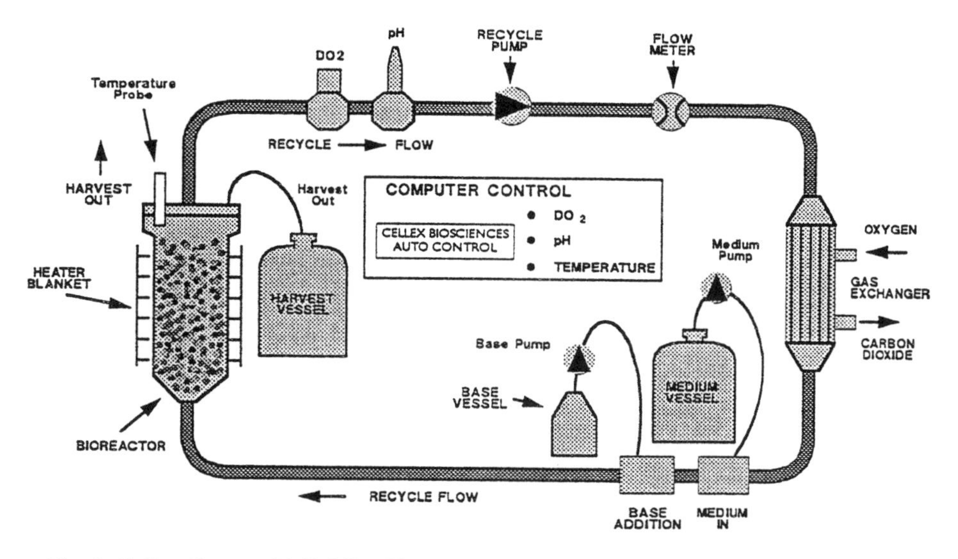

Fig. 2. Cellex System 20 fluidized bed reactor (FBR) recycle loop and process flowpath.

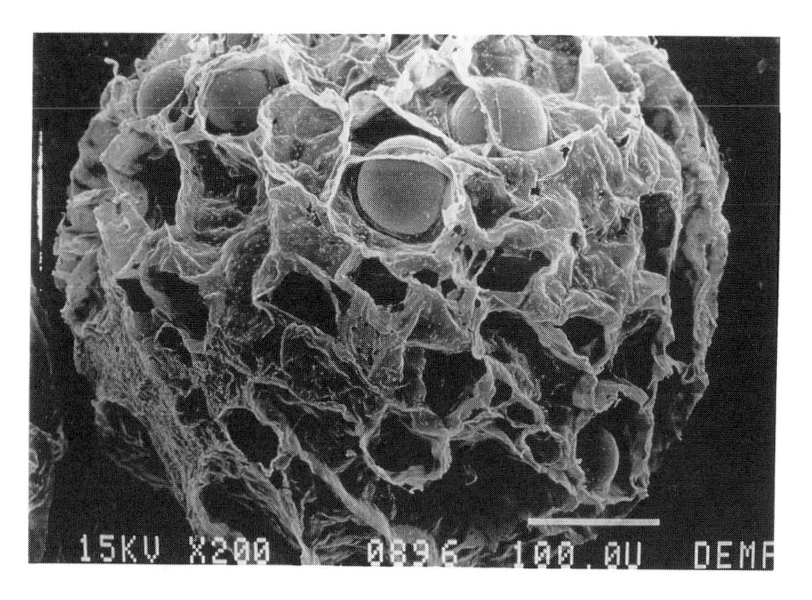

Fig. 3. A colonized microsphere. Bar = 100 µm.

2. Cytopilot™ Fluidized Bed Reactor
   All reactors (laboratory [2 L], pilot [20 L], production [80 L], and customized) are manufactured by Vogelbusch GmbH, Blechturmgasse 11, A-1050 Vienna, Austria.
3. Laboratory scale Cytopilot is distributed by Pharmacia Biotech AB, Björkgatan 30, S-751 82 Uppsala, Sweden.

Table 1
Advantages and Disadvantages of FBRs for Animal Cell Culture

Advantages
- Cell type—can be used to culture both anchorage-dependent and anchorage-independent cells. No lengthy suspension adaptation is necessary for anchorage-dependent cells, so faster process development is possible
- Protein quality—short residence times in the reactor mean that unstable proteins can be produced at high quality.
- High yield—high cell densities, typically achieved in perfusion culture, generally lead to increased product titers
- Scaleability—the Cellex System 20 is a benchtop reactor that can supply tens of liters of product stream a day. The process can be scaled up significantly simply by extending the length of the perfusion culture

Disadvantages
- Economics—generally, perfusion culture is an expensive mode of operation because of the increased volume of costly medium used
- Cell enumeration—difficult
- Regulatory issues—FBRs have never been used to date to manufacture therapeutics. This makes it a new technology that will need to be authorized by the regulatory bodies. A preferred route is generally one that has already been approved

2.1. Advantages and Disadvantages of Fluidized Bed Reactors for Animal Cell Culture

Table 1 summarizes some advantages and disadvantages of FBRs in animal cell culture.

3. Experimental Procedure

The exact procedure will vary with the make/model of the reactor. Below is a summary of the main steps involved in the setting up of a Cellex System 20 FBR.

3.1. FBR Preparation and Inoculation

1. Assemble the process loop.
2. Calibrate the pH probe and test the dissolved oxygen sensor.
3. Sterilize the assembly (by autoclaving at 121°C for 20 min).
4. Install the process loop in the System 20 cabinet, connecting the probes and heater jacket.
5. Attach the cell culture medium bottle, harvest bottle, and alkali bottle.
6. Fill the reactor and recycle loop with the culture medium (using the medium pump).
7. Set the recycle flow rate to 2.5 L/min.
8. Prime the base pump.

Table 2
Examples of Animal Cells Cultured in FBRs

Cell type	Ref.
HeLa	*1*
CHO	*2–4*
Hybridoma	*5–10*
Guinea pig keratinocyte	*6, 7*
Vero	*6, 7, 11*
MDBK	*12*
BHK	*9*

9. Calibrate the dissolved oxygen probe and bring the reactor to its operating conditions.
10. Add the microspheres and condition them with a medium flow of 5 mL/min for 24 h.
11. Switch the medium flow rate to zero.
12. Add the inoculum through a septum in the headplate (10^8 cells).
13. When the cells have attached to the microspheres (24–48 h), switch on the medium flow at a low flow rate to commence perfusion culture.

3.2. Process Monitoring

1. Take daily samples from the reactor and perform the following analyses:
   a. Cell number in suspension.
   b. Viability.
   c. Residual glucose concentration.
   d. Product concentration.
2. Freeze samples for further analyses such as lactate, glutamine, ammonia, amino acid concentrations, and lactate dehydrogenase activity.
3. To make daily estimates of cell growth on the beads, the glucose consumption rate is calculated. The medium flow rate can then be adjusted to maintain a constant residual glucose concentration. The optimal residual glucose concentration for a particular cell line must be determined empirically.
4. Microspheres from the reactor can be periodically sampled to measure directly the total cells in the system. Cell number and viability are determined via collagenase disruption of the microspheres.
5. As the cells colonize the microspheres, the fluidized bed height will expand, and beads may need to be removed to prevent their entry into the recycle loop.
6. The systems can be run in perfusion mode for several months.

4. Examples of Cells Grown in Fluidized Bed Reactors

Table 2 gives a list of some animal cell types, with references, cultured in FBRs.

Acknowledgments

The author would like to thank Receptor Technologies Limited for providing the photograph of the microsphere and Dr. Nigel Woods for his advice in writing this chapter.

References

1. Worden, M., Tremblay, M., Fahey, J., DeLucia, D., Runstadler, P. W., Pavlakis, G. N., Ciminale, V., and Campbell, M. (1990) Production of gp 120 HIV envelope protein by recombinant HeLa cells in collagen microspheres in a laboratory-scale fluidized bed bioreactor. *Verax Technical Bulletin*, Cellex Biosciences, 8500 Evergreen Boulevard, Minneapolis, MN 55433.
2. Ogata, M., Marumoto, Y., Oh-I, K., Shimizu, S., and Katoh, S. (1994) Continuous culture of CHO-K1 cells producing thrombomodulin and estimation of culture conditions. *J. Ferment. Bioeng.* **77,** 46–51.
3. Lüllau, E., Biselli, M., and Wandrey, C. (1994). Growth and metabolism of CHO cells in porous glass carriers, in *Animal Cell Technology: Products of Today, Prospects for Tomorrow* (Spier, R. E., Griffiths, J. B., and Berthold, W., eds.), Butterworth-Heinemann, Oxford, pp. 252–255.
4. Matthews, D. E., Piparo, K. E., Burkett, V. H., and Pray, C. R. (1994) Comparison of bioreactor technologies for efficient production of recombinant α-amidating enzyme, in *Animal Cell Technology: Products of Today, Prospects for Tomorrow* (Spier, R. E., Griffiths, J. B., and Berthold, W., eds.), Butterworth-Heinemann, Oxford, pp. 315–319.
5. Dean, R. C., Karkare, S. B., Ray, N. G., Runstadler, P. W., and Venkatasubramanian K. (1988) Large-scale culture of hybridoma and mammalian cells in fluidized bed bioreactors. *Ann. N.Y. Acade. Sci.* **606,** 129–146.
6. Looby, D. and Griffiths, J. B. (1989) Immobilisation of animal cells in fixed and fluidised porous glass sphere reactors, in *Advances in Animal Cell Biology and Technology for Bioprocesses* (Spier, R. E., Griffiths, J. B., Stephenne, J., and Crooy, P. J., eds.), Butterworths, Guildford, pp. 336–344.
7. Looby, D. and Griffiths, J. B. (1990) Immobilisation of animal cells in porous carrier culture. *Trends Biotechnol.* ***8,*** 204–209.
8. Rolef, G., Biselli, M., Dunker, R., and Wandrey, C. (1994) Optimization of antibody production in a fluidized bed bioreactor, in *Animal Cell Technology: Products of Today, Prospects for Tomorrow* (Spier, R. E., Griffiths, J. B., and Berthold, W., eds.), Butterworth- Heinemann, Oxford, pp. 481–484.
9. Kratje, R. B., Lind, W., and Wagner, R. (1994) Characterization of intra- and extracellular proteases in recombinant mammalian and hybridoma cells, in *Animal Cell Technology: Products of Today, Prospects for Tomorrow* (Spier, R. E., Griffiths, J. B., and Berthold, W., eds.), Butterworth-Heinemann, Oxford, pp. 679–682.
10. Reiter, M., Buchacher, G., Blüml, G., Zach, N., Steinfellner, W., Schmatz, C., Gaida, T., Assadian, A., and Katinger, H. (1994) Production of the HIV-1 neutralising human monoclonal antibody 2F5: stirred tank versus fluidized bed culture, in *Animal Cell Technology: Products of Today, Prospects for Tomorrow*

(Spier, R. E., Griffiths, J. B., and Berthold, W., eds.), Butterworth-Heinemann, Oxford, pp. 333–335.
11. Marique, Th., Malarme, D., Stragier, P., and Wérenne, J. (1994) On-line monitoring of growth and viral infection of VERO cells in collagen microspheres in a fluidized-bed bioreactor, in *Animal Cell Technology: Products of Today, Prospects for Tomorrow* (Spier, R. E., Griffiths, J. B., and Berthold, W., eds.), Butterworth-Heinemann, Oxford, pp. 369–371.
12. Marique, Th., Paul, Th., Blankaert, D., Texeira Guerra, I., and Wérenne, J. (1994) Collagenase production by a cell line in collagen microspheres in a fluidized-bed reactor and its modelization, in *Animal Cell Technology: Products of Today, Prospects for Tomorrow* (Spier, R. E., Griffiths, J. B., and Berthold, W., eds.), Butterworth-Heinemann, Oxford, pp. 245–247.

15

GPI-Anchored Fusion Proteins

Malcolm L. Kennard, Gregory A. Lizee, and Wilfred A. Jefferies

1. Introduction

Although most integral membrane proteins are bound to the lipid bilayer by a hydrophobic polypeptide transmembrane domain, a small functionally diverse group of proteins is uniquely anchored to the plasma membrane by the covalent attachment of a complex phospolipid anchor to the carboxyl terminus of the protein. This glycosylphosphatidyl inositol (GPI) anchor consists of a hydrophobic phospholipid, phosphatidylinositol, which attaches the carboxyl end of the protein to the outer lipid layer of the plasma membrane via a variable glycan chain and phosphoethanolamine *(1,2)*. The functional role of the anchor is not clearly established, and proposals range from cell motility to cell signaling *(1,3,4)*. The GPI-anchored protein can be released from the cell membrane by the action of a specific bacterial phospholipase, phosphatidylinositol phospholipase C (PI-PLC) *(5)*. The PI-PLC cleaves the anchor at the phosphodiester bond between the phophinositol group and the lipid portion of the anchor (**Fig. 1**). Diacyglycerol is left in the outer cell membrane, while the protein is released in a water-soluble form into the media with the hydrolyzed portion of the GPI anchor. Therefore, it is possible to harvest natural and recombinant GPI-anchored proteins in a controlled fashion by removing the cell growth medium and replacing it with a much smaller volume of PI-PLC solution *(6–8)*. This procedure yields a concentrated preparation of the desired protein that is relatively free of contaminating proteins contributed by the cells or the medium. This Chapter will discuss a method that creates GPI-anchored fusion proteins from proteins that are not normally GPI anchored such as those that are secreted or those that are attached to the cell surface by transmembrane sequences.

From: *Methods in Biotechnology, Vol. 8: Animal Cell Biotechnology*
Edited by: N. Jenkins © Humana Press Inc., Totowa, NJ

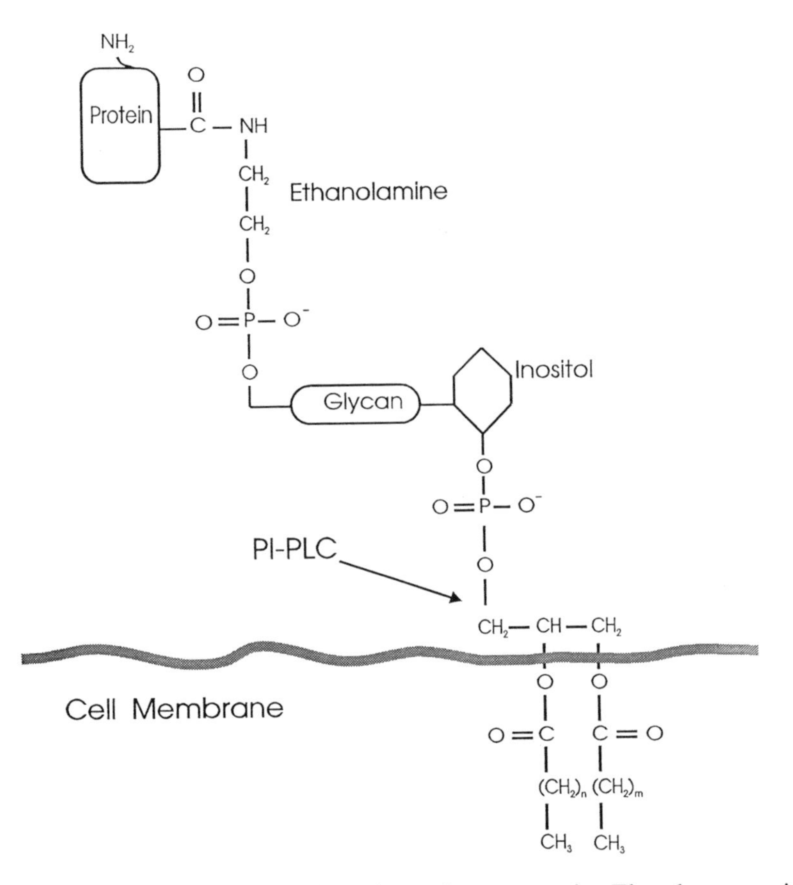

Fig. 1. Schematic of a GPI-anchored membrane protein. The cleavage site for PI-PLC is shown.

1.1. Post-Translational Attachment of the GPI Anchor

The GPI-anchored proteins and their GPI anchors are assembled separately and combined in the endoplasmic reticulum ***(9,10)***. The precursor protein is assembled with a pre-anchor sequence (PAS), which is a short COOH-terminal hydrophobic peptide sequence of 20–50 amino acids that contains a cleavage/attachment site for the GPI anchor attachment. The hydrophobic domain transiently anchors the precursor protein to the endoplasmic reticulum membrane positioning the cleavage/attachment site at an appropriate distance from the membrane for recognition by an unidentified transamidase enzyme. This enzyme cleaves part of the PAS and attaches the protein to a preassembled GPI anchor (*see* **Fig. 2**; **refs.** ***1***, ***11***, and ***12***). Site-directed mutagenesis and analysis of synthetic amino acid sequences ***(13–19)*** have established that the

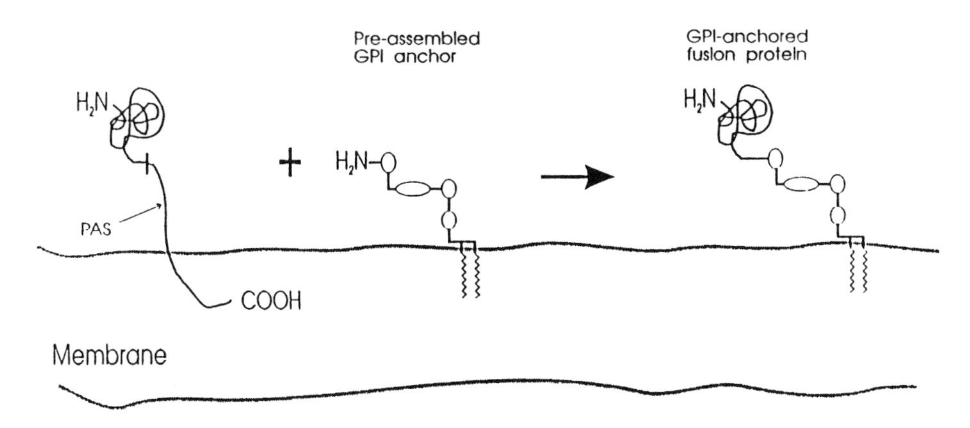

Fig. 2. Replacement of the pre-anchor sequence with a preassembled GPI anchor, which occures in the endoplasmic reticulum.

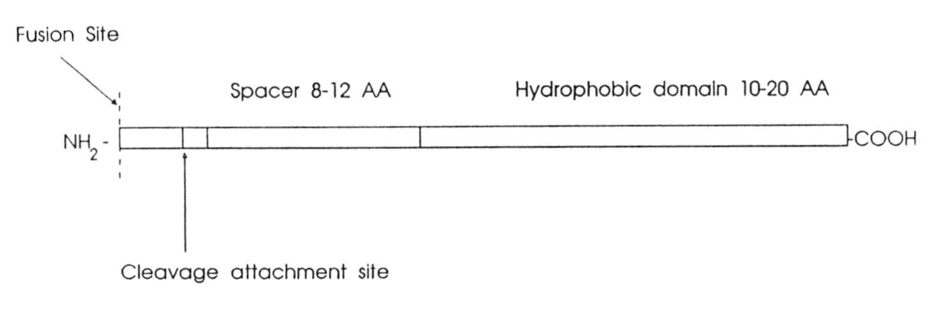

Fig. 3. Features of the pre-anchor sequence.

C-terminal GPI addition signal peptide or PAS consists of three parts that signal GPI anchoring (**Fig. 3**):

1. A cleavage/attachment domain that consists of a tripeptide made up of small amino acids (serine, aspartate, asparagine, alanine, cysteine, or glycine) at the first (GPI anchor attachment site) and the third positions. The second amino acid is less specific but also has to be small, nonhydrophobic, and nonaromatic. The regions found N-terminal of this cleavage/attachment site show no obvious homologies.
2. A spacer domain follows the cleavage/attachment site, which is moderately polar and consists of 8–12 amino acids. These sequences, from various GPI-anchored proteins, show no specific homologies to one another.
3. The PAS usually ends in a domain of 10–20 random hydrophobic amino acids ***(1,2,15)***. This hydrophobic sequence, as predicted by the Eisenberg algorithm ***(20)***, is not sufficient to act as a transmembrane sequence ***(21)***. In addition, in nearly all cases there is no typical cytoplasmic domain, as found in transmembrane proteins, immediately following the hydrophobic domain.

The hydrophilic sequences on either side of the cleavage/attachment site do not appear to play a role in the anchor attachment ***(17)***, although the spacer must be between 8 and 12 amino acids. The hydrophobic sequence, while absolutely necessary, can be easily replaced by any other hydrophobic sequence of similar length. This indicates that the overall degree of hydrophobicity rather than the primary sequence of the domain is important for its function ***(16,17,19)***.

1.2. GPI-Anchored Fusion Proteins

Although all GPI-anchored proteins are presumably processed via a common pathway, the complete lack of sequence homology of the PAS among the different GPI-anchored proteins indicates that the signal is of a general nature. Thus a fully functional PAS can be assembled from sequence elements totally unrelated to GPI anchoring as long as it possesses the three elements of the tripeptide cleavage/attachment site, spacer, and hydrophobic domains (**Fig. 3**). As a result, it has been possible to create synthetic PAS that can be fused to non-GPI-anchored proteins, which results in these proteins being targeted to the plasma membrane via a GPI linkage. For example, a synthetic peptide of 3 serines, 8 threonines and 14 leucines resulted in the GPI anchoring of the transmembrane protein CD46 ***(13)***. More commonly, chimeric versions of many type 1 integral transmembrane proteins or secreted proteins are rendered GPI anchored by attachment of PAS from existing GPI-anchored proteins ***(14,17,22–26)***. Although the coding sequence from any PAS can be used in most cases decay-accelerating factor (DAF), human placental alkaline phosphatase (HPAP), lymphocyte function-associated antigen (LFA)-3, and Thy-1 terminal carboxyl sequences are used (*see* **Table 1**). These GPI-anchored fusion proteins can then be released in a soluble form under controlled conditions by treatment of the cells with PI-PLC ***(6–8)***. For example, the carboxyl terminal 37 amino acids of DAF served as a PAS for the GPI attachment of a truncated form of glycoprotein D (gD-1) from herpes simplex virus type 1 ***(14)***.

Creation of GPI-anchored fusion proteins has been exploited by many workers who have had trouble expressing transmembrane and secreted proteins. The production of soluble PI-PLC-released proteins has many advantages:

1. It allows large quantities of protein to be produced at high concentrations and purity ***(6–8)***, which is a considerable improvement over the low levels of secreted proteins or low expression of surface proteins.
2. It allows controlled-release protein production ***(6)***.
3. Water-soluble proteins are produced that are free of detergents and other cellular components.
4. The presence of a hydrolyzed GPI anchor does not disrupt the conformation of the protein.

5. The process is especially useful in the production of monoclonal antibodies, in which cell surface proteins are poorly immunogenic.
6. Expression of the protein is not limited by expression of other surface proteins.
7. The process allows normally membrane-bound transmembrane proteins, such as receptors, to be studied independent of the cell membrane surface.

It should be noted that the fusion proteins are released with the hydrolyzed GPI anchor plus, in many cases, several amino acids from the parent PAS at their COOH terminus. However, it appears that the functional activity of these released fusion proteins is not impaired by the presence of the hydrolyzed GPI anchor, and they retain the full molecular interactive activities of their native counterparts ***(24,26)***. The presence of a common structure epitope, exposed after PI-PLC cleavage and commonly termed the crossreacting determinant (CRD) ***(1)***, may actually facilitate the purification of these released fusion proteins. For example, one could create an affinity column based on anti-CRD antibodies.

1.3. Construction and Expression of GPI-Anchored Fusion Proteins

Different methods have been used for preparing chimeric cDNAs, which include the PAS, that are able to direct GPI anchoring of the protein of interest. Usually a common restriction site is appropriately positioned at the 3' end of the cDNA encoding the protein of interest and 5' of the PAS, enabling direct fusion of the protein and PAS and insertion of the fused cDNA into the desired expression vector. Based on previous studies, there are vectors available with a built-in PAS with adjacent upstream subcloning sites into which the cDNA encoding the protein of interest can be subcloned ***(22,23)***. For example, the expression vector α+KH/HPAP20 ***(23)*** contains the carboxyl-terminal 45-amino-acid sequence of HPAP, which can be fused to the protein of interest using the cloning sites of the *Xho*I-*Not*I polylinker. Alternatively, polymerase chain reaction (PCR) amplification may be used to prepare the PAS sequence and the cDNA of the protein of interest sequentially. The PAS and protein sequences are fused by using oligos that overlap with the 5' end of the PAS and the 3' end of the protein of interest. At the extreme 5' end of the protein of interest and the 3' end of the PAS, oligos are used to create restriction sites that allow the fused protein to be readily subcloned into any eukaryotic expression vector that contains a specific selection marker such as G418 resistance (neomycin phosphotransferase gene). Before fusion of the protein of interest to the PAS, it may be necessary to modify the protein cDNA, especially in the case of transmembrane proteins, to delete the transmembrane and cytoplasmic regions prior to PCR amplification. The PAS cDNA clones can be prepared by the worker or obtained from other researchers, e.g., the plasmid pHPAP ***(24)*** that contains the HPAP PAS can be obtained from M. Davis (Stanford University, Palo Alto, CA).

Table 1
Summary of GPI-Anchored Fusion Proteins and Their Pre-anchor Sequences

Protein	Source of PAS (length amino acids)	Vector	Resistance marker	Cell line	Method of transfection	Ref.
CD46	Artificial sequence *(25)*	pSSFV-CD46	G418	CHO	Lipofection	***13***
Truncated herpes simplex glycoprotein D (gD-1)	Decay accelerating factor (DAF) *(37)*	pRSVcat	G418	CHO	Calcium phosphate coprecipitation	***14***
Human growth hormone (hGH)	DAF *(37)*	M13	MTX	COS	Calcium phosphate coprecipitation	***15,18,19***
i) High-affinity receptor for IgG (FcγRI) ii) CD2	Cell surface receptor molecule (LFA-3 or CD58) *(31)*	FcγRI-GPI (CDM)	MTX	COS	DEAE-dexran	***22***
i) Interleukin-1 receptor (IL-1RtI) ii) α-Subunit of interleukin-2 receptor (IL-2Rα) iii) E-selectin	Human placental alkaline phosphatase (HPAP) *(47)*	α+KH/HPAP20	G418	CHO	Electroporation	***23***
T-cell receptor heterodimer (TCR)	HPAP *(47)*	pBJ1–neo/A2 α/β-HPAP	G418	CHO	Electroporation	***24,27***
gD-1	DAF *(37)*	pMV6tk(neo)	G418	MDCK	Calcium phosphate coprecipitation	***25***

i) Human leukocyte adhesion molecule (ELAM-1) ii) α-Subunit of high-affinity receptor for IgE (FcεRIα) iii) Murine interleukin-1 receptor (IL1–R) iv) Human p70 subunit of the interleukin-2 receptor	IgG receptor FcGRIIIB (CD16) *(37)*	pBC12BI	G418	COS	DEAE-dextran	***26***
Murine MHC class II heterodimer	HPAP *(47)*	pBJi-Neo/MHCαβ	MTX	CHO	Electroporation	***28***
TCR	Thy-1 (CD90) *(31)*	pFRSV-TCRα/β/Thy-1	MTX	BW 5147 Thymoma	Electroporation	***29***
Cytokine macrophage colony stimulating factor (M-CSF)	DAF *(37)*	pM-CSF-GPI/REP4α	Hygromycin	KM-102 stromal cells	Lipofection	***30***
T-lymphocyte CD4	LFA-3 *(31)*	pMNCT4PI	G418	Murine T-cell hybridoma	Electroporation	***31***
T-lymphocyte CD4	DAF *(37)*	M13 CD4DAF	MTX	CHO	Calcium phosphate coprecipitation	***32***

Once the correct fusion constructs have been confirmed by DNA sequencing, the cell line of choice can then be transfected with the modified vector using standard techniques ***(33)***. Stable transfectants are isolated by their resistance to the antibiotic, and high expressing clones can be selected by flow cytometry using fluoresceinated monoclonal antibodies against the protein of interest. The selected cells can then be grown up to high density and the protein harvested under controlled conditions ***(6–8)***. Methods for harvesting GPI-anchored proteins from adherent or suspension cultures are described in Chapter 13. The recovered solubilized proteins can then be purified as necessary.

2. Materials

1. Nucleotides: deoxynucleotide triphosphate set (Boehringer-Mannheim [Laval, Canada], cat. no. 1277049).
2. Pfu polymerase (cloned) and buffer (Stratagene [Aurora, Canada], cat. no. 600154).
3. Thermal cycler: Perkin-Elmer (Mississauga, Canada) GeneAmp, model 9600.
4. Shuttle vector: pSV.SPORT 1 (GIBCO-BRL [Burlington, Canada], cat. no. 15388-010).
5. G418 resistance vector: pNEO (Pharmacia [Baie d'Urfe, Canada], cat. no. 22-4924-01).
6. Restriction enzymes (various suppliers).
7. Gel electrophoresis equipment: mini-horizontal electrophoresis system (Fischer [Nepean, Canada], cat. no. FB-MSV-965).
8. Electroporator: Gene Pulser II (Bio-Rad [Mississauga, Canada, cat. no. 165-2106); Capacitance Extender Plus (Bio-Rad, cat. no. 105-2108); Gene Pulser cuvettes (Bio-Rad, cat. no. 165-2088).
9. Gel purification kit: Qiaex II gel extraction kit 150 (Qiagen Inc. [Mississauga, Canada] cat. no. 20021).

3. Methods

A large number of variables need to be considered to create a recombinant fusion GPI-anchored protein.

1. Choice of PAS (e.g., from HPAP, DAF, LFA-3, or Thy-1).
2. Length of the PAS. Here it is best to minimize the number of amino acids upstream of the tripeptide cleavage/attachment site.
3. The protein of choice may need to be modified before fusion (e.g., removal of the cytoplasmic and transmembrane regions of cell surface proteins). This may affect the cell line chosen for the protein expression.
4. Choice of expression vector and resistance marker.
5. Choice of transfection method.
6. Choice of cell line and culture methods.

Since there are many routes for producing recombinant fusion GPI-anchored proteins, this section will summarize a general method based on creating an expression vector that contains the cDNA of the protein of interest fused to a PAS such as the 31-amino acid PAS from LFA-3 ***(22)***.

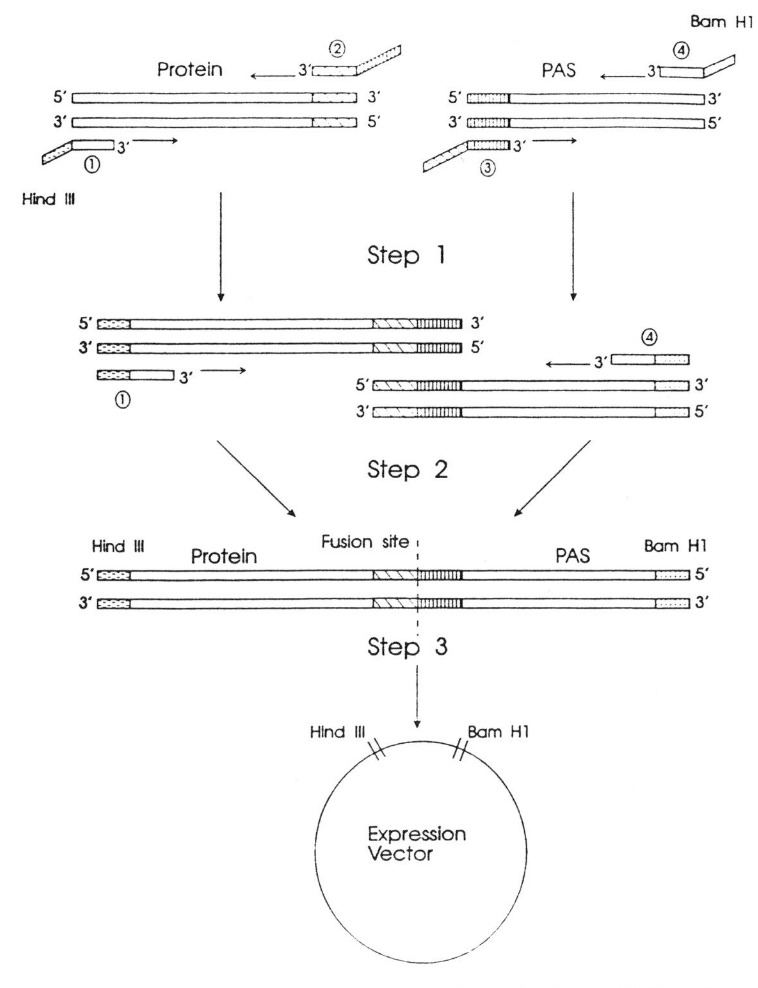

Fig. 4. Diagrammatic representation of the creation of a fusion GPI-anchored protein. Step 1: Amplification PCR of protein and PAS cDNA using oligonucleotide primers and the creation of restriction and fusion sites. Step 2: Combination/amplification PCR where modified protein and PAS fragments are fused using oligonucleotide primers 1 and 4. Step 3: Subcloning of resulting fusion sequence into expression vector using restriction sites.

3.1. Designing the Primers

This method requires the creation of four oligonucleotide primers (**Fig. 4**):

Primer 1. Contains 15–20 bp at its 3' end that is homologous to the 5' end of the gene encoding the protein of interest. The 5' end of the primer contains a restriction site (e.g., *Hin*dIII).

Primer 2. Contains 15–20 bp at its 3' end that will hybridize with the 3' end of the gene encoding the protein of interest. The 5' end of the primer contains a 15–20 bp hybridizing to the 5' end of the PAS.

Primer 3. Contains 15–20 bp at its 3' end that is homologous to the 5' end of the PAS. The 3' end of the primer contains 15–20 bp homologous to the 3' end of the gene encoding the protein of interest.

Primer 4. Contains 15–20 bp at its 3' end that will hybridize with the 3' end of the PAS. The 5' end of the primer contains a restriction site (e.g., *Bam*H1).

3.2. Preparation of the Fusion Sequence by PCR

1. This step involves two separate amplification polymerase chain reaction (PCR) reactions. The first involves using primers 1 and 2 with the gene encoding the protein of interest as a template. The second involves using primers 3 and 4 with the PAS as a template. The cDNA for the PAS may be obtained from other workers (**Table 1**; *see* **Note 1**).
   a. For each 50-µL reaction use: 5.0 µL 10× Pfu buffer, 4.0 µL 2.5 m*M* dNTPs, 0.1–0.5 µ*M* primers, 10–100 ng template DNA, and 0.5 µL Pfu polymerase. Make up to 50 µL with distilled water.
   b. Carry out amplification using a thermal cycler in 0.5-mL PCR tubes (*see* **Note 2**) and purify the DNA using agarose electrophoresis and ethidium bromide according to standard techniques (*see* **Note 3**; ref. *33*).
2. This involves combining the two PCR fragments from step 1 (modified protein and PAS cDNA) in a third combination/amplification PCR reaction. This reaction includes primers 1 and 4 and uses the two fragments as templates (**Fig. 4**). The 50-µL PCR reaction mix uses both fragments at 5–50 ng and 0.1–0.5 µ*M* primers 1 and 4. Carry out amplification and purification ***(33)***.

3.3. Screening for the Fusion Sequence

Since primers 1 and 4 were designed to include restriction sites, the fusion fragment should also contain these sites (e.g., *Hind*III at the 5' end and *Bam*H1 at the 3' end), which will allow for easy directional cloning into the expression vector of choice (*see* **Note 4**). Since these constructs, however, need to be be amplified and screened as bacterial clones prior to mammalian expression, it may be useful to subclone into a "shuttle-type" vector (such as pSV.SPORT 1; GIBCO-BRL), which contains sequences allowing it to exist and replicate inside both bacterial and mammalian cells (*see* **Note 5**).

1. Subclone the PCR fusion fragment into the vector of choice according to standard techniques ***(33)*** using the unique restriction sites.
2. Transform *Escherichia coli* with the clone using electroporation or a chemical method ***(31)***.
3. Grow the transformed bacteria in plates containing the antibiotic corresponding to the marker used in the cloning vector. Antibiotic-resistant colonies containing the complete recircularized plasmid can be selected.

4. Purify the plasmid DNA from positively identified bacterial clones and sequence the DNA to confirm correct DNA sequence, ligation sites, and fusion regions (e.g., using Sequenase Sequencing Protocol (US Biochemical, Indianapolis, IN).

3.4. Mammalian Cell Transfection and GPI-Anchored Protein Expression

Once the correct clone has been identified and sequenced it can be transfected into a mammalian cell line of choice. CHO cells have been found to be ideally suited to expressing GPI-anchored proteins.

1. Transfect the clone into CHO cells using electroporation, which has been found to be an effective method for CHO cell transfection, i.e., approx. 40 μg DNA/8 × 10^6 cells in 0.8 mL of growth medium at approx. 250 μF and 400 V. The cells and DNA are incubated on ice for about 10 min and then pulsed in a 0.4-cm gap cuvet.
2. Transfer the cells to T-flasks containing suitable growth medium e.g., Ham's F-12 medium (GIBCO-BRL) with 10% fetal bovine serum (GIBCO-BRL), at 37°C in a 5% CO_2 humidified atmosphere.
3. After 36–48 h, the medium is replaced with fresh media containing 1 mg/mL G418. The medium can be replaced every 2–3 days until resistant colonies appear.
4. Select individual high expressing cells using flow cytometry and fluoresceinated antibodies against the protein of interest.

4. Notes

1. If a *Bam*HI is present in the PAS, then either a *Bcl*I, *Bgl*II, or *Mbo*I site can be introduced, all of which have compatible overhangs with *Bam*HI.
2. A three-step cycling profile has been found to work well for this application. Choose an annealing temperature that is 5–10°C cooler than the melting temperature (*T*m) of the primer with the lowest *T*m. Generally 30–45 s at this temperature will suffice for proper annealing of the oligos to the template. The optimum polymerization temperature for Pfu polymerase is 72°C with extension times that depend on the length of the fragment (e.g., 30 s for 300–400 bp). If the lowest primer *T*m is below 72°C, it is advisable to polymerize at the same temperature as the annealing temperature or slightly higher. However, extension times will be longer. A denaturation step at 95°C for 20–30 s is typically included in every round of cycling.
3. DNA can be purified from agarose gels using a gel-purification kit such as Qiaex II. However, resuspend or elute the DNA in distilled water rather than *Tris*-EDTA, since the presence of EDTA may inhibit subsequent PCR reactions. An alternative to gel purification is to use a desalting column such as G-25 Sephadex. This is a quick and easy method but it requires larger volumes of pooled PCR fragments (>50 μL).
4. The fusion construct clone can be subcloned into a number of other vectors including pBluescript II SK/KS for mutagenesis and screening. After screening, the correct clone can be subcloned into a mammalian expression vector such as

pCDM8 (Invitrogen, San Diego, CA). This vector can be used to co-transfect cells together with a secondary vector containing a dominant selectable marker such as pNEO, which confers G418 resistance.

5. pSV.SPORT 1 does not contain a mammalian selection marker and can be used with pNEO to co-transfect mammalian cells.

Acknowledgments

Support was received from the Medical Research Council of Canada, Science Council of British Columbia and Synapse Technologies Inc.

References

1. Ferguson, M. A. J. and Williams, A. F. (1988) Cell-surface anchoring of proteins via glycosyl-phosphatidylinositol structures. *Annu. Rev. Biochem.* **57,** 285–320.
2. Low, M. G. (1989) The glycosyl-phosphatidylinositol anchor of membrane proteins. *Biochim. Biophys. Acta* **988,** 427–454.
3. Biovin, P. and Deluanay, J. (1991) La membrane du globule rouge. *XIe Congrès de la Societé Francaise d'Hématologie* **3(1),** 125–128.
4. Lisanti, M. P., Rodriguez-Boulan, E., and Saltiel, A. R. (1990) Emerging functional roles for the glycosyl-phosphatidylinositol membrane protein anchor. *J. Membr. Biol.* **117,** 1–10.
5. Sundler, R., Alberts, A. W., and Vagelos, P. R. 1(978) Enzymatic properties of phoshatidylinositol inositolphosphohydrolase from *Bacillus cereus. J. Biol. Chem.* **253,** 4175–4179.
6. Kennard, M. L., Food, M. R., Jefferies, W. A., and Piret, J. M. (1993) Controlled release process to recover heterologous glycosylphosphatidylinositol membrane anchored proteins from CHO cells. *Biotechnol. Bioeng.* **42,** 480–486.
7. Kennard, M. L. and Piret, J. M. (1994) Glycolipid membrane anchored recombinant protein production from CHO cells cultured on porous microcarriers. *Biotechnol. Bioeng.* **44,** 45–54.
8. Kennard, M. L. and Piret, J. M. (1995) Membrane anchored protein production from spheroid, porous, and solid microcarrier Chinese hamster ovary cell cultures. *Biotechnol. Bioeng.* **47,** 550–556.
9. Doering, T. A., Masterson, W. J., Hart, G. W., and Englund, P. T. (1990) Biosynthesis of glycosy-phosphatidylinositol membrane anchors. *J. Bio. Chem.* **265(2),** 611–614.
10. Bangs, J. A., Hereld, D., Krakow, J. L., Hart, G. W., and Englund, P. T. (1985) Rapid Processing of the carboxyl terminus of the trapanosome variant surface glycoprotein. *Proc. Natl. Acad. Sci. USA* **82,** 3207–3211.
11. Moran, P. and Caras, I. W. (1991) Fusion sequence from non-anchored proteins to generate a fully functional signal for glycophophatidylinositol membrane anchor attachment. *J. Cell Biol.* **115(6),** 1595–1600.
12. Boothroyd, J. C., Paynter, C. A., Cross, G. A., Bernards, A., and Borst, P. (1981) Variant surface glycoproteins of Trapanosoma brucei are synthetised with cleavable hydrophobic sequences at the carboxy and amino termini. *Nucleic Acids Res.* **9,** 4735–4743.

13. Coyne, K. E., Crisci, A., and Lublin, D. M. (1993) Construction of synthetic signals for glycosyl-phosphatidylinositol achor attachment. *J. Biol. Chem.* **268,** 6689–6693.
14. Caras, I. W., Weddell, G. N., Davitz, M. A., Nessenzweig, V., and Martin, D. W. (1987) Signal for attachment of a phospholipid membrane anchor in decay accelerating factor. *Science* **238,** 1280–1282.
15. Caras, I. W., Weddell, G. N., and Williams, S. R. (1989) Analysis of the signal for attachment of a glycophospholipid membrane anchor. *J. Cell Biol.* **108,** 1387–1396.
16. Moran, P., Raab, H., Kohr, W. J., and Caras, I. W. (1991) Glycophospholipid membrane anchor attachment. *J. Biol. Chem.* **266,** 1250–1257.
17. Moran, P. and Caras, I. W. (1991) Fusion sequence elements from non-anchored proteins to generate fully functional signal for glycophosphatidylinositol membrane anchor attachment. *J. Cell Biol.* **115,** 1595–1600.
18. Moran, P. and Caras, I. W. (1994) Requirements for glycophosphatidylinositol attachment are similar but not identical in mammalian cells and parsitic protozoa. *J. Cell Biol.* **125,** 333–343.
19. Caras, I. W. and Weddell, G. N. (1989) Signal peptide for protein secretion directing glycophospholipid membrane anchor attachment. *Science* **243,** 1196–1198.
20. Eisenberg, D., Schwarz, E., Kamaromy, M., and Wall, R. (1984) Analysis of membrane and surfae protein sequences with hydrophobic moment plot. *J. Mol. Biol.* **179,** 125–142.
21. Food, M. R., Rothenberger, S., Gabathuler, R., Haidl, G., Reid, G., and Jefferies, W. A. (1994) Transport and expression in human melanomas of a transferrin-like glycosylphosphatidylinositol-anchored protein. *J. Biol. Chem.* **269,** 3034–3040.
22. Harrison, P. T., Hutchinson, M. J., and Allen, J. M. (1994) A convenient method for the construction and expression of GPI-anchored proteins. *Nucleic Acids Res.* **22,** 3813,3814.
23. Whitehorn, E. A., Tate, E., Yanofsky, S. D., Kochersperger, L., Davis, A., Mortensen, R. B., Yonkovich, S., Bell, K., Dower, W. J., and Barrett, R. W. (1995) A generic method for expression and use of "tagged" soluble versions of cell surface receptors. *Bio/Technol*ogy **13,** 1215–1219.
24. Lin, A. Y., Devaux, B., Green, A., Sagerstrom, C., Elliott, J. F., and Davis, M. M. (1990) Expression of T cell antigen receptor heterodimers in a lipid-linked form. *Science.* **249,** 677–679.
25. Lisanti, M. P., Caras, I. W., Davitz, M. A., and Rodriguez-Boulan, E. (1989) A glycophospholipid membrane anchor acts as an apical targetting signal in polarized epithelial cells. *J. Cell Biol.* **109,** 2145–2156.
26. Scallon, B. J., Kado-Fong, H., Nettleton, M. Y., and Kochan, J. P. (1992) A novel strategy for secreting proteins: use of phosphatidylinositol-glycan specific phospholipase D to release chimeric phosphatidyinositol-glycan anchored proteins. *Bio/Technology* **10,** 550–556.
27. Devaux, B., Bjorkman, P. J., Stevenson, C., Grief, C., Elliot, J. F., Sagerstrom, C., Clayberger, C., Karensky, A. M., and Davis, M. M. (1991) Generation of mono-

clonal antibodies against soluble human T cell receptor polypeptides. *Eur. J. Immunol.* **21,** 2111–2119.

28. Wettstein, D. A., Boniface, J. J., Reay, P. A., Schild, H., and Davis, M. M. (1991) Expression of a class II major histocompatibility complex (MHC) heterodimer in a lipid linked form. *J. Exp. Med.* **174,** 219–228.
29. Slanetz, A. E. and Bothwell, A. L.M. (1991) Heterodimeric, disulphide-linked a/b T cell receptors in solution. *Eur. J. Immunol.* **21,** 179–183.
30. Weber, MC., Groger, R. K., and Tykocinski, M. L. (1994) A glycophosphatidylinositol-anchored cytokinecan function as a artificial cellular adhesin. *Exp. Cell Res.* **210,** 107–112.
31. Sleckman, B. P., Rosenstein, Y., Igras, V. E., Greenstein, J., and Burakoff, S. J. (1991) Glycolipid-anchored form of CD4 increases intracellular adhesion but is unable to enhance T cell activation. *J. Immunol.* **147,** 428–431.
32. Keller, G., Siegel, M. W., and Caras, I. W. (1992) Endocytosis of a glycophospholipid-anchored and transmembrane forms of CD4 by different endocytic pathways. *EMBO J.* **11,** 863–874.
33. Sambrook, J., Fritsch, E. F., and Maniantis, T. (1990) *Molecular Cloning: A Laboratory Manual,* 2nd ed. Cold Spring Harbor Laboratory Press, Cold Spring Harbor, NY.

16

Harvesting GPI-Anchored Proteins From CHO Cells

Malcolm L. Kennard and James M. Piret

1. Introduction

Glycosyl-phosphatidylinositol (GPI)-anchored proteins bound to the outer surface of cell membranes *(1,2)* can be selectively released in a soluble form by the action of a highly specific bacterial enzyme, phosphatidylinositol-phospholipase C (PI-PLC) *(3,4)*. This chapter describes a controlled release process *(5–7)* that increases both the concentration and purity of harvested GPI-anchored proteins by separating the harvesting of the protein from cell growth and protein expression. This approach has obtained product concentrations of up to 500 µg/mL at purities of 35% total protein *(5,7)*.

Recombinant proteins, expressed by mammalian cells are usually secreted into the culture at low concentrations and purities. In many cases it is difficult to recover more than a few micrograms per milliliter from the culture medium, which contains many contaminating proteins. To recover and purify the desired protein, lengthy and multiple purification steps are often necessary, with losses of protein at each stage. At laboratory and industrial scales these steps can represent the major cost of mammalian cell protein production. These normally secreted recombinant proteins can be expressed in a GPI-anchored form (*see* Chapter 12) to benefit from the controlled release method.

1.1. GPI Anchor

GPI-anchored proteins are covalently linked at their carboxyl end to glycosylphosphatidylinositol, via phosphoethanolamine and a variable glycan chain *(1,2)*. The hydrophobic glycolipid attaches the anchor to the cell membrane (*see* **Fig. 1**). The functional role of the anchor is not clear, and proposals range from cell motility to cell signaling *(8–10)*. The anchor is cleaved by PI-PLC at the phosphodiester bond between the phosphoinositol group and the

From: *Methods in Biotechnology, Vol. 8: Animal Cell Biotechnology*
Edited by: N. Jenkins © Humana Press Inc., Totowa, NJ

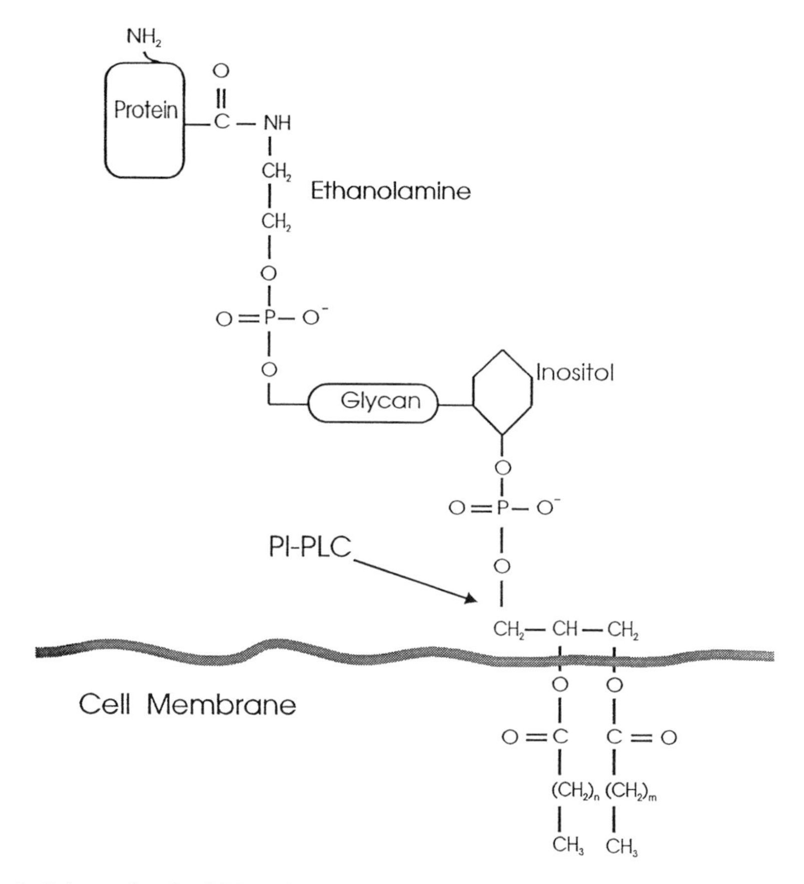

Fig. 1. Schematic of a GPI-anchored protein. The cleavage site for PI-PLC is shown.

lipid portion of the anchor. Diacyl-glycerol is presumably left in the membrane, while the protein (as well as the attached inositol glycan portion of the GPI anchor) is released in a soluble form into the medium.

1.2. GPI-Anchored Fusion Proteins

In the endoplasmic reticulum, GPI-anchored proteins are first assembled as precursor proteins with short COOH terminal hydrophobic pre-anchor sequences (PAS) of 20–37 amino acids that contain cleavage/attachment sites for the GPI anchor ***(10,11)***. The precursor protein is then post-translationally modified by replacing part of the PAS with a preassembled GPI anchor ***(1,12)***. Addition of a PAS to secreted proteins or replacement of the transmembrane regions of cell surface proteins with a PAS by recombinant DNA methods will result in GPI anchoring of the resulting fusion proteins to the plasma membrane ***(12–16)***. These GPI-anchored fusion proteins can then be released from the cell surface by PI-PLC under controlled conditions. The function of these released proteins is unaffected by the presence of the inositol glycan portion of the hydrolyzed GPI anchor and several amino acids from the PAS attached to their COOH termini ***(14,16)***.

1.3. Harvesting GPI-Anchored Proteins from CHO Cells

The controlled release method described in this chapter is based on harvesting from Chinese hamster ovary (CHO; cell line WTB, obtained from F. Maxfield, New York University, New York, NY) cells that have been transfected to expressed naturally occurring ***(5–7)*** or fusion GPI-anchored protein ***(12–16)***.

1.3.1. Harvesting from Suspension Cells

CHO cells can be cultured as a suspension culture by adapting the cells to grow in serum-free medium. By periodically replacing the growth medium with a much smaller volume of PI-PLC solution in phosphate-buffered saline (PBS), the GPI-anchored protein is released in a soluble form at high protein purity and concentration. The cells are then returned to fresh growth medium, where they re-express the harvested GPI-anchored protein for subsequent harvesting. This process can be repeated for many harvest cycles without deterioration in growth rate, viability, or cell-specific protein production ***(5)***.

1.3.2. Harvesting from Microcarriers

To avoid repeated washing and centrifugation of cells grown in suspension, adherent CHO cells can be immobilized on porous and solid microcarriers, such as Cultispher ***(17,18)*** and Cytodex ***(19,20)***. Cells grow in multilayers at far higher cell densities, 1×10^7 cells/mL of medium ***(17–19)***, than in suspension culture. Microcarriers settle rapidly under gravity, especially when densely populated with cells, which facilitates the multiple media replacement and cell washing steps. This type of high-cell-density microcarrier culture also allows the alternation between growth and harvest medium to be performed within a suitably designed bioreactor. Stable microcarrier cultures can be repeatedly harvested over many weeks in a similar manner to suspension cultures ***(6,7)***.

2. Materials

1. Cell line naturally expressing or transfected with GPI-anchored protein (*see* **Note 1**).
2. Serum-free medium for suspension culture such as CHO-S-SFM II (GIBCO-BRL [Burlington, Canada], cat. no. 12052).
3. Medium for adherent microcarrier culture or suspension spinner culture: Ham's F-12 (GIBCO-BRL, cat. no. 21700) supplemented with 2 m*M* L-glutamine (GIBCO-BRL, cat. no. 21051), 100 U/mL penicillin, and 100 μg/mL streptomycin (GIBCO-BRL, cat. no. 15070), and 10% newborn calf serum (NCS; GIBCO-BRL, cat. no. 1610; *see* **Note 2**).
4. Trypsin-EDTA: 10× solution (GIBCO-BRL, cat. no. 15050), used at 0.1% trypsin.

5. PBS: Ca/Mg free, sterile PBS (Sigma [Oakville, Canada], cat. no. D1480).
6. Enzyme solution: PI-PLC (Boehringer-Mannheim [Laval, Canada], cat. no. 1143-069; 50 U/mL) in PBS.
7. Microcarriers: porous microcarriers, Cultispher-G (Percell Biolytica [Lund, Sweden], cat. no. 9–000–080–001); solid microcarriers, Cytodex-1 (Pharmacia Biotech [Baie d'Urfe, Canada], cat. no. 17-0448-01).
8. T-flasks: 75 cm^2 and 150^2 (Nunc [Burlington, Canada], cat. no. 178891, 178883).
9. Spinner flasks: e.g., 500 mL (Bellco [Vineland, NJ], cat. no. 1965-00500).
10. Spinner flask magnetic stirrer (Bellco, cat. no. 7760–06005).
11. Siliconizing solution: 2% dimethyldichlorosilane in trichloethane (Pharmacia, cat. no. 17-1337-01).
12. Trypan blue stain (GIBCO-BRL, cat. no. 15250).

3. Methods

These methods apply to CHO cells but can be modified to deal with other GPI-anchored protein expressing mammalian cell lines.

3.1. Suspension CHO Culture

Adapt the CHO cells to the serum-free medium according to the manufacturer's instructions. Culture the cells (inoculated at 1–2 × 10^5 cells/mL) in 75-cm^2 T-flasks containing 25–50 mL of medium at 37°C in a 5% CO_2 humidified atmosphere. Serum-free medium is preferred since it reduces cell attachment to the T-flask surface. If larger volumes are required, transfer the culture to 500-mL spinners (*see* **Note 3**), inoculating at 2 × 10^5 cells/mL. Under these conditions the serum-free medium may be replaced by the considerably cheaper serum containing Ham's F-12. Monitor the cell density and viability using a hemocytometer and trypan blue exclusion. The cells should be harvested when the cell density has reached approx 2 × 10^6 cells/mL (*see* **Note 4**).

The following volumes refer to harvesting 25 mL of culture (approx 4 × 10^7 cells). All steps should be performed under aseptic conditions:

1. Check cell density and viability (preferably cells should be in late log phase, e.g., for CHO cells 1.5–2 × 10^6 cells/mL and >90% viability).
2. Pellet cells at 250*g* for 5 min.
3. Wash cells two times with 5 mL of PBS to remove medium proteins.
4. Resuspend cells in 1 mL of PI-PLC in PBS (30 mU/mL) and incubate at 37°C for 30 min with periodic agitation (*see* **Note 5**).
5. Pellet cells at 250*g* for 5 min and recover the enzyme solution containing the harvested protein (*see* **Note 6**).
6. Centrifuge the recovered protein solution at 10,000*g* for 15 min, filter through a 0.2-µm membrane, and store under sterile conditions.
7. Wash enzyme-treated cells two times in 15 mL of PBS to remove residual PI-PLC.
8. Resuspend cells in fresh medium and discard some cells if necessary to obtain a cell density of 0.5–1.0 × 10^6 cells/mL.

9. Culture cells for approximately 48 h to allow the cells to re-express the protein before repeating the protein harvesting (*see* **Note 7**).

The harvest cycle can be repeated as many times as needed as long as the culture remains viable (>90%) and contaminant free. If viability or expression falls, begin a new culture.

3.2. High-Density Microcarrier CHO Cell Culture

These methods are based on adherent CHO cells, which are initially cultured in T-flasks before being transferred to the microcarriers. Two types of microcarrier can be used: the porous microcarrier, Cultisper-G, which consists of crosslinked gelatin with a hydrated mean diameter of 200 μm (8×10^5 beads/g dry weight), or the solid microcarrier, Cytodex-1, which consists of a crosslinked dextran matrix with a hydrated mean diameter of 180 μm (6×10^6 beads/g dry weight).

The following steps apply to 250-mL cultures in 500-mL spinners (*see* **Note 3**). All steps should be performed under aseptic conditions:

3.2.1. Preparation of Cells

1. Prepare confluent cultures of adherent CHO cells in two 150-cm^2 T-flasks (*see* **Note 8**).
2. Wash the cells two times in 50 mL of PBS and then detach the cells by incubating with 10 mL of the dilute trypsin solution at room temperature for approx 2 min.
3. Spin the cells at 250*g* for 5 min, wash the pellet in 10 mL of PBS, and resuspend in fresh medium for spinner inoculation.

3.2.2. Preparation of the Spinners

1. Siliconize the spinner with dimethyldichlorosilane according to the manufacturer's instructions to minimize cell attachment to the glass surface of the spinner.
2. Wash the spinner thoroughly in distilled water.
3. Add 100 mL of distilled water and sterilize the spinner at 121°C for 30 min. Remove the water aseptically prior to inoculation.

3.2.3. Preparation of the Microcarriers

1. Rehydrate the microcarriers in Ca^{2+} and Mg^{2+} free PBS (1 g bead dry weight/100 mL) at room temperature for at least 1 h. Use approx 0.15 or 0.25 g bead dry weight/250 mL working volume for Cultispher-G and Cytodex-1, respectively.
2. Wash and resuspend the beads in fresh PBS and sterilize at 121°C for 30 min.
3. Remove the PBS and resuspend the beads in 30 mL of fresh Ham's F-12 medium for at least 1 h (approx 2 mL of suspended beads).

3.2.4. Seeding the Microcarriers and Preparing the Cultures for Harvesting

1. Transfer the beads to the spinner and seed the beads with approx 1×10^8 or 1.25×10^7 CHO cells for Cultispher-G and Cytodex-1, respectively. This will provide

an initial cell density of approx 500 cells/bead for Cultispher-G cultures and 5–10 cells/bead for Cytodex-1 cultures (*see* **Note 9**).
2. Place the spinner in the incubator at 37°C with 5% CO_2 humidified atmosphere. Provide gentle manual agitation (1 min every 30 min) for a 3-h period to distribute the cells evenly over the miocrocarriers. Attachment can be monitored by removing bead-free medium and determining the cell density.
3. After 3 h add further fresh medium to obtain a working volume of 250 mL and operate the spinner continuously at 50 rpm.
4. Monitor the glucose concentration. When the glucose concentration falls below 0.5 g/L, let the beads settle under gravity, decant the spent medium, and replace with fresh medium. The free and attached cell density may be monitored when necessary for process optimization (*see* **Note 10**).
5. Maintain the culture for 10 or 20 days for the Cultisper-G and Cytodex-1, respectively. The serum level in the medium can then be reduced to 2% and the cultures maintained for a further 10 days to adapt to the low serum medium (*see* **Note 11**) prior to harvesting the GPI-anchored protein.

3.2.5. Harvesting the Microcarrier Culture

1. Allow the beads to settle under gravity, decant spent medium, and transfer the beads to a 50-mL sterile centrifuge tube using a 10-mL glass pipette.
2. Wash the beads twice with 10 mL of PBS.
3. Resuspend the beads in 300 mU/mL PI-PLC solution in PBS at 2 mL bead/mL enzyme solution. Incubate at 37°C for 1 h with periodic agitation (*see* **Notes 5** and **12**).
4. Allow the beads to settle and recover the enzyme solution containing the harvested protein (*see* **Note 6**).
5. Wash enzyme treated cells three times (10 min/wash) with PBS (5 mL/mL of settled bead; *see* **Note 13**). Resuspend the beads in fresh medium and return to the spinner with fresh 250 mL of medium.
6. Culture the cells for approx 48 h to allow re-expression of the protein before harvesting the cells again (*see* **Note 7**).
7. Repeat harvest cycles as long as necessary. Monitor the rate of glucose consumption to ensure that glucose levels are maintained above 0.5 g/L. The immobilized and free cell densities may also be monitored if necessary to determine the stability of the culture (*see* **Notes 10** and **14**).

4. Notes

1. Once cells are transfected, high expressing clones can be readily selected using fluorescein-labeled antibodies against the expressed protein and flow cytometry *(5)*. This technique is extremely useful for optimizing the process and monitoring cell protein expression.
2. Although fetal calf serum may be required for some mammalian cell lines, the considerably cheaper newborn calf serum is found to be adequate for CHO cell culture. Contamination of the protein product by serum proteins is greatly reduced by the controlled release methodology.

3. The recommended efficient working volume for the 500-mL spinner is 200–250 mL of medium. Increased volumes reduce O_2 and CO_2 transfer, which could reduce the growth and viability of the cell culture.
4. Protein expression can vary considerably with growth phase ***(5,6)***. Thus it is useful to determine the maximum GPI-anchor expression by carrying out batch growth cultures prior to harvesting. Maximum expression usually occurs during exponential growth and can fall by >70% in the stationary and decline phases.
5. It may be necessary to optimize PI-PLC concentration and incubation times. These levels will depend on cell-specific expression, protein, and cell type. Volumes can be reduced and PI-PLC concentrations can be increased to obtain greater harvested protein concentrations. However, additional PI-PLC will increase costs and levels of contaminating enzyme.
6. The purity of the harvested GPI-anchored protein can reach 40% based on total protein ***(5–7)***. The contaminating proteins should consist mainly of PI-PLC and other naturally expressed GPI-anchored proteins.
7. CHO cells have been shown to be unaffected by PI-PLC treatment and completely recover their recombinant GPI-anchored protein expression within 48 h ***(5,6)***. This recovery time may have to be determined for other cell lines and may vary with different proteins.
8. If spinners or microcarriers are unavailable, the adherent cells can be repeatedly harvested while cultured in the T-flask. The confluent cells are washed twice in PBS and incubated with 20 mL of PI-PLC in PBS (30 mU/mL). The harvested protein is recovered, and then the cells are washed three times in PBS before adding fresh growth medium and reharvesting 2 days later. However this method of harvesting is not recommended, since the cultures are unstable and yield only low concentrations of protein (<4 μg/mL) ***(6)***.
9. The low cell-to-bead ratio is recommended for the Cytodex-1 beads to prevent cell and bead clumping and for more even coverage of cells over the beads. However, longer times will be required to establish stable cultures.
10. Cell density can be monitored by washing a known volume or number of beads with PBS for 5 min and then treating with trypsin solution at 37°C for up to 15 min with gentle vortexing. The gelatin beads dissolve, and the disaggregated cells can be counted ***(6)***.
11. When adherent cultures have been established on the microcarriers, they are at high density and at a low growth rate. Since the culture is stable and the cells grow very slowly, the serum content of the medium can be reduced without loss of protein productivity ***(6,7)***.
12. The harvested protein concentration can be increased by increasing the volume ratio of settled bead to enzyme solution and the PI-PLC concentration. However, loss of the harvested protein associated with the settled bead is increased. Furthermore, enzyme costs are increased and the protein is contaminated by higher PI-PLC concentrations.
13. Since a considerable volume of enzyme solution will remain with the decanted settled beads (up to 40% volume), the PBS washes will contain low levels of

harvested protein. It may be considered useful to save the first wash solution to recover this protein, which would be otherwise discarded.

14. Cells initially attach to the outer surface of the porous microcarrier and then slowly penetrate the porous interior. The interior becomes completely populated with cells, with multilayers of cells forming on the outside of the beads, giving the beads the appearance of cell aggregates or spheroids. If too many multilayers form, cells may die, forming necrotic cores within the beads due to lack of nutrients. Cultures on Cultispher-G are particularly prone to the formation of these necrotic cores *(6)*.

Acknowledgments

Financial support was provided by the Natural Sciences and Engineering Council of Canada and the British Columbia Health Research Foundation.

References

1. Ferguson, M. A. J. and Williams, A. F. (1988) Cell-surface anchoring of proteins via glycosyl-phosphatidylinositol structures. *Annu. Rev. Biochem.* **57,** 285–320.
2. Low, M. G. (1989) The glycosyl-phosphatidylinositol anchor of membrane proteins. *Biochim. Biophys. Acta* **988,** 427–454.
3. Sundler, R., Alberts, A. W., and Vagelos, P. R. (1978) Enzymatic properties of phoshatidylinositol inositolphosphohydrolase from *Bacillus cereus*. *J. Biol. Chem.* **253,** 4175–4179.
4. Taguchi, R., Asahi, Y., and Ikezawa, H. (1980) Purification and properties of phosphatidylinositol-specific phospholipase C of *Bacillus thuringiensis*. *Biochim. Biophys. Acta* **619,** 48–57.
5. Kennard, M. L., Food, M. R., Jefferies, W. A., and Piret, J. M. (1993) Controlled release process to recover heterologous glycosylphosphatidylinositol membrane anchored proteins from CHO cells. *Biotechnol. Bioeng.* **42,** 480–486.
6. Kennard, M. L. and Piret, J. M. (1994) Glycolipid membrane anchored recombinant protein production from CHO cell cultured on porous microcarriers. *Biotechnol. Bioeng.* **44,** 45–54.
7. Kennard, M. L. and Piret, J. M. (1995) Membrane anchored protein production from spheroid, porous, and solid microcarrier Chinese hamster ovary cell cultures. *Biotechnol. Bioeng.* **47,** 550–556.
8. Biovin, P. and Deluanay, J. (1991) La membrane du globule rouge. *XIe Congrès de la Société Française d'Hématologie* **3(1),** 125–128.
9. Lisanti, M. P., Rodriguez-Boulan, E., and Saltiel, A. R. (1990) Emerging functional roles for the glycosyl-phosphatidylinositol membrane protein anchor. *J. Membr. Biol.* **117,** 1–10.
10. Low, M. G. (1987) Biochemistry of the glycosyl-phosphatidylinositol membrane protein anchor. *Biochem. J.* **244,** 1–13.
11. Doering, T. A., Masterson, W. J., Hart, G. W., and Englund, P. T. (1990) Biosynthesis of glycosy-phosphatidylinositol membrane anchors. *J. Biol. Chem.* **265(2),** 611–614.

12. Moran, P. and Caras, I. W. (1991) Fusion sequence from non-anchored proteins to generate a fully functional signal for glycophophatidylinositol membrane anchor attachment. *J. Cell Biol.* **115(6),** 1595–1600.
13. Caras, I. W., Weddell, G. N., Davitz, M. A., Nussenzweig, V., and Martin, D. W. (1987) Signal attachment of a phospholipid membrane anchor in decay accelerating factor. *Science* **238,** 1280–1283.
14. Lin, A. Y., Devaux, B., Green, A., Sagerstrom, C., Elliott, J. F., and Davis, M. M. (1990) Expression of T cell antigen receptor heterodimers in a lipid-linked form. *Science* **249,** 677–679.
15. Lisanti, M. P., Caras, I. W., Davitz, M. A., and Rodriguez-Boulan, E. (1989) A glycophospolipid membrane anchor acts as an apical targeting signal in polarized epithelial cells. *J. Cell Biol.* **109,** 2145–2156.
16. Scallon, B. J., Kado-Fong, H., Nettleton, M. Y., and Kochan, J. P. (1992) A novel strategy for secreting proteins: use of phosphatidylinositol-glycan specific phospholipase D to release chimeric phosphatidyinositol-glycan anchored proteins. *Bio/Technol*ogy **10,** 550–556.
17. Mignot, G., Faure, T., Ganne, V., Arbeille, B., Pavirani, A., and Romet-Lemonne, J. L. (1990) Production of von Willebrand factor by CHO cells cultured in macroporous microcarriers. *Cytotechnol*ogy **4,** 163–171.
18. Nikolai, T. J. and Hu, W. S. (1992) Cultivation of mammalian cells on macroporous microcarriers. *Enz. Microb. Technol.* **14,** 203–208.
19. Croughan, S. M., Hamel, J. P., and Wang, D. I.C. (1988) Effects of microcarrier concentration in animal cell culture. *Biotechnol. Bioeng.* **32,** 975–982.
20. Tao, T., Ji, G., and Hu, W. (1988) Serial propagation of mammalian cells on gelatin coated microcarriers. *Biotechnol. Bioeng* **32,** 1037–1052.

17

Hematopoietic Cells for Cellular and Gene Therapy

I. Basic Assay Techniques

Sigma S. Mostafa, Diane L. Hevehan, Todd A. McAdams, E. Terry Papoutsakis, and William M. Miller

1. Introduction

1.1. Hematopoietic Cell Culture: Challenges

Mature hematopoietic cells (blood cells) found circulating in the body are derived from stem cells in the bone marrow and have a variety of functions ranging from oxygen transport and blood clotting to such immune defense mechanisms as combating bacterial and viral infections **(Fig. 1)**. Ex vivo expansion of hematopoietic cells is a rapidly growing area with many potential applications in bone marrow transplantation, gene therapy, and the production of blood products. Many challenges unique to hematopoiesis must be overcome in developing productive hematopoietic cell culture systems. Complexities arise from the inherent heterogeneity and variability of the cell population that are both cell-source and time dependent. Cultures differ not only in the proportion of cell types of separate lineages, but also in the proportion of cells occupying various differentiation stages within a specific lineage **(Fig. 1)**. Growth of hematopoietic cells is further complicated by rigorous medium and growth factor requirements *(1)*. A number of assays are available to distinguish cells present at each differentiation stage for different lineages during culture. This chapter focuses on several fundamental hematopoietic cell assays, along with the challenges, advantages, and disadvantages associated with each method.

1.2. Hematopoietic Cell Culture: Assays

Much of our insight into hematopoietic cell culture is owed to the development of the colony-forming unit (CFU-C) assay, an in vitro culture tech-

From: *Methods in Biotechnology, Vol. 8: Animal Cell Biotechnology*
Edited by: N. Jenkins © Humana Press Inc., Totowa, NJ

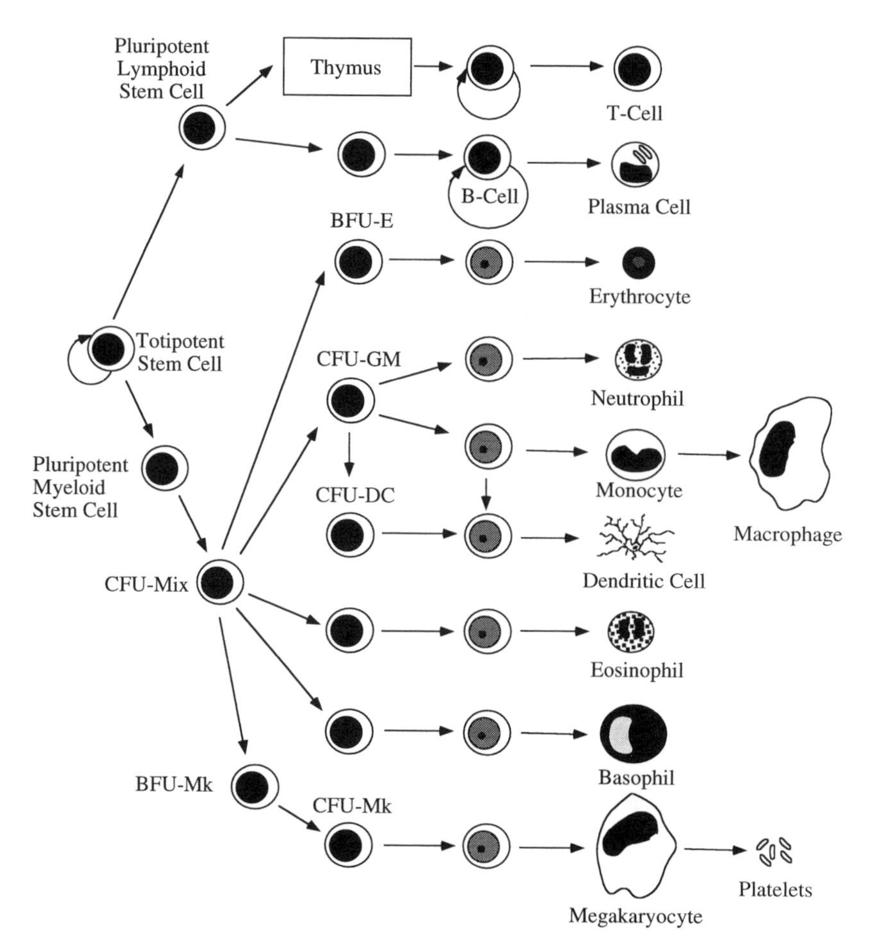

Fig. 1. Structure of the hematopoietic system.

nique used to detect progenitor cells with colony-forming potential. A small number of cells are inoculated in semisolid medium under optimal conditions. After 14 days, cultures are scored for the presence of colonies arising from single cells based on certain criteria, such as colony color and size. We describe a standard colony assay which is used to identify granulocyte (CFU-G), macrophage (CFU-M), granulocyte–macrophage (CFU-GM), erythroblast (BFU-E), and multipotent (CFU-Mix) colony-forming cells **(Fig 1)**. Colonies derived from dendritic cell precursors (CFU-DC, which are not described elsewhere in this chapter) are identified by their distinct morphology when cultured in media containing granulocyte–macrophage colony–stimulating factor (GM-CSF) and tumor necrosis factor-α *(2)*. Colony assays for megakaryocyte progenitors (BFU-MK and CFU-MK) are less standardized and more difficult to perform, and they consist of a culture phase followed by an immunostaining phase for the antigen CD41.

The CFU-C assay selects for precursor cells that are already committed to one or several specific lineages. To assess the content of more primitive cells, the long-term culture-initiating-cell (LTC-IC) assay is used ***(3)***. Hematopoietic cells are supported on an adherent layer of bone marrow stromal cells and maintained in culture for 5–8 wk, during which time colony-forming cells (CFC) differentiate into mature cells. Any remaining CFC which were derived from early progenitors are then detected via the CFU-C assay. Although LTC-IC comprises only a subset of stem cells, the LTC-IC assay is the closest in vitro estimate, thus far, of the stem cell content. In vivo models of hematopoiesis remain the only true means of quantitating totipotent human stem cells. For this purpose, a number of animal and xenograft systems, including immunodeficient mice and fetal sheep, have been developed to measure reconstitution of hematopoietic activity ***(4,5)***.

Microscopic determination of blood cell lineage and stage of maturation is possible through the use of morphological and immunocytochemical stains. Heterogeneity in hematopoietic cell populations can also be identified and quantitated using flow cytometry. This technology allows for monitoring of cell-surface antigens including CD34, which is a marker of hematopoietic stem and progenitor cells ***(6)***. One major drawback of flow cytometry is that there are no unique markers for specific CFC such as BFU-E and CFU-GM. Thus, BFU-E and CFU-GM cannot be quantitated via flow cytometry. The usefulness of flow cytometry is also limited for any case in which cells are present at very low frequencies.

2. Materials

Filtered, deionized, cell-culture grade water is used. Unless otherwise indicated, chemical reagents are available from Sigma (St. Louis, MO) and all plastic and glassware are available from Fisher (Itasca, IL).

2.1. Cell Counting

2.1.1. Coulter Counting of Nucleated Cells (for **Subheading 3.1.1.***)*

1. Coulter counter (Hialeah, FL), Coulter sample vials, 0.22-μm filters.
2. Cetrimide (hexadecyltrimethylammonium bromide) solution:

Cetrimide	240 g
NaCl	66.7 g
EDTA	2.93 g
Water	8 L

Stir overnight to dissolve solid particles. Filter cetrimide solution into 1-L bottles and store at room temperature.

2.1.2. Hemacytometer Counting of Viable and Nonviable Cells (for **Subheading 3.1.2.**)

1. Neubauer phase hemacytometer slide (0.1 mm deep; Hausser Scientific, Horsham, PA), light microscope (Olympus model BHTU Biological microscope, Lake Success, NY, D plan 10× lens).
2. Trypan Blue solution:

Trypan Blue dye	0.125 g
Water	50 mL

2.1.3. Benzidine Staining of Hemoglobin-Containing Cells (for **Subheading 3.1.3.**)

1. Light microscope, hemacytometer *(see above)*.
2. 30% H_2O_2.
3. Benzidine dihydrochloride stock solution: 0.2% (v/v) benzidine dihydrochloride in 0.5 *M* acetic acid.

2.2. CFU-C Assay

2.2.1. Stock Methylcellulose (for **Subheading 3.2.1.**)

1. Heated stir plate, large stir bar, 1-L graduated cylinder, 2-L Erlenmeyer flask, 50-mL centrifuge tubes.
2. Methylcellulose (GIBCO, Grand Island, NY).
3. 2X IMDM (Iscove's Modified Dulbecco's Medium) solution:

IMDM powder	(to make 1 L of 1X IMDM solution)
Sodium bicarbonate	3.02 g
Water	~490 mL
Gentamycin sulfate (50 mg/mL solution)	1 mL
2-Mercaptoethanol (2-ME)	12.7 µL

Dissolve IMDM powder and bicarbonate in 450 mL water. Rinse the IMDM bottle with 10 mL of water, add this wash liquid, gentamycin, and 2-ME, and then bring the volume up to 500 mL. Filter the media with a 0.22-µm filter into a sterile storage bottle.

2.2.2. Methylcellulose Media (for **Subheading 3.2.2.**)

1. Fetal bovine serum (FBS) (Hyclone, Logan, UT or StemCell Technologies, Vancouver, BC, Canada), crude bovine serum albumin (BSA) (Leptalb 7, Intergen, Purchase, NY).
2. Granulocyte colony-stimulating factor (G-CSF, Neupogen; Amgen, Thousand Oaks, CA), GM-CSF (Leukine; Immunex, Seattle,WA), and erythropoietin (Epo, Epoetin Alfa Procrit, Ortho Biotec, Raritan, NJ) are approved for clinical use and are available at hospital pharmacies. Stem cell factor (SCF), interleukin-3 (IL-3), and interleukin-6 (IL-6) are available from R&D Systems (Minneapolis, MN), PeproTech (Rocky Hill, NJ), and other cytokine suppliers.

2.2.3. CFU-C Plating/Scoring (for **Subheadings 3.2.3.** and **3.2.4.**)

1. Vortexing mixer, Eppendorf positive displacement repeat pipetter, stereo dissecting microscope, humidified 37° incubator at 5% CO_2.
2. 17 × 100 mm polypropylene culture tubes, 12.5-mL and 5.0-mL Eppendorf Combitip Biopur pipets, gridded 35-mm culture dishes, 150 × 15-mm culture dishes.

2.3. Morphological and Immunocytochemical Staining (for **Subheading 3.3.**)

2.3.1. Morphological Staining

1. Shandon cytocentrifuge (Shandon, Pittsburgh, PA), light microscope, disposable filter cards (Shandon), coated glass slides (75 × 25-mm Shandon), slide coverslips metal holders (Shandon), plastic funnels (Shandon).
2. Wright-Giemsa stain (EM Diagnostic Systems, Gibbstown, NJ).
3. Accumount (VWR/Scientific Products, Itasco, NY).

2.3.2. Immunocytochemical Staining

1. Shandon cytocentrifuge (Shandon), light microscope, disposable filter cards (Shandon), coated glass slides (75 × 25-mm Shandon), slide coverslips metal holders (Shandon), plastic funnels (Shandon).
2. KCl, KH_2PO_4, NaCl, $Na_2HPO_4 \cdot 7H_2O$, 30% H_2O_2 solution, hematoxylin (Biomeda Corp., Foster City, CA), Crystal mount (Biomeda Crop.), Accumount (VWR/Scientific Products), acetone.
3. Purified freeze-dried primary antibody (mouse IgG specific for desired antigen).
4. Streptavidin–horseradish peroxidase (HRP) kit (KPL, Gaithersburg, MD): contains normal goat serum, biotinylated goat antimouse IgG (secondary antibody), streptavidin–peroxidase.
5. Peroxidase chromogen kit (Biomeda): contains concentrated chromogen, buffer, and substrate.
6. Phosphate-buffered saline (PBS) solution:

Water	4 L
KCl	0.8 g
KH_2PO_4	0.8 g
NaCl	32 g
$Na_2HPO_4 \cdot 7H_2O$	8.64 g

Add salts to water and adjust pH to 7.35 while stirring. Sterile filter into 1-L bottles and store at room temperature.

2.4. Flow Cytometry (for **Subheading 3.4.**)

1. Centrifuge, flow cytometer, vortexing mixer, 12 × 75-mm polystyrene tubes.
2. BSA (Fraction V, Sigma), cacodylate acid sodium salt, EDTA, ethyl alcohol, $KHCO_3$, NaCl, NaOH, NH_4Cl, paraformaldehyde, propidium iodide (PI), sodium azide.

3. Control fluorescent antibodies (Becton-Dickinson, San Jose, CA): FITC isotype control, PE isotype control, FITC CD45, PE CD45.
4. Experiment-specific fluorescent antibodies (Becton-Dickinson): for example, FITC CD15, PE CD11b.
5. PAB solution:

Sodium azide	0.1% (w/v)
BSA	0.5% (w/v)
PBS	1 L

Dissolve sodium azide and BSA in PBS. Stir over low heat until dissolved. May need to stir overnight. Filter and store at 4°C and use within 6 months of preparation.
6. Ammonium chloride, red-blood-cell lysing solution:

Water	1 L
NH_4Cl	8.29 g
$KHCO_3$	1.09 g
EDTA	0.037 g

Dissolve salts into water while stirring. Store at room temperature and use within 1 week of preparation.
7. Paraformaldehyde solution:

Cacodylate acid sodium salt	21.4 g
Water	2 L
Paraformaldehyde	20.0 g
NaCl	15.0 g

Dissolve cacodylate acid sodium salt in water. Adjust pH to 7.2 using 1.0 *N* NaOH. Add paraformaldehyde, stirring on a hot plate in a fume hood until dissolved. Add NaCl and ensure it is fully dissolved. Filter sterilize and store at 4°C and use within 12 months of preparation.
8. Propidium iodide (PI) solution:

Ethanol (reagent grade)	1 mL
Propidium iodide (PI)	0.002 g
PBS	9 mL

Dissolve PI in ethanol and then add PBS. Store at 4°C in 15-mL centrifuge tube wrapped in aluminum foil.

3. Methods

3.1. Cell Counting Methods

The most widely used methods for counting cells in the laboratory entail use of the Coulter counter and the hemacytometer.

3.1.1. Coulter Counting/Cetrimide Protocol for Nucleated Cells

Beyond determining the number of particles, Coulter counters also distinguish between particle sizes, allowing cell or nuclei size distributions to be obtained. Cells are suspended in a conducting liquid (cetrimide solution) and are forced through an orifice with electrodes on each side. As the cells pass through the orifice, the impedance of electrical current, which is a function of cell volume, is measured. Because cetrimide lyses the cell membrane but leaves the nuclear membrane intact, only nucleated cells are counted. This is a significant advantage in cell samples that contain (non-nucleated) red blood cells (e.g., fresh peripheral blood apheresis products or umbilical cord blood), which one would generally like to exclude.

1. Collect sample. Add 10–25 µL cell suspension to 10 mL cetrimide solution placed in a plastic sample vial (blank is 10 mL cetrimide only).
2. Cap the diluted cell suspension, mix gently by inversion, and place on the sampling stand.
3. Set the counter to aspirate 2 mL cell suspension through the orifice.
4. Calculate cell concentration as follows:

$$\frac{\text{Cells}}{\text{mL (original)}} = \frac{\text{Cell count displayed}}{\text{2 mL mixture}} \times \frac{\text{10 mL mixture}}{\text{0.01–0.025 mL cell suspension (original)}}$$

Notes:

1. Coulter counting does not differentiate between viable and nonviable cells.
2. If cell count displayed is below 1000, then a larger cell sample should be used, keeping the total volume in the vial at 10 mL.
3. If cell count displayed is above 10,000, then the sample should be diluted and repeated ***(7)***.
4. Measurements are unreliable if cells form aggregates or are unhealthy, or if the suspension is too concentrated.
5. A discernible nuclei peak should be visible. Samples with heavy counts in small-diameter channels (<4 µm) and no visible peak are not accurate cell nuclei counts ***(8)***. Warming the sample vial to 37°C may help by reducing crystallization of cetrimide solution.
6. In our experience, the presence of two distinct peaks on a nuclei size distribution is often indicative of an actively proliferating cell population ***(8)***.
7. In our experience, gating between 2.712 and 7.567 µm is a useful range for cell nuclei in most types of hematopoietic cultures.

3.1.2. Hemacytometry/Trypan Blue

Hemacytometer counts are used to microscopically assess cell number and viability. Trypan Blue dye penetrates nonviable cells and stains intracellular proteins, whereas the intact membrane of a living cell serves as a barrier to the dye. The concentration of cells in a cell suspension is determined by placing the suspension in a hemacytometer slide and then enumerating cells using a microscope following standard procedures ***(9)***. Cell clumping is easily visualized and can be accounted for, in contrast to Coulter counting.

Notes:

1. Single red blood cells often do not look obviously "red" under the microscope and must be identified by their discoid morphology, which can be difficult to discern under certain lighting conditions.
2. The viability of thawed cells may drop gradually with time, so a viability estimation should be performed as close to culture initiation as possible.
3. Large aggregates may not enter the chamber, which can lead to errors in cell density estimates.
4. Errors may arise if the sample is not mixed immediately before being transferred to the chamber.

3.1.3. Benzidine Staining

This assay employs hemoglobin (Hb) as a marker to monitor commitment to the erythroid lineage. Hb synthesis begins in the late basophilic normoblast stage of erythropoiesis and extends through the subsequent stages of erythrocyte formation.

1. Prepare working benzidine solution: add 4 µL 30% H_2O_2 to 1.0 mL stock benzidine solution.
2. Dilute working benzidine solution 10× with cell sample; e.g., add 10 µL working benzidine solution to 90 µL cell suspension (4 µL:36 µL minimum).
3. Place approximately 40 µL of stained sample onto a clean hemacytometer slide and do a cell count ***(9)***. Brightly stained blue cells indicate cells containing Hb. For accuracy, at least 200 total cells should be counted.
4. Calculate fraction benzidine positive cells (number of blue-stained cells as a fraction of total cells).

Notes:

1. Working benzidine solution must be prepared fresh each time.
2. Cell samples should be counted within 10 min of staining or positive cells may become brown-black in color and more difficult to accurately count.
3. Cells do not stain well for benzidine after they have been deposited on slides using the Shandon cytocentrifuge.
4. It may be necessary to centrifuge harvested culture samples before staining in order to count an adequate numbers of cells; 5 min at 800*g* in an Eppendorf centrifuge is adequate.
5. In most samples, there is approximately 5–10% nonspecific background staining.
6. Very mature polymorphonuclear granulocytes (typically found in fresh mononuclear cell samples) may stain weakly positive with benzidine ***(10)***.
7. Due to the presence of red blood cells (RBCs) at early times (first 6 days) in mononuclear cell (MNC) cultures, it is difficult to determine the number of benzidine-positive nucleated cells.

3.2. CFU-C Assay

There are two critical components of this assay. First, methylcellulose media is prepared using appropriate cytokines and optimal concentrations of serum with the aim of promoting maximal colony formation. Second, culture conditions and plating parameters are adjusted so that individual colonies can be identified.

3.2.1. Stock Methylcellulose (2.1%) Protocol

To prepare a total volume of 1.6 L:

1. Autoclave 800 mL water (with a large stir bar) in an Erlenmeyer flask for 20 min on the liquid cycle.
2. Immediately place flask on a preheated stir plate.
3. While water is still hot, stir in methylcellulose powder (33.2 g).
4. Sterilize the solution by bringing just to boiling three times using the heated stir plate. Be careful to avoid overflow during boiling by watching closely and removing flask from heat source when boiling approaches.
5. Let the solution cool to room temperature while continuously stirring.
6. Slowly add 800 mL cold (4°C) 2X IMDM with 2-ME (*see* **Subheading 2.**) while stirring.
7. Stir the stock methylcellulose solution for at least 48 h at 4°C to allow complete polymerization.
8. Aliquot methylcellulose using a sterile disposable pipet into 50-mL centrifuge tubes and store at –20°C for up to 6 months. Thawed tubes may be stored at 4°C.

3.2.2. Methylcellulose Media (1.1%) Protocol

To prepare a total volume of 200 mL:

1. Thaw 2.1% methylcellulose stock in a 37°C water bath for 45 min.
2. To each of four 50-mL centrifuge tubes add the following:

Methylcellulose stock	27.5 mL
FBS (heat inactivated)	15.0 mL
BSA (Letpalb-7)	3.33 mL
2X IMDM	3.75 mL

3. Mix tubes by shaking vigorously, then centrifuge for 20 min at 700*g* to precipitate any debris present in the solution.
4. Pour the contents of the tubes aseptically into a 200-mL sterile bottle.
5. Add the following cytokines, mix thoroughly, and store the medium at 4°C:

Cytokine	Suggested stock concentration	Final medium concentration	Volume (μL)
rh IL-3	15 μg/mL	5 ng/mL	66.7
rh GM-CSF	1×10^6 U/mL	200 U/mL	40
rh G-CSF	1×10^6 U/mL	150 U/mL	30
rh EPO	1×10^4 U/mL	3 U/mL	60
rh IL-6	30 μg/mL	10 ng/mL	66.7
rh SCF	50 μg/mL	50 ng/mL	200

Notes:

1. Different lots of FBS and BSA differ substantially in their capacity to support hematopoietic cells in the CFU-C assay. Therefore, either prescreened components (StemCell Technologies) must be purchased or the investigator should use the CFU-C assay to screen and select the most appropriate lots of these components. Samples from different lots and different companies may be evaluated. Our lab has achieved good, consistent results using various lots of Hyclone FBS for several years.
2. The medium should be used within approximately 2 months. After that, it is still usable, but performance may decline.
3. Viscosity of the medium declines over time for unknown reasons. Although it is still able to support the CFU-C assay, scoring of plates (see below) may become more difficult as colonies can get displaced during handling of the plates.

3.2.3. Plating Procedures

1. Preequilibrate media: add 3 mL of methylcellulose media to each of an appropriate number of 17 × 100-mm polypropylene culture tubes via a 12.5-mL Eppendorf Combitip pipet and an Eppendorf repeat pipetter. Place the tubes in a 37°C incubator for 1 h.
2. Inoculate (5–10) × 10^3 MNC/mL or (1–2.5) × 10^3 $CD34^+$ cells/mL into the methylcellulose. Cell concentrations within the colony assay are based on obtaining a desired 100 colonies/plate.
3. Vortex tubes thoroughly and, using a 5-mL Eppendorf Combitip pipet and repeat pipetter, plate 1-mL aliquots in duplicate, labeled 35-mm gridded culture dishes. Use a separate pipet for each tube. Rotate dish to ensure a complete distribution of the viscous medium.
4. Place dishes into a larger petri dish (150 × 15 mm) containing 10–15 mL sterile water on the bottom to prevent evaporation from the culture dishes.
5. Incubate undisturbed for 14 days at 37°C under fully humidified conditions in an atmosphere of 5% CO_2, and 5% O_2, 90% N_2.
6. Score colonies containing a minimum of 50 cells as CFU-G, CFU-M, CFU-GM, BFU-E, or CFU-Mix using a stereo dissecting microscope.

Notes:

1. After the technique is well-practiced, one can often reduce the volume of methylcellulose medium from 3.0 to 2.75 mL/sample to save on costs.
2. Plating cells at too high a density may result in an overcrowded culture dish, leading to depletion of nutrients, to the accumulation of toxic by-products, and to reduced pH. If the pH drops too low, full differentiation of BFU-E will not occur and the colonies may appear light pink or white, making them much more difficult to distinguish from CFU-GM.
3. A CFU-C assay may be performed in a 20% O_2 (rather than 5% O_2) atmosphere, but the overall number of colonies may be reduced.
4. Adequate mixing is necessary to give a homogeneous cell suspension. Sample reproducibility will suffer from a lack of mixing.

3.2.4. Scoring Procedures (more details in **Ref. 10**)

1. G colonies: small, white, usually tightly packed collection of cells sharp in contrast. However, where differentiation has occurred, small cells may form a halo or become entirely dispersed.
2. M colonies: large, opaque cells usually uniformly dispersed, but in some cases may be closely packed.
3. GM colonies: tight center of small G cells with a surrounding halo of M and G cells.
4. BFU-E colonies: multicentric colonies of small, tightly packed cells, red or pink in color, and homogeneous in size.
5. Mixed colonies: may appear as single, multicolored colonies encompassed by a heterogeneous population of cells. The G, M, and E lineages may all be represented in a mixed colony.

Notes:

1. It is very important not to overlook colonies at the circumference of culture dishes. Mixing cultures in a centrifugal motion distributes a large number of CFC on the periphery. In addition, the meniscus effect leads to a greater number of cells in the thicker regions at the edges.
2. Cultures left out of the incubator longer than 30 min or on the microscope light source longer than 5 min may begin to dry up, resulting in the medium becoming opaque and making it difficult to score colonies.
3. Plucking and phenotyping of colonies may be necessary to distinguish among CFU-G, CFU-M, and CFU-GM. Therefore, many investigators report the sum of these three colony types as CFU-GM.

3.3. Morphological and Immunocytochemical Staining

Microscopic visualization of cell morphology or cell-surface antigens is possible through the use of various cell-staining reagents. Wright-Giemsa is the most commonly used morphological stain for blood cells *(11)*. Methylene Blue and Eosin are the primary components of the Wright-Giemsa stain. Eosin is an acidic dye that reacts with basic cellular components such as hemoglobin and certain cytoplasmic granules and colors them various shades of orange-red. Acidic cellular components such as nuclear DNA are stained a range of violet-blue by the basic dye Methylene Blue. Technicians with the proper training can distinguish cells at various differentiation stages within different lineages based on their appearance subsequent to Wright-Giemsa staining.

Immunocytochemistry takes advantage of the affinity of a specific primary antibody to label cells expressing a particular antigen. The labeling molecule can be directly attached to the primary antibody or it can be part of a secondary antibody that binds to the primary antibody. Enzymes, radioisotopes, gold colloids, and fluorophors are the commonly used labeling molecules. In the procedure presented here, a biotin–streptavidin–peroxidase system is used to visualize cells positive for the selected primary antibody.

3.3.1. Morphological Staining

1. Each slide should contain approximately 1.5×10^4 cells in 100–500 μL of sample volume in the cytocentrifuge funnel. For samples with very high or very low cell density, dilution or concentration of the sample will be required, respectively.
2. Assemble the four components needed for the cytocentrifuge slide preparation in the following order: the metallic holder, a coated glass slide, a filter card, and the plastic funnel. Add sample carefully into the funnel.
3. Centrifuge using the cytocentrifuge at 850 rpm for 5 min.
4. Pour Wright-Giemsa stain in a staining container and immerse the slides for 1 min.
5. Rinse the slides in deionized water for 1 min and air-dry.
6. Add one drop of Accumount on top of cells and place a coverslip on top.
7. Evaluate slides using a bright field microscope for the presence of the cell types of interest.

Note:

1. Once the centrifugation is done, hold slides up to room light for a quick check to make sure that cells are present and check under the microscope to make sure cells are not too sparse or closely packed. If the distribution is undesirable, repeat centrifugation with more or fewer cell per slide. Add the cover slip at an angle to minimize trapping air bubbles in the Accumount.

3.3.2. Immunocytochemical Staining

1. Follow **steps 1** through **3** from **Subheading 3.3.1.**
2. Immediately after centrifugation, soak the slides in pure acetone for 1 min to fix the cells and air-dry overnight. Dried slides may be stored up to 1 month before immunostaining.
3. Soak the dry slides first in PBS for 5 min and then in a 15 : 1 solution of PBS and 30% H_2O_2 for 10 min to block endogenous peroxidase. Wash again with PBS for 5 min.
4. Dry the top and bottom of each slide using a tissue, avoiding the circular area containing the cells.
5. Add normal goat serum to the slides and incubate for 15 min to prevent nonspecific binding.
6. Flick the slide to remove excess serum and dry off the slide (according to **step 4**).
7. Add 100 μL of primary antibody (mouse IgG) solution to each slide and incubate for 1 h.
8. Shake off the primary antibody and rinse the slide with PBS three times.
9. Dry off the PBS as before and add the secondary antibody (goat antimouse conjugated with biotin). Incubate for 30 min.
10. Shake off secondary antibody, rinse the slide with PBS three times, dry off PBS, and incubate with streptavidin–peroxidase for 30 min.
11. Shake off streptavidin–peroxidase and rinse the slides with PBS three times.
12. Make fresh peroxidase–chromogen solution according to manufacturer's instruction.

13. Dry off PBS. Add chromogen solution to each slide and incubate for 2–10 min. In order to minimize nonspecific staining, view one of the stained slides under the microscope to evaluate the extent of color development. When positive cells can be adequately distinguished, stop incubation and rinse the slides with deionized water.
14. Add hematoxylin, a nuclear counterstain, to a staining container and immerse the slides in this solution for 1 min. Check slides under the light microscope to verify that nuclei have stained purple/blue. If nuclei staining is too light, use a longer staining period.
15. Rinse slides with distilled water and air-dry.
16. Add one drop of Crystal mount to cells and dry overnight.
17. Add one drop of Accumount over dried Crystal mount and add cover slip. Dry overnight.
18. Count at least 200 cells per slide to determine the percentage of positive cells.

Notes:

1. Be sure to include a positive and a negative control slide for each experiment while developing the technique. For positive control, use a sample that has abundant cells expressing the antigen of interest. For negative control, use the experimental sample but do not add the primary antibody.
2. When adding the normal goat serum, secondary antibody, streptavidin–peroxidase, and the chromogen solution, add only enough to cover the region of the slide containing the cells. Usually this amounts to one to two drops per reagent per slide. Use of the special hydrophobic-coated slides is very helpful to prevent reagents from covering the entire surface of the slide.
3. The slides must be kept humidified during the primary and secondary antibody incubation steps. For example, lay the slides on wetted paper towels and put them inside a slide box during incubation.
4. Each particular primary antibody requires titration to determine the optimal concentration for the assay. In general, dilutions ranging from 1 : 100 to 1 : 500 should be tested, with 1 : 400 working well in most cases.
5. Notes from **Subheading 3.3.1.** also apply.

3.4. Flow Cytometry

Flow cytometry allows simultaneous multiparametric analysis of cell-surface antigens, cell size, and cell granularity. The light-scattering properties of individual cells are measured by forward and side scatter, which, in turn, relate to cell size and granularity, respectively. Antibodies attached to fluorescent dyes such as fluorescein (FITC) and phycoerythrin (PE) can also be detected by a flow cytometer. The availability of lineage-specific cell-surface antigens that are expressed at defined stages of cell differentiation makes it possible to study subpopulations within a given lineage of cells. For example, within the leukocyte population, monocyte precursors express CD11b antigen first and then CD15 antigen, whereas granulocyte precursors show the opposite pattern of antigen expression. Also, the level of CD15 expression is higher on mature granulocytes as compared to mature monocytes.

For each batch of samples, three control tubes are needed. Tube 1 is the isotype control containing FITC and PE antibodies of the same isotype as the primary antibodies (usually IgG). In this control, sample cells are incubated with an antibody that has the same isotype as the primary antibody but does not have a relevant epitope specificity. Cells with an Fc receptor bind to these antibodies so that the percentage of cells positive for control 1 should be subtracted from the percentage positive for each of the respective primary antibodies. Tubes 2 and 3 containing FITC CD45 and PE CD45, respectively, are for fluorescence compensation. The purpose of compensation is to account for one fluorochrome's fluorescence interfering with detection in the other fluorescence channel. For a more detailed description of flow cytometry fundamentals and protocols, *see* **refs. *12*** and ***13***.

3.4.1. Cell Staining

1. Add 10^5–10^6 cells per test tube.
2. Centrifuge at 750*g* for 3 min and pour off the supernatant.
3. Resuspend pellet in 1 mL PAB. Repeat **step 2**.
4. Add 10–20 µL appropriate antibody (~1 mg/mL) to pellet, vortex, and incubate at 4°C for 15 min.
5. Add 3 mL of NH_4Cl RBC lysing solution to each tube and shake tube for 1 min. Repeat **step 2**.
6. Wash twice with PAB as in **step 3**.
7. To fix cells for analysis at a later time, resuspend pellet in paraformaldehyde solution. Wrap the test tube with aluminum foil and store in a 4°C refrigerator until analysis (within 5 days at most). To do live cell analysis, resuspend the cells in 1 mL PAB.

Note:

1. Antibody titration is essential for each new batch and type of antibody. The optimal antibody titer is not necessarily one that produces the highest percentage of positive cells since non-specific binding may give false positive results. The criteria for optimal concentration is maximum signal-to-noise ratio (distance between positive and negative peaks on a one-color histogram) for the antibody being titrated.

3.4.2. Flow Cytometric Data Acquisition (As Performed on a Becton-Dickinson FACScan Flow Cytometer)

1. Flow cytometer settings:

Excitation:	488 nm
Data acquisition type:	List Mode
FITC (green) emission (FL1):	530 ± 15 nm (short band-pass filter)
PE (orange) emission (FL2):	575 ± 15 nm (short band-pass filter)
PI (bright red) emission (FL3):	>620 nm (long band-pass filter)
Amplification:	Forward and side scatter (linear), all fluorescence (logarithmic)
Number of total events acquired:	$(1–3) \times 10^4$

2. Adjust settings to get cells on scale:
   a. Run tube 1, which is the isotype control.
   b. Display forward vs side scatter dotplot.
   c. Adjust forward scatter amplifier and side scatter detector to get the cell population centered on the screen. May need to raise the forward scatter threshold to leave debris out of the plot.
   d. Set a gate on the lymphocyte population at the lower left corner of the plot. (Lymphocyte fraction diminishes very quickly in our cultures. Therefore, this step may not be applicable after day 1 of culture.)
   e. Turn on the gate.
3. Adjust fluorescence parameters (use tube 1 data from **step 2a**):
   a. Display FL1 vs FL2 plot.
   b. Any positive event for either FL1 or FL2 indicates nonspecific staining. Adjust FL1 and FL2 gains to bring the cell population to the double-negative (lower left) quadrant.
4. Set fluorescence compensation:
   a. Run tube 2, which contains FITC CD45. CD45 is expressed by all MNC.
   b. Display FL1 vs FL2 plot.
   c. Any positive event for FL2 is due to FITC emission overlap into the PE range. Adjust FL2-%FL1 until the cell population lies in the FL2 negative region.
   d. Run tube 3, which contains PE CD45.
   e. Display FL1 vs FL2 plot.
   f. Any positive event for FL1 is due to PE emission overlap into FITC range. This is usually a minor problem compared to FITC bleeding into PE. Adjust FL1-%FL2 to bring cell population to FL1 negative quadrants.
   g. If doing live-cell analysis, display FL2 vs FL3 plot.
   h. Any positive event for FL3 is due to PE (FL2) bleeding into PI range. Adjust FL3-%FL2 to bring the cell population in the FL3 negative quadrants.
   i. Record parameter settings for future reference.
5. Propidium iodide addition and data acquisition:
   a. For live-cell analysis, add 10 μL propidium iodide solution to each of the tubes.
   b. Clear the gate to collect all events from the experimental samples.
   c. Run tube 1 again.
   d. Acquire data and save to a disk.
   e. Run and save data for the rest of the samples.

Note:

1. Fixed cells should be analyzed as soon as possible. Prolonged storage leads to increased autofluorescence. Due to the variability among samples from different cell sources, a separate set of controls (tubes 1–3) should be used.

3.4.3. Data Analysis

1. Read saved file from disk.
2. Gate out the PI-labeled dead cells (high FL3 expression) and evaluate only the viable population.

3. Display dotplot of FL1 vs FL2 to quantify the fractions of cell populations positive for CD15 or CD11b.
4. Print graph and associated statistics.

Note:

1. Plotting fluorescence data against forward or side scatter may provide some unique information. For example, displaying CD45 antigen florescence level in combination with side scatter has been used to identify eight major types of blood cells (lymphocytes, monocytes, myeloblasts, segmented neutrophils and bands, etc.) ***(14)***. Additional information on flow cytometry protocols can be found in **refs.** ***6*** and ***12–14***).

Acknowledgments

This work was supported by NIH Grant RO1 HL48276. SSM was partially supported by a Predoctoral Biotechnology Training Grant (NIH GM 08449). TAM was partially supported by a U.S. Army Breast Cancer Training Grant (USAMRMC Grant DAMD 17-94-J-4466).

References

1. McAdams, T. A., Miller, W. M., and Papoutsakis, E. T. (1996) Hematopoietic cell culture therapies (Part I): cell culture considerations. *Trends Biotechnol.* **14,** 341–349.
2. Young, J. W., Szabolcs, P., and Moore, M. A. (1995) Identification of dendritic cell colony-forming units among normal human CD34+ bone marrow progenitors that are expanded by c-*kit*-ligand and yield pure dendritic cell colonies in the presence of granulocyte/macrophage colony-stimulating factor and tumor necrosis factor alpha. *J. Exp. Med.* **182,** 1111–1119.
3. Sutherland, H. J., Lansdorp, P. M., Henkelman, D. H., Eaves, A. C., and Eaves, C. J. (1990) Functional characterization of individual human hematopoietic stem cells cultured at limiting dilution on supportive marrow stromal layers. *Proc. Natl. Acad. Sci. (USA)* **87,** 3584–3588.
4. Zanjani, E. D., Almeida-Porada, G., and Flake, A. W. (1995) Retention and multilineage expression of human hematopoietic stem cells in human-sheep chimeras. *Stem Cells* **13,** 101–111.
5. Szilvassy, S. J., Humphries, R. K., Lansdorp, P. M., Eaves, A. C., and Eaves, C. J. (1990) Quantitative assay for totipotent reconstituting hematopoietic stem cells by a competitive repopulation strategy. *Proc. Natl. Acad. Sci. (USA)* **87,** 8736–8740.
6. Sutherland, D. R., Anderson, L., Keeney, M., Nayar R., and Chin-Yee, I. (1996), The ISHAGE guidelines for CD34+ cell determination by flow cytometry. *J. Hematother.* **5,** 213–226.
7. Darling, D. C. and Morgan, S. J. (1994) *Animal Cells: Culture and Media*. John Wiley & Sons, Chichester, pp. 90–118.
8. Collins, P. C., Papoutsakis, E. T., and Miller, W. M. (1998) Nuclei size distribution as a predictor of hematopoietic cell proliferation, submitted.

9. Doyle, A. and Newell, D. G. (1994) Core techniques, in *Cell and Tissue Culture: Laboratory Procedures* (Doyle A., Griffiths, J. B., and Newell, D. G., eds.), John Wiley and Sons, Chichester, pp. 4B:1. 1–4B:1. 5.
10. Metcalf, D. (1984) Techniques for the scoring and analysis of clonal hemopoietic cultures, in *Clonal Culture of Hemopoietic Cells: Techniques and Applications*, Elsevier, Amsterdam, pp. 73–93.
11. Shafer, J. A. (1991) Preparation and interpretation of peripheral blood smears, in *Hematology: Basic Principles and Practice* (Hoffman R, Benz E. J., Jr., Shattil, S. J., Furie, B., Cohen H. J., and Silberstein, L. E., eds.), Churchill Livingstone, New York, pp. 2202–2209.
12. Jackson, A. L. and Warner, N. (1986) Preparation, staining, and analysis by flow cytometry of peripheral blood leukocytes, in *Manual of Clinical Laboratory Immunology* (Rose, N. R., Friedman, H., and Fahey, J. L., eds.), American Society of Microbiology, Washington, pp. 226–235.
13. Bender, J. G., Unverzagt, K., Walker D. E., Lee W., Van Epps D. E., Smith, D. H., Stewart, C. C, and To, L. B. (1991) Identification and comparison of CD34-positive cells and their subpopulations from normal peripheral blood and bone marrow using multicolor flow cytometry, *Blood* **77,** 2591–2596.
14. Stelzer, G. T., Shults, K. E., and Loken, M. R. (1993) CD45 gating for routine flow cytometric analysis of human bone marrow specimens. Ann. NY Acad. Sci. **677,** 265.

18

Hematopoietic Cells for Cellular and Gene Therapy

II. Expansion Protocols

Todd A. McAdams, E. Terry Papoutsakis, and William M. Miller

1. Introduction

Hematopoietic culture is a rapidly growing area of mammalian cell culture with broad applications. As discussed in Chapter 17, many different types of cells can be produced of multiple lineages and multiple states of differentiation within each lineage. In this chapter, we illustrate the basic techniques and principles of hematopoietic cell culture by providing protocols for hematopoietic cultures of three types: retroviral transduction of $CD34^+$ cells for gene therapy, and expansion of colony-forming unit–granulocyte monocyte (CFU-GM) progenitors and neutrophil postprogenitors for transplantation therapies.

There are potentially several clinical applications for the use of ex vivo expanded hematopoietic cells in a variety of transplantation therapies. One application is the use of hematopoietic cells as vectors for the delivery of therapeutic genes to cure genetic disorders. By transducing long-lived hematopoietic stem cells with the therapeutic gene of interest, it may be possible to provide long-term correction of single-gene genetic defects such as ADA (adenosine deaminase) deficiency and sickle cell anemia. A second possible clinical benefit of hematopoietic cell cultures is to expand donor-cell populations to enable the collection and use of smaller samples (e.g., from bone marrow aspirates or umbilical cord blood) for adult transplants. A third use of hematopoietic cell expansion is the production of large numbers of progenitor (CFU-GM) and/or postprogenitor cells of the granulocyte lineage in order to decrease the time to short-term neutrophil engraftment. It is currently not clear

From: *Methods in Biotechnology, Vol. 8: Animal Cell Biotechnology*
Edited by: N. Jenkins © Humana Press Inc., Totowa, NJ

which cell population(s) (long-term culture-initiating cells [LTC-IC], CFU-GM, or postprogenitors) are primarily responsible for short-term engraftment. The production of megakaryocyte progenitors and postprogenitors has the potential to speed platelet engraftment, but these cultures are beyond the scope of this chapter. Clinical trials to date ***(1,20)*** have generally shown no toxicity due to culture-expanded cells, but also little clinical benefit. Whether the lack of clinical benefit so far is due to an insufficient quantity of cells or some type of engraftment deficiency has yet to be determined. The expansion protocols described in this chapter have not been tested for the maintenance of engraftable stem cells and are intended only as a supplement to unmanipulated donor populations.

2. Materials

Filtered, deionized, cell-culture grade water is used. Unless otherwise indicated, chemical reagents are available from Sigma (St. Louis, MO) and supplies are available from Fisher (Itasca, IL).

2.1. Depletion of Red Blood Cells (for Subheading 3.1.)

1. Centrifuge (≥ 300*g*).
2. 50-mL centrifuge tubes.
3. Histopaque 1077.
4. Phosphate-buffered saline (PBS) (*see* Chapter 17, **Subheading 2.3.2.**), Iscove's modified Dulbecco's medium (IMDM).

2.2. Retroviral Infection of CD34+ Peripheral Blood Cells (for Subheading 3.3.)

1. Incubator (37°C, 5% CO_2, 95% air), centrifuge (≥ 300*g*).
2. 50-mL centrifuge tubes, T-225 flasks.
3. Stem cell factor (SCF; Peprotech, Rocky Hills, NJ), flt3 ligand (FL; Peprotech), interleukin-3 (IL-3; Peprotech), interleukin-6 (IL-6; Peprotech), protamine sulfate (Lymphomed, Deerfield, IL), fetal bovine serum (FBS) and bovine serum albumin (BSA) (*see* Chapter 17, **Subheading 2.2.2.**), L-glutamine, IMDM, 2-mercaptoethanol (2-ME).
4. Retroviral supernatant (*see* **ref. *4***).

2.3. Expansion of CFU-GM Progenitors from Cord Blood

1. Incubator (37°C, 5% CO_2, 95% air).
2. Teflon fluorinated ethylene polypropylene (FEP) culture bags (American Fluoroseal, Columbia, MD).
3. Granulocyte colony-stimulating factor (G-CSF; Amgen, Thousand Oaks, CA), erythropoietin (Epo; Ortho Biotech, Raritan, NJ), SCF, IL-3, IL-6, FL, serum-free culture medium (*see* **Subheading 4.1.**).

2.4. Expansion of Neutrophil Postprogenitors from Mobilized Peripheral Blood

1. Incubator (37°C, 5% CO_2, 95% air).
2. Teflon FEP culture bags (American Fluoroseal).
3. SCF, IL-3, IL-6, G-CSF, FBS, horse serum (HS), autologous serum or plasma, serum-free culture medium (*see* **Subheading 4.1.**).

3. Methods

3.1. Depletion of Red Blood Cells (RBC) and Polymorphonuclear (PMN) Granulocytes

Most sources of hematopoietic cells used in culture contain large numbers of RBC and mature PMN granulocytes that do not contribute to culture expansion and will increase nutrient uptake and acid production by the culture. For these reasons, it is desirable to deplete samples of RBC and/or isolate the mononuclear cells (MNC). A number of procedures have been reported and compared *(2)*, including density gradient centrifugation using Ficoll/Histopaque or Percoll, sedimentation through gelatin, methylcellulose, or starch, and ammonium chloride lysis. Gelatin sedimentation and ammonium chloride lysis deplete samples of approximately 95% of RBC, and are recommended if maximal recovery of MNC is necessary. The technique that provides the cleanest sample (removing maximal RBC and PMN granulocytes) and is most commonly used is Ficoll/Histopaque density gradient centrifugation. Due to the removal of PMN granulocytes, nucleated cell recoveries are typically 25–50% using this procedure.

1. Determine the number of 50-mL centrifuge tubes that will be needed and the volume of cell sample (i.e., cord blood [CB], bone marrow [BM] aspirate or mobilized peripheral blood [PB] apheresis product) to be added to each tube. Approximately 5 mL of blood or 2×10^7 nucleated cells can be adequately separated in one tube.
2. Add the appropriate volume of sample (5 mL or less) to each centrifuge tube, then fill up the tube to 35 mL with PBS.
3. Fill a 10-mL pipet with approximately 13 mL of prewarmed (37°C) Histopaque 1077. Place the tip of the pipet at the bottom tip of the centrifuge tube, and very slowly and gently begin dispensing Histopaque at the bottom of the centrifuge tube so that a layer of Histopaque forms below the blood/PBS.
4. Without mixing layers, place the tubes in the centrifuge and spin for 20 min at 300*g*.
5. Upon removal from the centrifuge, the majority of RBC should be in a pellet at the bottom of the tube, and a hazy white layer (containing MNC) should be visible at the interface between the Histopaque and the PBS/plasma. Use a 10-mL pipet to extract this hazy white layer, approaching it from the top. The liquid extracted with the cell layer should be PBS/plasma with as little Histopaque as possible.

6. Collect each 10 mL of MNC fraction into a fresh 50-mL centrifuge tube and add 30 mL of culture medium (IMDM or other serum-free medium) to wash the cells.
7. Centrifuge for 10 min at 300*g*.
8. Aspirate and discard the supernatant, being careful not to disturb the cell pellet. The cell pellet often contains residual RBC, giving it a slightly red color.
9. Resuspend the cell pellet in 5 mL of culture medium (IMDM or actual culture medium).

3.2. Selection of $CD34^+$ Cells from Mononuclear Cells

The use of $CD34^+$ cell culture has certain advantages for experimental purposes. The effects of culture conditions on the proliferation of hematopoietic progenitors are easier to detect without the presence of large numbers of accessory cells, due to the greater expansion ratios achieved with $CD34^+$ cells at low inoculum densities. However, both MNC and $CD34^+$ cells have potential advantages that must be considered when choosing the starting population for hematopoietic cultures with potential clinical use ***(1)***.

A number of products are commercially available for the isolation of $CD34^+$ cells. See **ref. *3*** for a review of $CD34^+$ cell selection techniques. Follow manufacturers instructions for the isolation of $CD34^+$ cells. We recommend the "MiniMACS" system (Miltenyi Biotech, Auburn, CA) for its high recovery, high yield, and its ease of use. In our experience, it is very important to strictly adhere to the expiration date of reagents for the MiniMACS system to avoid significant reductions in the yield of $CD34^+$ cells. The manufacturer's instructions recommend the use of a nylon mesh column to avoid cell clumps that might clog the separation column. In our experience with non-cryo-preserved PB MNC, this step is not necessary unless the sample contains large, visible clumps. When using CB MNC or any previously cryo-preserved cells, the nylon mesh column should always be used. We also recommend removing cells from the MiniMACS buffer and resuspending selected cells in a hematopoietic culture medium within approximately 1 hour of selection.

3.3. Retroviral Infection of $CD34^+$ Peripheral Blood Cells

The majority of colony-forming cells (CFU-C) and LTC-IC cultured using this protocol should be transduced by the retroviral vector. Primitive stem cells, as assessed by growth in immunodeficient mice, are also likely to be transduced by this protocol. Actual transduction efficiencies may vary considerably from experiment to experiment due to variations in the quality of the cells used. This protocol was adapted from those described by Nolta et al. ***(4)***.

1. Prepare 100 mL of transduction medium: IMDM with 30% FBS, 1% BSA, 0.1 m*M* 2-mercaptoethanol, 2 m*M* L-glutamine, 8 μg/mL protamine sulfate, 20 ng/mL IL-6, 100 ng/mL SCF, 100 ng/mL FL, and 10 ng/mL IL-3. Preequilibrate medium in

incubator (37°C, 5% CO_2, 95% air) for at least several hours before inoculating cells to achieve proper pH and temperature.
2. Inoculate cells (1×10^5 $CD34^+$ cells/mL) into a combination of 25 mL transduction medium plus 25 mL vector-containing supernatant and place in T-225 flask. Place flask into incubator with 5% CO_2 in air (*see* **Subheading 4.5.**).
3. After 24 h, pipet the culture into a 50-mL centrifuge tube, centrifuge for 10 min at 300g, and discard supernatant.
4. Resuspend pellet in 25 mL transduction medium plus 25 mL vector-containing supernatant and return to T-225 flask.
5. Repeat **steps 3** and **4** two more times (on d 2 and 3)
6. On day 4, harvest culture by removing contents from T-225 flask. Assay for CFU-C and primitive progenitors as required, according to the vector and selectable marker being transduced.

3.4. Expansion of CFU-GM Progenitors from Cord Blood

The CB cells cultured using this protocol should undergo considerable expansion of CFU-C (primarily CFU-GM). Actual expansion values may vary considerably from experiment to experiment, due to variations in the quality of the CB cells.

1. Prepare 1 L of serum-free medium (*see* **Subheading 4.1.**) containing cytokines at the following concentrations: 50 ng/mL SCF; 50 ng/mL FL; 1.5 ng/mL IL-3; 10 ng/mL IL-6; 150 U/mL G-CSF; 0.1 U/mL Epo (*see* **Subheading 4.4.**). Preequilibrate medium in incubator (37°C, 5% CO_2, 95% air) for at least several hours before inoculating cells to achieve proper pH and temperature.
2. Inoculate cells (2×10^5 MNC/mL or 2×10^4 $CD34^+$ cells/mL; *see* **Subheading 4.2.**) into 500 mL of preequilibrated medium and place into 1-L culture bag (*see* **Subheading 4.6.**). Place culture bag into incubator with 5% CO_2 and 20% O_2 gas phase (*see* **Subheading 4.5.**).
3. On day 6, feed the culture by adding 500 mL of preequilibrated medium containing cytokines to the culture bag (*see* **Subheading 4.3.**).
4. On day 10, harvest culture by removing contents from culture bag and perform total cell count and CFU-C assay (Chapter 17, **Subheading 3.2.**). Additional characterization via flow cytometry, morphological staining, and immunostaining, may be carried out if desired (*see* Chapter 17). In order to calculate CFU-GM fold expansion, a CFU-C assay must also be performed on the original cell sample.

3.5. Expansion of Neutrophil Postprogenitors from Mobilized Peripheral Blood (PB) Apheresis Products

Cells cultured using this protocol should undergo considerable expansion of neutrophil postprogenitor cells ($CD15^+/CD11b^-$ promyelocytes and $CD15^+CD11b^+$ metamyelocytes and myelocytes) and CFU-GM. Actual expansion values may vary substantially from experiment to experiment due to variations in the quality of the PB cells. The expansion potential of mobilized PB cells may depend on the age and disease state of the patient, amount of prior chemotherapy, and mobilization regimen.

1. Prepare 1 L of serum-free medium (*see* **Subheading 4.1.**) containing cytokines at the following concentrations: 50 ng/mL SCF; 1.5 ng/mL IL-3; 10 ng/mL IL-6; 150 U/mL G-CSF (*see* **Subheading 4.4.**). For increased yield of neutrophil postprogenitors, it is recommended that a source of serum be added to the medium. For clinical protocols, 2–5% autologous serum or plasma may be used. For experimental purposes, either 10% FBS or 12.5% FBS/12.5% HS may be added (*see* **Subheading 4.1.**). Preequilibrate medium in incubator (37°C, 5% CO_2, 95% air) for at least several hours before inoculating cells to achieve proper pH and temperature.
2. Inoculate cells (1×10^5 MNC/mL or 1×10^4 $CD34^+$/mL; *see* **Subheading 4.2.**) into 500 mL of preequilibrated medium and place into a 1-L culture bag (*see* **Subheading 4.6.**). Place culture bag into incubator with 5% CO_2 and 20% O_2 gas phase (*see* **Subheading 4.5.**).
3. On day 6, feed the culture by adding 500 mL of preequilibrated medium containing cytokines to the culture bag (*see* **Subheading 4.3.**).
4. On day 12, harvest culture by removing contents from culture bag and perform total cell count and CFU-C assay. Additional characterization via flow cytometry, morphological staining, and immunostaining, may be carried out if desired (*see* Chapter 18). In order to calculate neutrophil postprogenitor expansion, flow cytometry, immunostaining, or morphological analysis must also be performed on the original cell sample.

4. Notes

4.1. Culture Medium

In general, serum-containing (FBS and HS) formulations are superior for the expansion and maturation of the granulocyte/macrophage lineage ***(5)***. Serum-free media promote greater progenitor and total cell expansion of the erythroid and megakaryocytic lineages, primarily because transforming growth factor-β (TGF-β, which is typically found in serum) is inhibitory to the expansion of these lineages ***(6)***. Serum-free media have supported excellent expansion of highly purified primitive hematopoietic cells stimulated by multiple cytokines.

A number of serum-free hematopoietic media have been described ***(7)*** and can be assembled from individual components. However, in our experience ***(5)***, the adequate performance of self-assembled media depends significantly on the screening of multiple lots of major components such as transferrin and bovine serum albumin. Accordingly, we and others have found the use of commercially available serum-free hematopoietic media to be preferable. Excellent results have been obtained using X-VIVO 20 (BioWhittaker, Walkersville, MD) ***(8)*** and X-VIVO 10 (BioWhittaker) + 1% human serum albumin (Baxter, McGaw Park, IL) ***(9)***. Other available serum-free media include X-VIVO 15 (BioWhittaker), StemPro-34 SFM (Life Technologies, Gaithersburg, MD),

QBSF-59 (Quality Biological, Inc. , Gaithersburg, MD), and QBSF-60 (Quality Biological, Inc.)

4.2. Inoculation Density

In static cultures of MNC and $CD34^+$ cells, expansion ratios and the kinetics of expansion are dependent on cell inoculation density *(10,11)*. In general, lower inoculation densities lead to greater total cell expansion and greater depletion of colony-forming cells. Higher-density cultures achieve greater total cell and CFU-C numbers, but a lower total cell expansion ratio when compared to lower density cultures. Variability in expansion potential has been shown to decrease when cultures are performed at high cell densities *(11)*. In addition, when BM MNC are cultured at sufficiently high density, they become essentially stroma independent in terms of LTC-IC maintenance and expansion *(11)*.

If rapid, maximal expansion of total cells is desired, low inoculation densitites (5×10^4–10^5 MNC/mL, 10^3–10^4 $CD34^+$ cells/mL) should be used. If slower expansion kinetics, reduced sample variability, and maintenance of more primitive cells are preferable, then higher cell densities ([2–10] $\times 10^5$ MNC/mL; [2–5] $\times 10^5$ $CD34^+$ cells/mL) should be used. If high cell densities are to be used, however, other factors such as the frequency of medium exchange required to prevent a decrease in pH should be considered.

4.3. Feeding Protocols

Frequent feeding has been found to improve the performance of human stroma coculture systems *(12)*. Haylock et al. have shown that the total cell yield for $CD34^+$ cell cultures is greatly increased by frequent medium exchange *(10)*. Koller et al. have demonstrated that the benefits of frequent medium exchange are greater in cultures with greater metabolic demand, such as those at high cell densities and in the presence of a stromal layer *(11)*. In contrast, Sandstrom et al. noted that the benefits of perfusion were greatest on samples with the least expansion potential *(13)*, suggesting that the benefits of perfusion were due to removal of an inhibitory factor.

One must balance the advantage of frequent feeding against the increased medium requirement and culture manipulation required. We have recommended one feeding by volume doubling at the midpoint of the culture because it provides a significant increase in culture performance with a minimum of additional culture manipulation. Additional feedings, especially toward the end of the culture period, when cell numbers are rapidly increasing, will provide additional benefits to culture performance (especially for high inoculation densities). G-CSF has shown to be fairly unstable in hematopoietic culture *(14)* and, thus, it may be beneficial to provide additional feedings (either G-CSF or medium + G-CSF) when performing cultures using this cytokine.

4.4. Cytokines

Hematopoietic cytokines constitute a large and growing family of glycoprotein factors that are indispensable in hematopoietic cultures. In general, the greater the number of positive-acting cytokines, the greater the total cell and committed progenitor numbers. The optimal time of harvest to collect cells at the peak of a particular type of CFU-C must be determined empirically with each new cytokine mixture. In general, erythroid CFU-C peak earlier (days 4–7) than granulocyte, macrophage, and megakaryocyte CFU-C (days 10–14). The cytokine concentrations suggested have been used with success in our laboratory and are based on literature values for single cytokine titrations in CFU-C assays. Ideally, cytokine levels should be titrated for each individual hematopoietic culture system.

4.5. Oxygen and pH Levels

In vivo measurements of oxygen tension in bone marrow have yielded values equivalent to medium in equilibrium with a gas phase containing 5% oxygen ***(15)***. Because of this observation, many experiments have been performed comparing the performance between ex vivo hematopoietic cultures carried out under gas phases containing 5% and 20% oxygen. The size and number of colonies detected in methylcellulose colony assays (a measure of progenitor cell differentiation) were larger at low (5%) oxygen ***(16)***. Additional studies in our laboratory have demonstrated that the production of mature cells and progenitors of different lineages is dependent on the oxygen tension ***(17)***. Given this information, the choice of a 5% or 20% gas-phase oxygen tension can be made depending on the goal of the culture and the likelihood of a particular culture configuration to develop oxygen limitations.

The effects of pH on hematopoietic cell culture have not been widely addressed. We have recently examined the effects of pH on human hematopoietic cell differentiation ***(18)***. In the granulocyte/macrophage lineage, differentiation from progenitor cells to mature cells is optimal between pH 7.2 and 7.4. Macrophage progenitors are slightly more sensitive to acidic conditions than are granulocytic progenitors. In the erythroid lineage, the rate and extent of differentiation increases with rising pH from 7.1 to 7.6. Hemoglobin content and other markers of erythroid development show that erythroid cells differentiate at a greater rate at high pH (7.6) than at a standard pH (7.35). Measuring and controlling pH is thus an important consideration in hematopoietic cell culture.

4.6. Culture Vessels

In **Subheadings 3.4.** and **3.5.**, we have suggested the use of 1-L culture bags as a large-scale static culture system likely to prove useful in clinical protocols. Based on our studies of the effects of biomaterials on hematopoietic culture *(19)*, we recommend the use of Teflon FEP culture bags (American Fluoroseal). Recently, we have had success with polystyrene Nunc Cell Factories (Nalge Nunc International, Rochester, NY), which are available in sizes that will accommodate 250, 500, and 2500 mL final culture volume. The protocols described above can be scaled down using the same cell density to a number of static culture vessels such as well plates or T-flasks. Suggested medium inoculation volumes for other vessels are as follows: 45 mL T-225, 15-mL T-75, 1 mL 24-well plate, 0.1-mL 96-well plate. In all cases, feeding at day 6 still involves doubling the initial volume.

Acknowledgments

This work is supported by NIH Grant R01 HL48276. TAM was partially supported by a U.S. Army Breast Cancer Training Grant USAMRMC Grant DAMD17-94-J-4466.

References

1. McAdams, T. A., Winter, J. N., Miller, W. M., and Papoutsakis, E. T. (1996) Hematopoietic cell culture therapies (Part II): clinical aspects and applications. *Trends Biotechnol.* **14,** 388–395.
2. Denning-Kendall, P., Donaldson, C., Nicol, A., Bradley, B., and Hows, J. (1996) Optimal processing of human umbilical cord blood for clinical banking. *Exp. Hematol.* **24,** 1394–1401.
3. de Wynter, E. A., Coutinho, L. H., Pei, X., Marsh, J. C. W., Hows, J., Luft, T., and Testa, N. G. (1995) Comparison of purity and enrichment of $CD34^+$ cells from bone marrow, umbilical cord and peripheral blood (primed for apheresis) using five separation systems. *Stem Cells* **13,** 524–532.
4. Nolta, J. A., Smogorzewska, E. M., and Kohn, D. B. (1995) Analysis of optimal conditions for retroviral-mediated transduction of primitive human hematopoietic cells. *Blood* **86,** 101–110.
5. Sandstrom, C. E., Collins, P. C., McAdams, T. A., Bender, J. G., Papoutsakis, E. T., and Miller, W. M. (1996) Comparison of whole-serum-deprived media for ex vivo expansion of hematopoietic progenitor cells from cord blood and mobilized peripheral blood mononuclear cells. *J. Hematother.* **5,** 461–473.
6. Dybedal, I. and Jacobsen, S. E. (1995) Transforming growth factor β (TGF-β), a potent inhibitor of erythropoiesis: neutralizing TGF-β antibodies show erythropoietin as a potent stimulator of murine burst-forming unit erythroid colony formation in the absence of a burst-promoting activity. *Blood* **86,** 949–957.
7. Sandstrom, C. E., Miller, W. M., and Papoutsakis, E. T. (1994) Review: serum-free media for cultures of primitive and mature hematopoietic cells. *Biotech. Bioeng.* **43,** 706–733.
8. Loudovaris, M., Qiao, X., Smith, S., Unverzagt, K., Hazelton, B., Schwartzberg, L., and Bender, J. G. (1994) In vitro proliferation and differentiation of human CD34+ cells into neutrophil progenitors and megakaryocytes. *Exp. Hematol.* **22,** 817.

9. Williams, S. F., Lee, W. J., Bender, J. G., Zimmerman, T., Swinney, P., Blake, M., Carreon, J., Schilling, M., Smith, S., Williams, D. E., Oldham, F., and Van Epps, D. (1996) Selection and expansion of peripheral blood $CD34^+$ cells in autologous stem cell transplantation for breast cancer. *Blood* **87,** 1687–1691.
10. Haylock, D. N., To, L. B., Makino, S., Dowse, T. L., Juttner, C. A., and Simmons, P. J. (1995) Ex vivo expansion of human hemopoietic progenitors with cytokines, in *Hematopoietic Stem Cells: Biology and Therapeutic Applications* (Levitt, D. and Mertelsmann, R., eds.), Marcel Dekker, Inc., New York, pp. 491–517.
11. Koller, M. R., Manchel, I., Palsson, M. A., Maher, R. J., and Palsson, B. Ø. (1996) Different measures of ex vivo human hematopoiesis culture performance optimized under vastly different conditions. *Biotech. Bioeng.* **50,** 505–513.
12. Schwartz, R. M., Palsson, B. Ø., and Emerson, S. G. (1991) Rapid medium perfusion rate significantly increases the productivity and longevity of human bone marrow cultures. *Proc. Natl. Acad. Sci. (USA)* **88,** 6760–6764.
13. Sandstrom, C. E., Bender, J. G., Papoutsakis, E. T., and Miller, W. M. (1995) Effects of $CD34^+$ cell selection and perfusion on ex vivo expansion of peripheral blood mononuclear cells. *Blood* **86,** 958–970.
14. Unverzagt, K. L., Martinson, J. A., Loudovaris, M., Smith, S. L., Lee, W., Berger, C., Qiao, X., McNiece, J., and Bender, J. G. (1996) Proliferation and differentiation of CD34+ cells into neutrophil precursors. Evaluation of feeding schedules and function. *Exp. Hematol.* **24,** 1075.
15. Ishikawa, Y. and Ito, T. (1988) Kinetics of hemopoietic stem cells in a hypoxic culture. *Eur. J. Haematol.* **40,** 126–129.
16. Broxmeyer, H. E., Cooper, S., and Gabig, T. (1989) The effects of oxidizing species derived from molecular oxygen on the proliferation in vitro of human granulocyte-macrophage progenitor cells. *Ann. NY Acad. Sci.* **554,** 177–184.
17. LaIuppa, J. A., Papoutsakis, E. T., and Miller, W. M. (1995) Oxygen tension alters the effects of cytokines on the megakaryocyte, erythrocyte, and granulocyte lineages. *Exp. Hematol.* **26,** 835–843.
18. McAdams, T. A., Miller, W. M., and Papoutsakis, E. T. (1997) Variations in culture pH affect the cloning efficiency and differentiation of progenitor cells in ex vivo haematopoiesis. *Br. J. Haematol.* **97,** 889–895.
19. LaIuppa, J. A., McAdams, T. A., Papoutsakis, E. T., and Miller, W. M. (1997). Culture materials affect *ex vivo* expansion of hematopoietic progenitor cells. *J. Biomed. Mater. Res.* **36,** 347–359.
20. Bertolini, F., Battaglia, M., Pedrazzoli, P., Da Prada, G. A., Lanza, A., Soligo, D., et al. (1997) Megakaryocytic progenitors can be generated ex vivo and safely administered to autologous peripheral blood progenitor cell transplant recipients. *Blood* **89,** 2679–2688.

19

Cytotoxicity Testing Using Cell Lines

Lorraine D. Buckberry

1. Introduction

Commercially exploitable compounds are being produced using modern biotechnology for use as food additives, chemotherapeutic agents, and pesticides. Traditionally, animal testing has always played an important role in the safety evaluation of such agents. However, financial and ethical considerations, together with an increased awareness of the limitations of animal models in relation to human metabolism, now warrant the development of alternative testing methods. Therefore, it is fitting that the potential of biotechnology should provide mammalian cell systems for in vitro testing. The ultimate aim of in vitro toxicity testing is the replacement of animals in testing protocols, but in the short term, procedures are refined to reduce the numbers of animals required. This "three Rs" philosophy of reduction, refinement, and replacement was first proposed by Russell and Burch as early as 1959 *(1)*, and is now recognized in the UK Animals in Scientific Procedures Act, 1986 and EC Directive 86/609/ECC *(2)*.

Toxicity testing encompasses a wide range of causes and effects such as mutagenicity, carcinogenicity, teratogenicity, and acute and chronic cytotoxicity. Whereas a detailed treatment of these areas is beyond the scope of this chapter, the fundamental requirements of in vitro cytotoxicity testing remain constant:

1. An assay system should give a reproducible dose-response curve over a concentration range that includes the in vivo dose.
2. There should be a linear relationship between cell number and assay response, and the resultant dose-response curve should relate predictively to the in vivo effect of the same compound (ideally the same range of concentrations should give the same effect) *(3)*.

From: *Methods in Biotechnology, Vol. 8: Animal Cell Biotechnology*
Edited by: N. Jenkins © Humana Press Inc., Totowa, NJ

2. Designing Toxicity Studies

2.1. Culture Method

Five types of culture methods are commonly employed in toxicity studies:

1. Primary cells and organ tissues are used in toxicity testing. This type of culture method is beyond the scope of this chapter, but information on these techniques can be obtained from **refs.** ***4*** and ***5***.
2. Spheroid cultures are being used for penetration assays and solid tumor modeling. The end point of spheroidal cultures is taken to be spheroid size, as determined by cloning.
3. Suspension cultures can be used in long- and short-term assays for drug sensitivity but are mainly applied to systems using tumor biopsy material, as no growth is required and clonal selection is minimized, allowing rapid results to be obtained.
4. The main application of clonogenic growth in soft agar is also in the growth of primary tumor cells. Clonogenic assays have the advantage of minimizing the growth of anchorage-dependent stromal cells, and the response is determined in cells with a high capacity for self-renewal, such as tumor stem cells.
5. Monolayer culture is the most widely used culture method employed in cytotoxicity testing. This method offers the greatest flexibility in terms of exposure, recovery, and quantification of drug effect. Exposure times of between 1 and 24 h with or without a recovery period are commonly used. The end point is based on cell number as an indication of survival or metabolic capacity. Monolayer culture requires few cells compared with other techniques and so can easily be used for microtitration assays. Microtiter plate (96-well) assays allow wide concentration ranges to be tested. Any toxic effects can be easily quantified spectrophotometrically using a microtiter plate reader and hence can be automated relatively easily ***(3–6)***.

2.2. Assay Design

2.2.1. Exposure to Study Compound

The concentration of the test compound should be dictated by the exposure level experienced in vivo. If kinetic data are available on the in vivo clearance of a compound, peak plasma concentration and plasma clearance curves can be used to estimate the most appropriate concentration range for testing, which must be adjusted to give a dose-response curve. The sensitivity required over the chosen concentration range will also influence the choice of assay. In vivo kinetic data show that the maximum exposure to most compounds following acute exposure via intravenous injection occurs after 1 h, and therefore an exposure period of 1 h is adopted in most studies. This is an adequate approach for compounds such as cell cycle-specific drugs. However, in cases of chronic exposure to environmental agents and non-cycle-specific drugs, longer exposure times may give more relevant data. It has been shown that inhibition dose (ID_{50}) values (the test compound concentration required to reduce a response to half that of the control) can decrease on prolonged exposure to certain agents ***(3)***; however, the rate at which the test compound enters the cells alters the assay response with short exposure times. This is because the time taken for penetration accounts for a large percentage of the total exposure time. Ultimately, exposure times of 1 or 24 h are usually adopted as common testing periods, and it is vital that the testing period be kept constant throughout a given study ***(3, 7)***.

2.2.2. Recovery Period

A recovery period must be instituted when metabolic parameters are used as an index of the test compound's effect, as this allows recovery from any metabolic disfunction unrelated to cell death. Sublethally damaged cells can recover over this period, and any delayed cytotoxicity may occur. Omitting a recovery period can lead to under- or overestimation of the level of cell killing observed. However, if the recovery period is too long, cell killing can be underestimated due to overgrowth of a resistant population ***(3,6)***.

2.2.3. Use of Controls Appropriate to the Mode Toxin of Application

Many potential toxins are highly lipophilic and therefore can only be added to culture medium in an organic solution. In this case, the solvent may cause toxicity in its own right and so should not exceed a total concentration of >1% (v/v) in the medium. Certain oils may dissolve into plastic culture vessels, resulting in toxic compounds leaching from the plastic. Particulate samples, which may not dissolve in an organic solvent, may be applied as suspensions. However, these routes of application may influence the resultant data; therefore the use of blank controls, in which the cells are treated with media containing solvent alone, are vitally important. The reliability of the assay can be monitored by using positive control compounds chosen on the basis of their ability to demonstrate the effect being tested ***(3,6)***.

2.3. Choice of Cell Type

The choice of cell type will be dictated by the type of compound to be tested ***(7)***. For general toxicity screening against a reference compound, fast growing robust fibroblast or epithelial lines are often employed (e.g., LLC-PK1, MDCK, CHO, V79, or HeLa cells), although the genetic instability of cells of this type may compromise the results in long-term studies. Cell lines commonly used for toxicity studies are commercially available from commercial cell culture suppliers. However, more specialized lines may be obtained from the American Type Culture Collection (12301 Parklawn Drive, Rockville, MD 20852–1776;

Tel: [301] 881-2600, [800] 638-6597; Fax: [301] 231-5826) or the European Collection of Animal Cell Culture (CAMR, Porton Down, Salisbury, SP4 0JG, England; Tel: [01980] 610391; Fax: [0980] 611 315).

Studies directed toward specific cellular functions may preclude the use of the transformed cell lines described above due to their dedifferentiated status. Many compounds undergo metabolic transformations into an activated derivative catalyzed by P450-dependent enzymes. Although the use of established, well-characterized immortalized lines has obvious advantages, the problem of the maintenance of normal levels of cytochrome P450 in cultures (particularly hepatocytes) cannot be overlooked.

In primary cell lines, levels of certain forms of cytochrome P450 decrease rapidly in the first 24 h of culture, whereas the levels of other multiple forms increase. However, a number of strategies have been employed to maintain a normal level of P450 population in the cell. These include media manipulation, such as hormone supplements and the omission of various amino acids, and coculture of hepatocytes with rat liver epithelium.

A range of specialist assay protocols including those for cell isolation are available through INVITTOX, The ERGATT/FRAME Data Bank of *In Vitro* Techniques in Toxicology. A current protocol list can be obtained from INVITTOX, Russell and Burch House, 96–98 North Sherwood Street, Nottingham, NG1 4EE, UK; Tel: (0115) 958-4740; Fax: (0115) 950-3570; E-mail: invittox@frame-uk.demon.co.uk.

3. Basic Screening Tests

Toxicity is a complex event in vivo and exhibits a wide spectrum of effects from cell death to metabolic aberrations, which result in functional changes in cells rather than death. While simple determination of cell death may initially give useful information regarding the gross toxicity of a compound, the complexity of toxic processes may involve minor metabolic changes or an alteration in cell-cell signaling, which may not necessarily result in cell death. All assays oversimplify the processes that they measure, and gross tests of cytotoxicity should be supplemented with more specific tests of metabolic aberration.

3.1. Assays Based on Cell Viability or Cell Survival

Traditionally, cytotoxicity assays have been based on indices of cell viability and survival. Viability is an instantaneous parameter thought to be predictive of survival. It can be determined by colony growth, net change in population size, mass, or gross metabolic activity. Survival is usually defined by determination of reproductive capacity via clonogenicity. Metabolic parameters may be used when a population is allowed a suitable recovery period following exposure. Viability and survival assays have been developed with quantifiable end points. The ability of a compound to induce an inflammatory or allergic response is more difficult to determine, although assays in this area are developing rapidly, particularly as alternatives to the Draise eye test.

Cytotoxicity assays can be divided into three groups, as defined by Freshney ***(5)***:

1. Those that infer viability by determination of a change in membrane permeability or metabolism.
2. Those that determine viability through absolute long-term survival, as quantified by the capacity to regenerate (clonogenicity).
3. Those that determine survival in an altered state by expressing a genetic mutation or malignant transformation.

3.1.1. Viability Assays

Viability assays determine the proportion of viable cells remaining following a toxic insult. The abilities of cells to exclude vital dyes such as eosin, nigrosin, trypan blue, and neutral red, or to include vital dyes (such as neutral red and diacetyl fluorescein) or isotopes (such as ^{51}Cr) have been widely used as indices of cell viability. Application of these tests to cell survival is limited, and comparisons should be made with clonogenic survival following two or three doubling times after removal of the test substance, as this gives an indication of long-term survival.

3.1.1.1. Dye Exclusion Methods

Viable cells exclude a number of vital dyes. A cell suspension is combined with dye and examined using a hemocytometer and optical microscope. The number of unstained cells is expressed as a percentage of the total population. Staining for viability assessment is more suited to suspension than monolayer cultures, as dead cells detach from the monolayer.

3.1.1.1.1. Materials

1. Calcium- and magnesium-free phosphate-buffered saline solution (PBS).
2. Trypsin (0.25% w/v) prepared in GIBCO (Gaithersburg, MD and Paisley, Scotland) solution A.
3. Cell growth media.
4. Stain (0.4% w/v trypan blue).
5. Neubauer microscope (magnification ×100).

3.1.1.1.2. Method. Following a toxic insult carried out with appropriate recovery periods, blanks, and controls (*see* **Subheading 2.2.3.**):

1. A cell suspension is prepared by trypsinization at a density of approx 10^6 cell/ml.
2. The cell suspension (20 µL) is then mixed with trypan blue (20 µL).
3. The viable cells (that will be white due to successful exclusion of the dye) are then counted using an improved Neubauer hemocytometer and expressed as a

percentage of the total cell count (blue nonviable cells + white cells) (*see* **Note 1**).

3.1.1.2. Dye Uptake Methods

These methods can be adapted for use with flow cytometers and microtiter plates. Neutral red is taken up into viable cells in culture, and this property can be used in a microtiter plate dye uptake assay. The method that follows is based on that of Borenfreund and Puerner ***(8)***.

3.1.1.2.1. Specialist Equipment

1. Flat bed shaker.
2. Microtiter plate reader with a 540-nm filter.

3.1.1.2.2. Materials

1. Cells growing in a 96-well tissue culture plate (for a 24-h exposure period, seed the cells at approximately 2×10^4 cells/mL and incubate overnight to allow adherence and recovery from trypsinization).
2. Neutral red stock solution (2 mg/mL cell growth medium).
3. Neutral red medium, 0.05 mg/mL (neutral red stock solution/medium, 1:39); incubate overnight at 37°C to allow any neutral red crystals to form and as a sterility check.
4. Neutral red destain solution (glacial acetic acid/ethanol/distilled water, 1:50:49).
5. PBS (calcium and magnesium free) (GIBCO).

3.1.1.2.3. Method

1. Cells are grown in columns 2–11 of a 96-well tissue culture plate and incubated to a point in the exponential growth phase at which they are challenged with the test compound. Cells in columns 1 and 12 experience slightly different growth conditions and so are not used for testing.
2. The concentration range of the test compound should be such that the lowest concentration kills none of the cells and the highest kills >90% of the cells. This can best be achieved by preparing a fivefold serial dilution to give eight concentrations.
3. Cell culture medium (190 μL) containing the highest test compound concentration is placed in each well (A–H) of column 3 on a 96-well plate. The wells of the other columns are filled with 150 μL of cell culture medium. Cell culture medium (40 μL) containing the test compound is transferred to the wells (A–H) of column 4 using a multichannel pipet and mixed. The transfer of 40 μL of media from column to column is repeated until a 21% serial dilution over a range of eight concentrations has been generated. Tests should be carried out using eight replicate wells at each concentration (columns 3–10). Column 2 should contain media alone and any solvent to drive the test compound into solution added prior to addition to the media; column 1 will be used as an instrument blank (**Fig. 1**).
4. Three hours before the end of the test period medium is aspirated from the cells and replaced with 150 μL of neutral red medium in each well. The incubation period is then completed.

	1	2	3	4	5	6	7	8	9	10	11	12
A	Instrument Blank											
B		Culture Medium Only	Top Dose	⇨	⇨	⇨	⇨	⇨	⇨	⇨	Lowest Dose	
C			Top Dose	⇨	⇨	⇨	⇨	⇨	⇨	⇨	Lowest Dose	
D			Top Dose	⇨	⇨	⇨	⇨	⇨	⇨	⇨	Lowest Dose	
E			Top Dose	⇨	⇨	⇨	⇨	⇨	⇨	⇨	Lowest Dose	
F			Top Dose	⇨	⇨	⇨	⇨	⇨	⇨	⇨	Lowest Dose	
G			Top Dose	⇨	⇨	⇨	⇨	⇨	⇨	⇨	Lowest Dose	
H												

Fig. 1. The loading of samples in a 96-well plate for microtitration assays.

5. The neutral red medium is removed and the wells washed with warm PBS (37°C) to wash out any excess stain.
6. Destain (150 µL/well) is added to fix the cells and to elute the neutral red in the cells into solution.
7. The plates are gently shaken until the color in the wells is homogeneous.
8. The absorbance of each well is determined at 540 nm against a reference well, which does not contain cells.
9. Absorbances should correlate linearly with cell number over the range 0.2–1.0 OD units. The absorbance observed at each test compound concentration is expressed as a percentage of the absorbance observed with unchallenged cells. A graph of concentration against percent inhibition of growth (100% of control absorbance) is plotted. A sigmoid or exponential curve may be obtained from which the ID_{20}, ID_{50}, and ID_{80} values can be calculated (**Fig. 2**). The curve is fitted using a polynomial regression equation available on most curve fitting packages. The ID_{20}, ID_{50}, and ID_{80} values can be calculated by substituting 20, 50, and 80 into the equation for the curve (*see* **Notes 2–4**).

3.1.1.3. Fluorescence Method

Viable cells also take up diacetyl fluorescein, which can subsequently be hydrolyzed to fluorescein, causing viable cells to fluoresce green. Nonviable cells can be visualized using ethidium or propidium bromide, which causes them to fluoresce red. Viability is expressed as the number of cells that fluoresce green expressed as a percentage of total cells counted (which includes both the green and red fluorescent cells **ref. *9***.

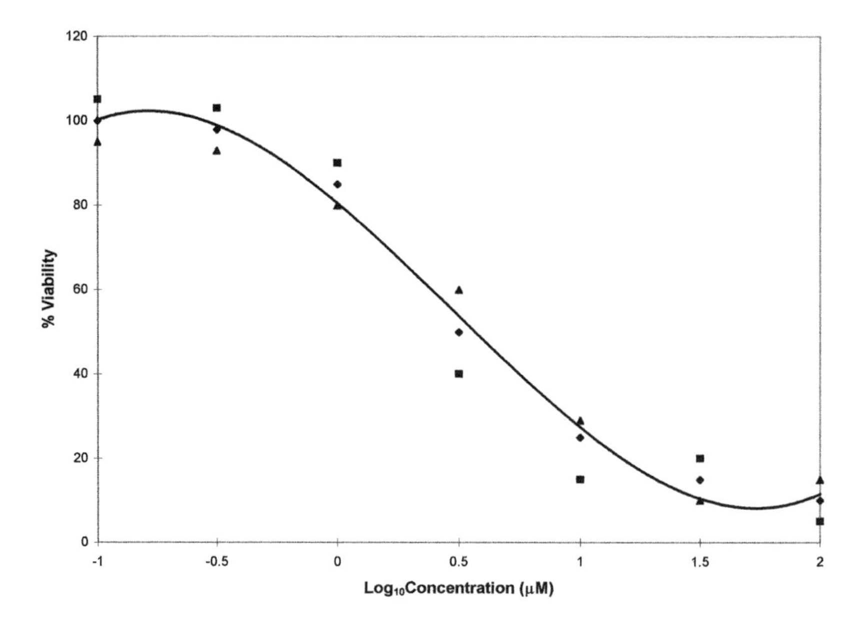

Fig. 2. Plot of percentage cell viability against the concentration of test compound ($\log_{10}$).

3.1.1.3.1. Specialist Equipment

1. Fluorescence microscope, equipped with a light source and filters to detect fluorescein (excitation 450/590 nm, emission 515 nm) and propidium iodide (excitation 488 nm, emission 615 nm).
2. Flow cytometer.

3.1.1.3.2. Materials

1. Dye mixture: fluorescein diacetate (10 μg/mL), and propidium iodide (500 μg/mL) in Hanks' balanced salt solution.

3.1.1.3.3. Method

1. A cell suspension is prepared by trypsinization.
2. Dye mixtures and the cell suspension are combined in a ratio of 1:10 (v/v) to give final dye concentrations of 1 μg/mL (fluorescein) and 10 μg/mL (propidium iodide).
3. The cells are then examined by fluorescence microscopy or flow cytometry.
4. Cells fluorescing green are viable and cells fluorescing red are nonviable.

3.1.1.4. Dye Release Methods

Dye release assays are exemplified by the neutral red release assay ***(10,11)***. A confluent monolayer of cells is preloaded with dye, and the medium is then replaced with fresh dye-free medium. The cells are then exposed to a test substance for 1 min and washed to remove liberated dye. A fixing agent is then added and optical density measurements taken from the resulting solutions in the wells of the plate. Owing to the short exposure time this test is best employed to test rapidly occurring effects directed at the cell membrane and not those compounds that require activation.

3.1.1.4.1. Specialist Equipment. As for **Subheading 3.1.1.2** dye uptake method.

3.1.1.4.2. Materials. As for **Subheading 3.1.1.2** dye uptake method.

3.1.1.4.3. Method. Cells are grown to confluence in a 96-well tissue culture plate.

1. Culture medium is replaced with 150 μL of neutral red medium in each well, and the plate is incubated for 3 h at 37°C.
2. The neutral red medium is removed and replaced with 150 μL of fresh media, prewarmed to the incubation temperature, to halt dye uptake.
3. The culture medium is replaced with PBS (200 μL), which is then removed and replaced with the test compound solution (50 μL).
4. After a contact time of 1 min the test sample is removed and replaced with PBS (200 μL).
5. To release the neutral red, the PBS is replaced with neutral red destain (100 μL).
6. Plates are shaken until well color becomes homogenous and the absorbance of the solutions is then determined in each well at 540 nm.
7. It is assumed that under control conditions, the live cells do not release neutral red. Calculate the ID_{20}, ID_{50}, and ID_{80} values as for neutral red uptake.

3.1.1.5. Isotope Release Methods

Reduced $^{51}Cr^{3+}$ is taken up by viable cells and undergoes oxidation to $^{51}Cr^{2+}$, which cannot travel across viable membranes because of covalent binding of chromate to amino acids residues in proteins *(3)*. Dead cells subsequently release the $^{51}Cr^{2+}$. Following application of the test substance and an $Na_2{}^{51}CrO_4$ label, aliquots of medium are γ-counted for ^{51}Cr release. Toxicity is expressed as a percentage of total (-counts for cells and medium; this does not give an absolute viability figure but is useful for comparison between substances. Also, chromate leakage can be high in viable cells; therefore, test substance exposure is functionally limited to a maximum of 4 h. Owing to the problems of handling radioisotopes, dye release methods are usually more practical.

3.1.2. Long-Term Survival

Often, there is a delay of several hours before cells exhibit responses to toxic insult; a different type of assay is required in these cases taking into account the detachment of dead cells. These assays often measure compromised metabolic or proliferative capacity to infer long-term damage rather than short-term toxicity, which may be reversible.

3.1.2.1. Metabolism Assays

Alterations in metabolism (e.g., changes in processes like glycolysis, enzyme activities, and the ability to incorporate labeled precursors such as [^{3}H]thymidine) have all been used to measure response to toxic stimuli. However, the viability data obtained from this type of assay should not be interpreted as specific to the metabolic parameter tested. Toxin-induced changes in respiration and glycolysis can be determined using manometry by determination of oxygen use and carbon dioxide production, respectively; however, this is extremely time consuming. A 3-(4,5-dimethylthiazol-2–yl)-2,5-diphenyltetrazolium bromide (MTT)-based assay is faster, expressing cell number as an end point for viability ***(12)***. MTT is taken up into viable cells and reduced by succinate dehydrogenase to the formazan derivative, which is purple in color. On solubilization the formazan can be quantified colorometrically.

3.1.2.1.1. Materials

1. Trypsin (0.25% w/v), EDTA (1 m*M*) in PBS (GIBCO).
2. MTT (50 mg/mL cell culture medium) sterilized by filtration (Sigma, St. Louis, MO).
3. Glycine (0.1 *M*), and sodium chloride (0.1 *M*) adjusted to pH 10.5 with sodium hydroxide (1 *M*).
4. Dimethylsulfoxide (DMSO).

3.1.2.1.2. Method

1. *See* Method, **Subheading 3.1.2.3., Steps 1–3.**
2. The medium is removed at the end of the exposure period of incubation and replaced with 150 µL of fresh media prewarmed to the incubation temperature.
3. Cells are grown for a period sufficient for the number of cells to increase two- to threefold.
4. Medium is replaced with MTT-medium solution and incubated for 4–8 h.
5. Solution is replaced with DMSO (200 µL) to dissolve the remaining crystals.
6. Plates are shaken until the well color becomes homogenous and the absorbance of the solutions in each well determined at 570 nm.
7. It is assumed that under control conditions no MTT-formazan is formed. Calculate the ID_{20}, ID_{50}, and ID_{80} values as for neutral red uptake (**Subheading 3.1.1.2.**) (*see* **Notes 5** and **6**).

3.1.2.2. Protein Content

Protein content is a useful measure of total cellular material as an estimation of cell number, although it will vary with cell size and stage in the cell cycle. While the absorbance of protein at 280 nm can be determined directly, colorimetric assays are more sensitive. Standard approaches have been based on the Lowry and Bradford methods ***(13)***; however, a major disadvantage of these techniques is the formation of a ring of dried protein above the cell culture, which results in poor assay reproducibility. Incorporation of radiolabeled amino acids such as [^{3}H]leucine or [^{35}S]methionine has been used as a more direct index of protein synthesis (*see* **Notes 7** and **8**). Coomasssie brilliant blue R-250 dye binding ***(13)*** is outlined below.

3.1.2.2.1. Materials

1. PBS pH 7.4.
2. Fixative (glacial acetic acid/ethanol/water, 1:49:50, v/v/v).
3. Stock stain solution (Coomasssie brilliant blue R-250 stain (0.05%, w/v, Sigma) in ethanol/water, 25:63, v/v).
4. Working stain solution is made by diluting the stock solution in glacial acetic acid solution, 88:12 (v/v)
5. Wash solution (ethanol/glacial acetic acid/water, 10:5:85, v/v/v).
6. Desorbing solution (potassium acetate, 1 *M*, in ethanol/water, 7:3, v/v).

3.1.2.2.2. Method. Cells are grown in a 96-well tissue culture plate and exposed to the test compound as described previously in **Subheading 3.1.2.1.**

1. After exposure to the test compound, remove the medium and wash the wells with warm PBS (37°C).
3. Add fixative (150 µL/well) and shake on a flat bed shaker for 20 min.
4. Remove the fixative, add stain (150 µL/well), and shake on a flat bed shaker for 20 min.
5. Remove the stain and fill each well with wash solution (250 µL).
6. Aspirate the wash solution and repeat **steps 5** and **6**.
7. Add desorbing solution (150 µL/well) and shake vigorously on a flat bed shaker until a homogeneous solution is observed.
8. Determine the absorbance of each well at 577 nm with a 404-nm reference blank. The absorbance observed at each test compound concentration is expressed as a percentage of the absorbance observed with unchallanged cells. A graph is plotted of protein concentration against the percent inhibition of growth (relative to the absorbance at zero toxin concentration). A sigmoid or exponential curve may be obtained from which the ID_{20}, ID_{50}, and ID_{80} values can be calculated.

3.2. Irritancy Assays

A number of in vitro methods are currently undergoing validation as alternatives for the EU/Home office Draize eye test ***(14–16)***. Effects on intracellular metabolism can be reflected by a decrease in extracellular acidification rate as determined by a silicon microphysiometer, which can in turn be used as a measure of eye irritation potential. Toxin membran-olytic activity and its ability to cause protein denaturation in mammalian erythrocytes has also been used as a measure of potential ocular irritancy ***(16,17)***. The neutral red bioassay using BALB/C 3T3 cells may also be used to assess potential irritancy ***(14)***.

4. Conclusions

Clearly, all the approaches discussed above have major advantages and disadvantages. Batteries of assays offer a partial solution to the inherent problems involved in extrapolating toxicity data from one assay alone. For example, The FRAME cytotoxicity test ***(13)*** uses a combination of Coomasssie brilliant blue R-250 protein staining with the neutral red release assay to give an estimation of gross protein content and membrane integrity. Coomassie brilliant blue R-250 staining is performed on cells tested by the neutral red method, following removal of neutral red destain prior to the addition of the kenacid blue stain.

5. Relevance and Evaluation of Results

Even with the use of human cell lines, it will never be possible to predict the type or level of toxicity caused by a compound in vivo from in vitro studies. At best, only qualitative extrapolations can be made.

The validity of a cytotoxicity index is dependent on the degree of linearity between cell number and the chosen toxicity parameter. In clonogenic assays, a linear relationship may not occur at low cell numbers due to the dependence of growth on conditioning factors, while nutritional deficiencies may affect growth if the cell number is too high. In cytotoxicity assays, density-dependent inhibition of growth may occur while the sensitivity of the assay may cause a loss of linearity at the lower end of the assay range.

6. Notes

1. It is not practical to use a hemocytometer for cell counting to determine viability at a range of concentrations to generate a titration curve, because of the time taken to carry out the counting.
2. Neutral red is preferentially taken up into the lysosomes/endosomes of the cell. Therefore any compound having a localized effect on these organelles will give artificially low or high viability data. Neutral red can precipitate out in visible needle-like crystals. Therefore, it is advisable to centrifuge neutral red medium prior to use to remove any crystals that have formed on overnight incubation and to inspect treated wells carefully prior to analysis.
3. When using viability assays the following factors must be considered: it may require several days before lethally damaged cells lose their membrane integrity and during this time the "surviving" cells may continue to proliferate; and some lethally damaged cells may undergo early disintegration so that they are not present to be stained with dye at the end of the testing period. These factors may then cause an underestimation of the cell death when the result of the assay is expressed as percentage viability.
4. However, the use of optical density determination as opposed to cell counts does allow the assay to be employed for microtitration studies and is also less prone to

operator error. Loaded cells that are adherent are easily detached by violent pipetting. After dye loading, cells must be exposed to the test compound within 2 h. If the test substance is PBS insoluble the sample should be shaken vigorously before addition to the wells, and additional replicates should be used. Care should be taken when timing the removal and addition of the test substance. The order of addition and removal should be constant to ensure a consistent length of exposure. When seeding cells into microtiter plates, it is vital to ensure that the cells are well dispersed (i.e., that no clumping has occurred following trypsinization) as this will lead to inconsistent results.

5. The MTT-formazan dye is unstable. The addition of glycine buffer (25 mL) to each well may increase the stability; however, it is advisable to read the plate immediately on addition of the DMSO.
6. When seeding cells into microtiter plates it is vital to ensure that the cells are well dispersed (i.e., that no clumping has occurred following trypsinization) as this will lead to inconsistent results.
7. The incorporation of radiolabeled nucleotides such as [^{3}H]thymidine and [^{3}H]uridine into DNA and RNA, respectively, can be employed as a measure of the ability to proliferate. In assays performed without a recovery period the observed changes may relate to changes in the nucleotide pool rather than DNA synthesis or toxin-related inhibition of pyrimidine biosynthesis ***(5)***.
8. It is acknowledged that overestimation of viability may occur through the detection of respiring cells that have lost the capacity to proliferate prior to death. If any dark blue rings of stain remain on the well walls after washing these should be wiped away. If this is not the possible the plate should be discarded.

Acknowledgments

The author would like to thank Nigel Jenkins for his advice and guidance during the production of this chapter and her colleagues at De Montfort University for allowing her to bounce ideas off them during its production.

References

1. Russell, W. M. S. and Burch, R. L. (1959) *The Principles of Humane Experimental Technique,* Methuen, London.
2. Fentem, J. and Balls, M. (1992) *In vitro* alternatives to toxicity testing in animals. *Chem. Ind.* 207–211.
3. Wilson, A. P. (1989) Cytotoxicity and viability assays, in *Animal Cell Culture: A Practical Approach* (Freshney, R. I., ed.), IRL, Oxford, pp. 183–216.
4. Freshney, R. I. (1994) *Culture of Animal Cells, A Manual of Basic Techniques,* Wiley-Liss, New York.
5. Freshney, R. I. (1989) *Animal Cell Culture: A Practical Approach*, IRL, Oxford.
6. Benford, D. J. and Hubbard, S. A. (1987) Preparation and culture of mammalian cells, in *Biochemical Toxicology: A Practical Approach* (Snell, K. and Mullock, B., eds.), IRL, Oxford, pp. 57–82.

7. Riddell, R. J., Panacer, D. S., Wilde, S. M., Clothier, R. H., and Balls, M. (1986) The importance of exposure period and cell type in *in vitro* cytotoxicity testing. *ATLA* **14,** 86–92.
8. Borenfreund, E. and Puerner, J. A. (1985) Toxicity determined *in vitro* by morphological alterations and neutral red absorption. *Toxicol. Lett.* **24,** 119–124.
9. Rotman, B. and Papermaster, B. W. (1966) Membrane properties of living animal cells as studied by enzymatic hydrolysis of fluorogenic esters. *Proc. Natl. Acad. Sci. USA* **55,** 134–141.
10. Reader, S. J., Blackwell, V., O'Hara, R., Clothier, R. H., Griffin, G., and Balls, M. (1989) A vital dye release method for assaying the short term cytotoxic effects of chemicals and formulations. *ATLA* **17,** 28–33.
11. Reader, S. J., Blackwell, V., O'Hara, R., Clothier, R. H., Griffin, G., and Balls, M. (1990) Neutral red release from pre-loaded cells as an in vitro approach to testing for eye irritancy potential. *Toxicol. In Vitro*, **4,** 264–266.
12. Iselt, M, Holtei, W., and Hilgard, P. (1989) The tetrazolium dye assay for rapid *in vitro* assessment of cytotoxicity. *Arzneimittel-forschung* **39,** 747–749.
13. Clothier, R. H., Atkinson, K. A., Garle, M. J., Ward, R. K., and Willshaw, A. (1995) The development and evaluation of in vitro tests by the FRAME alternatives laboratory. *ATLA* **23,** 75–90.
14. Spirlmann, H., Balls, M., Brand, M., Doring, B., Holzhutter, H. G., Kalweit, S., Klecak, G., Eplattenier, H. L., Liebsch, M., Lovell, W. W., Maurer, T., Moldenhauer, F., Moore, L., Pape, W. J. W., Pfanenbecker, U., Potthast, J., Desilva, O., Steiling, W., and Willshaw, A. (1994) EEC COLIPA project on in vitro phototoxicity testing—first results obtained with Balb/c 3T3 cell phototoxicity assay. *Toxicol. in Vitro* **8,** 793–796.
15. Harbell, J. D., Tsai, Y. C., Maibach, H. I., Gay, R., Miller, K., and Mun, G. C. (1994) An *in vivo* correlation with 3 in vitro assays to assess skin irritation potential. *J. Toxicol. Cut. Ocul. Toxicol.* **13,** 171–183.
16. Lewis, M., McCall, J. C., Botham, P. A., and Trebilock, R. (1994) A comparison of 2 cytotoxicity tests for predicting the ocular irritancy of surfactants. *Toxicol. in Vitro* **8,** 867–869.
17. Christian, M. S. and Diener, R. M. (1996) Soaps and detergents—alternatives to animal eye irritation tests. *J. Am. Coll. Toxicol.* **15,** 1–44.

V

Product Evaluation Protocols

20

Measuring the Folding Dynamics of Recombinant Proteins Secreted from Mammalian Cells

Neil J. Bulleid and Adrian R. Walmsley

1. Introduction

The biotechnologist faced with the challenge of producing significant quantities of recombinant protein will endeavor to identify a host cell for protein production that will synthesize authentic, biologically active proteins at relatively high yields. Due to the complex folding pathway and posttranslational modifications required for some proteins to elicit full biological activity, the choice of host cell is limited to mammalian cells (for review, *see* **ref. *1***). The restrictions do not end here, as some cell lines have a greater capacity to carry out specific modifications than others, thus requiring an evaluation of host cells to fold, modify, and secrete the model protein ***(1–3)***. This chapter will summarize the main steps in carrying out the evaluation process at a practical level. The protocols are applicable to most proteins but some optimization of conditions will be necessary depending on the protein being studied.

The focus of this chapter is on measuring the folding dynamics of proteins, hence the choice of expression system used will not be covered here. However, the level of expression does have a bearing on such parameters as labeling time and rates of secretion. Generally, stable cell lines will be used, but valuable information can be gained from transient transfection, which has the advantage that information regarding folding kinetics, rates of secretion, and authenticity of protein produced can be obtained in a relatively short time, allowing several cell lines to be screened.

2. Materials

1. A ^{35}S-labeled methionine and cysteine mix was obtained from NEN (DuPont UK, Stevenage, UK, or Amersham, Little Chalfont, UK).

From: *Methods in Biotechnology, Vol. 8: Animal Cell Biotechnology*
Edited by: N. Jenkins © Humana Press Inc., Totowa, NJ

2. *N*-ethylmaleimide (NEM), phenylmethylsulphonyl fluoride (PMSF), Triton X-100, sodium azide, and cycloheximide were obtained from Sigma (Poole, UK).
3. Antipain, chymostatin, leupeptin, pepstatin, endoglycosidase H, and soybean trypsin inhibitor were obtained from Boehringer Mannheim (Lewes, UK).
4. Protein A Sepharose was obtained from Pharmacia (St. Albans, UK).

3. Methods

3.1. Detection and Radiolabeling

To study the folding and secretion of a particular protein expressed within mammalian cells it is essential to be able to detect specifically the protein of interest. To achieve this, an antibody specific to the protein of interest that can immunoprecipitate must be used. To study the kinetics of folding and transport, the target protein is radiolabeled so it can be monitored by sodium dodecyl sulfate-polyacrylamide gel electrophoresis (SDS-PAGE) analysis followed by autoradiography. Thus, the first stage in analysis is to carry out a cell labeling experiment followed by immunoprecipitation to determine the optimum conditions for pulse chase (For a practical guide to the use of antibodies, *see* **ref.** ***4***.) The ideal is a labeling time that is as short as possible while still allowing detection. This will typically be between 5 and 10 min but will obviously depend on the levels of protein expression and the number of methionine and cysteine residues present in the protein, as these are the radiolabeled amino acids used. A typical protocol would be as follows:

1. Cells are grown to 90% confluence in 100-mm dishes, washed twice with methionine- and cysteine-free medium, and then preincubated in the same medium for 20 min at 37°C. Each dish is labeled with 20 µCi of ^{35}S-labeled methionine and cysteine/mL in a final volume of 5 mL of prewarmed medium for various times at 37°C.
2. After the cells have been labeled, the dishes are transferred to ice and washed twice with ice-cold phosphate-buffered saline (PBS) containing 20 mM NEM to prevent formation or rearrangement of disulfide bonds in the protein.
3. The cells are lysed with 750 µL/dish of ice-cold lysis buffer (25 mM Tris-HCl, pH 7.4, containing 0.5% Triton X-100, 50 m*M* NaCl, 2 m*M* EDTA, 20 m*M* NEM, 1 m*M* PMSF, 10 µg/mL each of antipain, chymostatin, leupeptin, and pepstatin, and 10 µg/mL soybean trypsin inhibitor). To ensure complete lysis cells may be scraped from the dish using a cell scraper. Lysates are adjusted to 1.5 mL with lysis buffer and then spun for 15 min at 12,000*g* at 4°C to pellet nuclei and cell debris.
4. Cell lysates (typically between 300 and 750 µL) are diluted to 1 mL with immunoprecipitation buffer (50 m*M* Tris-HCl, pH 7.4, containing 1% Triton X-100, 150 m*M* NaCl, 2 m*M* EDTA, and 0.02% w/v sodium azide). Samples are preincubated with 50 µL protein A Sepharose (10% w/v in PBS) with mixing for 1 h at 4°C to preclear the samples of protein A binding components. After 30 s of centrifugation at 12,000*g* to pellet the protein A Sepharose, the supernatant is

removed to a fresh tube and incubated with 50 μL protein A Sepharose; antibody is then added (the amount needs to be determined empirically to ensure saturation of binding) followed by a further incubation for 15 h at 4°C. The complexes are washed three times with fresh immunoprecipitation buffer and then resuspended in SDS-PAGE sample buffer (0.25 *M* Tris-HCl, pH 6.8, containing 2% w/v SDS, 20% v/v glycerol, and 0.004% w/v bromophenol blue) and where necessary dithiotreitol (DTT) added to 50 m*M*, if samples are to be analyzed after reduction of disulfide bonds (*see* **Note 1**).

3.2. Kinetics of Secretion

Once the labeling conditions have been optimized to ensure a detectable signal in the shortest labeling time, the kinetics of secretion can be monitored. It is generally the case that proteins that do not fold correctly are not secreted from mammalian cells so a measure of secretion also indicates that the protein has folded correctly. The passage of proteins through the secretory pathway can be monitored by two techniques, namely, the appearance of protein in the culture medium and the acquisition of complex modifications to the oligosaccharide side chain. The former approach is not applicable to membrane proteins, whereas the latter approach is obviously only applicable to glycoproteins and indicates that the protein has been transported from the endoplasmic reticulum to the Golgi apparatus. Modification to the oligosaccharide side chain can most conveniently be measured by the sensitivity to digestion with endoglycosidase H. High mannose-containing oligosaccharides such as those found on proteins within the ER are sensitive to digestion, whereas trimmed and modified oligosaccharides (as occurs in the Golgi apparatus) are resistant to digestion. The secretion kinetics can be followed by carrying out a pulse-chase experiment as described below:

1. Cells are grown to 90% confluence in 100-mm dishes, washed twice with methionine- and cysteine-free medium, and then preincubated in the same medium for 20 min at 37°C. Each dish is labeled with 20 μCi of ^{35}S-labeled methionine and cysteine/mL in a final volume of 5 mL of prewarmed medium at 37°C. Label cells using the optimized labeling conditions as determined previously.
2. Labeling is terminated by adding 5 mL of pre-warmed medium with 5 m*M* each of unlabeled methionine and cysteine. Cycloheximide (500 μ*M*) is usually included in the chase medium to block completion of labeled nascent chains. Cells are incubated in the chase medium for various lengths of time (usually up to 3 h but perhaps longer depending on the protein).
3. Lyse cells and immunoprecipitate cell lysate as above. The medium from each chase time is also immunoprecipitated to determine whether protein has been secreted. Usually 1 mL of medium is diluted with 0.5 mL of immunoprecipitation buffer to a final volume of 1.5 mL and immunoprecipitation carried out as described above.

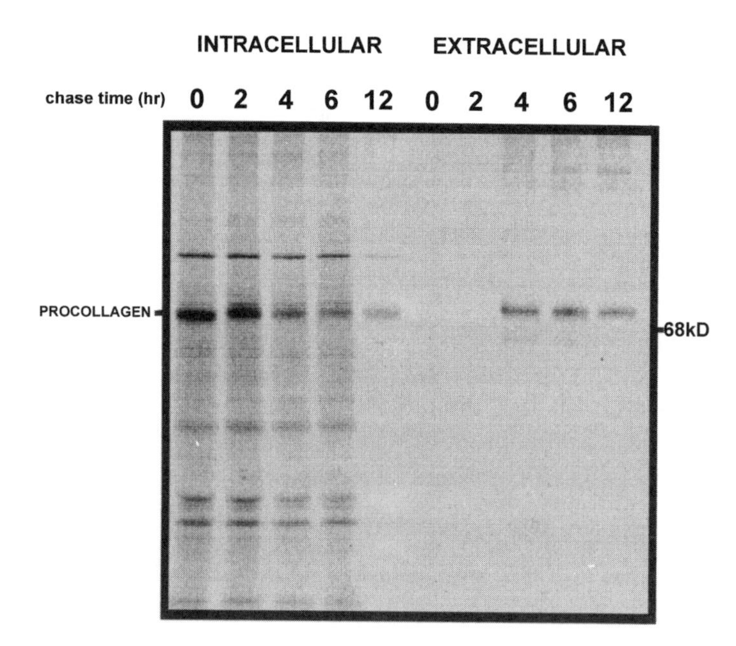

Fig. 1. Pulse-chase analysis of intracellular and extracellular procollagen synthesized by a stable cell line. Cells were labeled for 15 min with 20 μCi/mL [^{35}S]-methionine and chased in the presence of excess cold methionine for the times indicated. Intracellular procollagen chains were immunoprecipitated from cell lysates using a polyclonal antibody raised against type III procollagen. Extracellular procollagen chains were immunoprecipitated from the medium. Samples were separated under reducing conditions by SDS-PAGE, and dried gels were analyzed by autoradiography.

4. For endoglycosidase H digestion, after the immunoprecipitates have been washed the final pellet is resuspended in 15 μL of 50 m*M* Tris-HCl buffer, pH 8.0, containing 1% w/v SDS and boiled for 5 min. An equal volume of endoglycosidase H digestion buffer (150 m*M* sodium citrate, pH 5.5 ,containing 0.5 m*M* PMSF and 0.02% w/v sodium azide) is added to each sample, which is then incubated for 16 h with 1 mU endoglycosidase H at 37°C. The reaction is terminated by adding an equal volume of SDS-PAGE sample buffer containing 50 m*M* DTT (*see* **Note 2**).

A secretion time-course for procollagen expressed in an HT1080 cell line is illustrated in **Fig. 1**. Here the appearance of immunoprecipitated procollagen in the medium coincides with a decrease in the amount of procollagen in the cell lysate, indicating secretion.

3.3 Kinetics of Folding

To ascertain the folding dynamics of the expressed protein, you need to have an assay that determines conformation. Simply running the immunoprecipitate

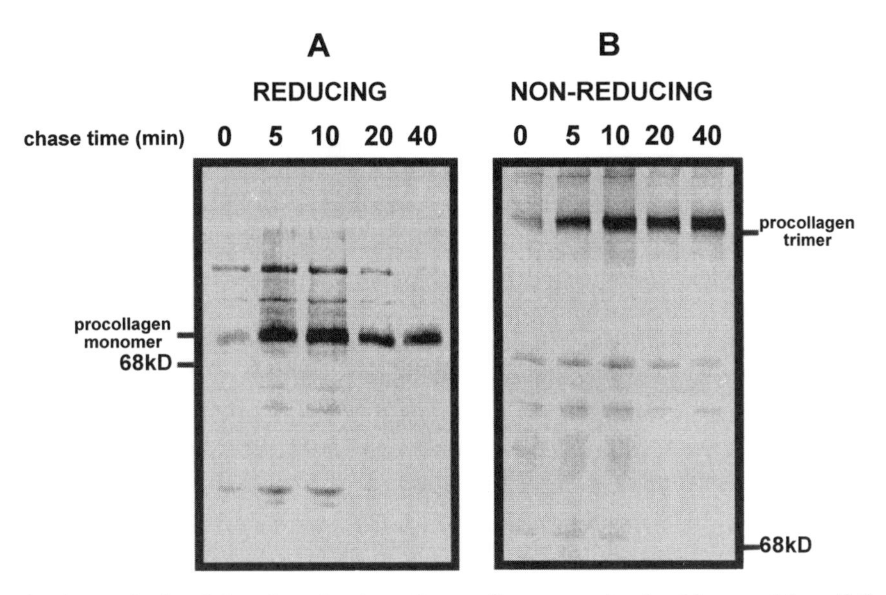

Fig. 2. Analysis of the trimerization of procollagen synthesized by a stable cell line. Labeling of procollagen and immunoprecipitation was carried out as described in **Fig. 1**. Samples were separated under reducing **(A)** or nonreducing **(B)** conditions by SDS-PAGE, and dried gels were analyzed by autoradiography.

under nonreducing conditions will give you some indication of whether disulfide bonds have formed *(5–8)*. A faster mobility relative to the reduced protein indicates the formation of intrachain disulfide bonds, as the protein has a lower hydrodynamic volume *(9)*, whereas slower migrating products would indicate the formation of one or more interchain disulfides. This approach is illustrated in **Fig. 2**, which monitors disulfide bond formation in procollagen by separating immunoprecipitated products either before or after reduction. The appearance of a higher molecular weight species when the material is separated without prior reduction indicates that interchain disulfide bonds have formed and that the polypeptide chains have folded correctly and assembled into trimeric structure stabilized by covalent bonds.

Another convenient way of assessing folding is to use a conformation-specific antibody. Certain antibody epitopes only mature once the protein has folded to assume a native conformation. If such an antibody exists for your protein then a simple immunoprecipitation will indicate folding. A pulse-chase time-course could be carried out as described in **Subheading 3.2.** followed by immunoprecipitation with both conformation- and non-conformation-specific antibodies to determine the relationship between protein synthesis and folding. Other methods of assessing the folding state of proteins involve acquisition of resistance to digestion with proteases, which often occurs as a protein folds to its native state. One example of this approach is the use of trypsin digestion to demonstrate the folding of a collagen triple helix ***(10)***. Alternatively binding to specific ligands would also be indicative of correct folding, for example, the binding of a protease to its inhibitor. An example of such an interaction occurs when tissue-type plasminogen activator binds to its inhibitor ***(11)***. Such a complex could then be co-immunoprecipitated with antibodies directed against the ligand.

An alternative approach to studying the folding dynamics of recombinant proteins is to determine the effect of exogenously added reagents that perturb the normal folding pathway. These are often reversible, so that folding can be prevented during the labeling period, for example, and then initiated posttranslationally, thereby effectively synchronizing folding. A reducing agent such as DTT, usually used at 5 m*M*, which prevents the formation of disulphide bonds, is an example of a reversible agent ***(12,13)***. The *N*-linked glycosylation inhibitor tunicamycin is an example of an irreversible agent that will provide information on the effect of a particular modification, in this case glycosylation, on folding ***(14)***. This inhibitor blocks the addition of *N*-glycans by interfering with the synthesis of the lipid-linked precursor oligosaccharide.

Once a time-course for folding and secretion has been established for a specific protein using the pulse-chase methodology, comparisons can be made regarding expression of a particular protein in different cell types or within the same cell type at different stages of culture. This provides valuable information that will enable the best host cell type to be identified for the expression of biologically active recombinant protein.

4. Notes

1. For the use of protein A Sepharose, if the chosen antibody is of a class that does not bind to protein A (for example, if it was raised in rat) then you may need to add a secondary antibody such as rabbit antirat IgG to ensure quantitative binding. Most preparations of protein A Sepharose have a binding capacity of 100 μg IgG/50 μL 10% suspension of protein A Sepharose. Serum concentrations of IgG are typically 10–15 mg/mL.
2. For the use of cycloheximide, an increase in the amount of radiolabeled protein may occur for a few minutes after addition of the chase medium and cycloheximide due to completion of labeled nascent chains. Therefore, it is not unusual for the first time point of the chase to have more radiolabeled protein than at time zero ***(5)***.

References

1. Bendig, M. M. (1988) The production of foreign proteins in mammalian cells. *Genet. Engin.* **7,** 91–127.

2. Jackson, C. M. and Nemersom, Y. (1980) Blood coagulation. *Annu. Rev. Biochem.* **49,** 765–811.
3. Parekh, R. B., Dwek, R. A., Thomas, J. R., Opdenakker, G., and Rademacher, T. W. (1989) Cell-type-specific and site-specific N-glycosylation of type I and type II human tissue plasminogen activator. *Biochemistry* **28,** 7644–7662.
4. Harlow, E. and Lane, D. (1988) *Anitibodies: A Laboratory Manual,* Cold Spring Harbor Laboratory, New York.
5. Braakman, I., Hoover-Litty, H., Wagner, K. R., and Helenius, A. (1991) Folding of influenza hemagglutinin in the endoplasmic reticulum. *J. Biol. Chem.* **114,** 401–411.
6. de Silva, A., Braakman, I., and Helenius, A. (1993) Posttranslational folding of vesicular stomatis virus G protein in the ER: involvement of covalent and non-covalent complexes. *J. Biol. Chem.* **120,** 647–655.
7. Braakman, I., Helenius, J., and Helenius, A. (1992) Role of ATP and disulphide bonds during protein folding in the endoplasmic reticulum. *Nature* **356,** 260–262.
8. Bulleid, N. J., Wilson, R., and Lees, J. F. (1996) Type III procollagen assembly in semi-intact cells: chain association, nucleation and triple-helix folding do not require formation of inter-chain disulphide bonds but triple-helix nucleation does require hydroxylation. *Biochem. J.* **317,** 195–202.
9. Goldenberg, D. P. and Creighton, T. E. (1984) Gel electrophoresis in studies of protein folding and conformation. *Anal. Biochem.* **138,** 1–18.
10. Bruckner, P. and Prockop, D. J. (1981) Proteolytic enzymes as probes for the triple-helical conformation of procollagen. *Anal. Biochem.* **110,** 360–368.
11. Spengers, E. D. and Kluft, C. (1987) Plasminogen activator inhibitors. *Blood* **69,** 381–387.
12. Braakman, I., Helenius, J., and Helenius, A. (1992) Manipulating disulfide bond formation and protein folding in the endoplasmic reticulum. *EMBO J.* **11,** 1717–1722.
13. Allen, S., Naim, H. Y., and Bulleid, N. J. (1995) Intracellular folding of tissue-type plasminogen activator: effect of disulphide bond formation on N-linked glycosylation. *J. Biol. Chem.* **270,** 4797–4804.
14. Elbein, A. D. (1987) Inhibitors of the biosynthesis and processing of N-linked glycosylation. *Annu. Rev. Biochem.* **56,** 497–534.

21

Control of Proteolysis in Cell Culture

Use of Inhibitors and Engineered Cell Lines

Maria G. Castro, Simon Windeatt, and Marcelo J. Perone

1. Introduction

Mammalian cells are increasingly used as hosts for the expression of recombinant proteins. This has been facilitated by the development of better methods of gene transfer and the use of strong promoters. The advantage of using mammalian cells as hosts for the expression of recombinant proteins from higher eukaryotes stems from the knowledge that the signals for synthesis, posttranslational processing, and secretion of these proteins are correctly and efficiently recognized in mammalian cells. If large amounts of protein will be required for purification purposes, the culture system should be scaled up to produce large amounts of cells or supernatant. The most commonly used systems are roller bottle production, growth in spinners, and fermenters.

The purification of recombinant proteins from mammalian cells can be significantly hampered by unwanted proteolysis. The disruption of cells during purification and analytical procedures often releases proteinases, which can cause proteolysis. Protocols have to be designed to minimize any endogenous proteolytic activity as it is impossible to remove all putative proteinases present. **Table 1** provides suggestions that may help to prevent unwanted proteolysis during sample preparation.

1.2. Inhibition of Proteolysis in Tissue Culture

1.2.1. Use of Proteinase Inhibitors

A widely used and successful approach toward preventing unwanted proteolysis during protein isolation from animal cells in culture is to include proteinase inhibitors during sample preparation, characterization, and purification

From: *Methods in Biotechnology, Vol. 8: Animal Cell Biotechnology*
Edited by: N. Jenkins © Humana Press Inc., Totowa, NJ

Table 1
Common Techniques for the Prevention of Proteolysis in Samples

Component	Notes
Low temperature	The activity of proteinases is minimized. Samples should be stored frozen at –80°C
Time	There is less chance of proteolysis occurring if shorter procedure times are used. For longer techniques such as chromatography or dialysis, the inclusion of proteinase inhibitors is recommended
pH and buffers	The use of a pH that is not optimum for proteinase activity will reduce proteolysis. Usually, above neutral pH, lysosomal enzymes will not be active. Some buffers, e.g., phosphate, have been shown to have a stabilizing effect ***(13)***
Activators omitted	Exclude any proteinase activators from the extraction buffer such as divalent cations (Ca^{2+} and Mg^{2+}) or by employing chelating agents (EDTA)
Added protein	The use of proteins such as bovine serum albumin to provide an alternative substrate for the proteinases can provide protection, although additional steps need to be included for the removal of the added protein
Stabilizing substances	The inclusion of 10% DMSO or 15–35% glycerol can help prevent unwanted proteolysis during preparation and storage. Reducing agents such as DTT (0.1–1 m*M*) or mercaptoethanol (1–10 m*M*) will diminish inhibitor effectiveness and activate cysteine proteinases, but they ensure that free thiol groups are not oxidized

procedures. This section will explain which inhibitors are available, how they are prepared, and what to consider before deciding on the best inhibitors (**Table 2**). The best results are usually obtained using a cocktail of several inhibitors. The most commonly used are phenylmethanesulphonyl fluroide (PMSF), EDTA, benzamidine, leupeptin, tosyl lysyl chloromethyl ketone (TLCK), tosyl phenylalanyl chloromethylketone (TPCK), pepstatin, and either soybean trypsin inhibitor (SBTI) or aprotinin at the final concentrations indicated in Table 2 (*see* **Note 1**).

1.2.2. Removal of Proteinases in Tissue Culture

Choosing cell lines that have low endogenous proteinase activity can avoid some of the problems related to unwanted proteolysis. There are a variety of mechanisms by which the level of proteinases can be controlled:

Table 2
Proteolytic Inhibitors

Proteinase class	Inhibitor	Stock solution	Final concentration
Metallo-proteinases	EDTA	0.5 *M* in water, pH 7.0	5 m*M*
	1,10-phenanthroline	0.1 *M* in DMSO	0.1 m*M*
General	3,4-DCI	5 m*M* in DMSO	0.1 m*M*
serine proteases	SBTI or Aprotinin	0.1 mg/mL in aqueous solution	1 µg/mL
	PMSF	0.1 *M* in 100% ethanol	1 m*M*
Trypsin-like serine proteases	Leupeptin	0.1 *M* in water	1–100 m*M*
	TLCK	10 m*M* in aqueous solution	10–100 µM
	Benzamidine	0.1 *M* in aqueous solution	1 m*M*
	Antipain	0.1 *M* in water	1–100 m*M*
Chymotrypsin-like serine proteases	TPCK	10 m*M* in methanol or ethanol	10–100 µM
Cysteine proteases	Iodoacetic acid	1 *M* in water	10 m*M*
	E-64	1 m*M* in water	0.02 m*M*
Aspartic proteases	Pepstatin	1 m*M* in methanol	1 µM

1. The stationary phase of mammalian cell cultures could trigger cell starvation and therefore stimulate cellular proteinase synthesis and activity if harvesting of the cells is delayed.
2. Gentle procedures for the disruption and subcellular fractionation of mammalian cells can prevent rupture of organelles such as lysosomes and vacuoles, which contain a large proportion of proteolytic enzymes and therefore minimize proteolysis.
3. Harsh detergents such as Triton X-100 should be avoided where possible and isotonic solutions (e.g., 0.25 *M* sucrose) should be used for homogenization to prevent organelle rupture. Another alternative is to use digitonin at 0.8 mg/mL to perforate the plasma membrane without disturbing the membranes surrounding the organelles. Differential centrifugation can then be used to remove the fraction containing the lysosomes and vacuoles.

4. Procedures should be carried out at 4°C to minimize the risk of organelle rupture and proteolytic activity.
5. Proteinases can be removed by incorporating specific affinity techniques into the purification procedure. Many of the adsorbents used for this purpose are immobilized proteinase inhibitors or substrates. The choice of procedure will depend on the proteinase to be removed. The main disadvantage of using this technique is the high cost, which therefore limits the scaling up of these procedures.

1.2.3. Monitoring Proteolytic Activity in Mammalian Cell Cultures

Various techniques can be employed to evaluate the synthesis, purification, and proteolysis of a given recombinant protein from mammalian cell lines. Detection methods that allow determination of the presence of a protein within cells or in their conditioned medium include immunohistochemistry, radioimmunoassay, or sodium dodecyl sulfate-polyacrylamide gel electrophoresis (SDS-PAGE) followed by coomassie blue or silver staining. Western blotting can also be used to evaluate proteolytic activity since it allows the identification of the desired proteins using specific antibodies. Proteins separated using SDS-PAGE can be transferred to a nitrocellulose membrane or other hydrophobic membranes. The membrane is incubated with an appropriate dilution of the primary antibody against the protein of interest. The membrane and protein-antibody complexes are visualized by a color reaction catalyzed enzymatically.

To evaluate proteolytic activity, incubation of conditioned medium from cell lines expressing proteases together with their respective substrates may be employed, following detection of the products by high-performance liquid chromatography, fluorography, and radioimmunoassay.

Radiolabeling followed by immunoprecipitation and SDS-PAGE followed by autoradiography are commonly used to assess the integrity of recombinant proteins generated within mammalian cell lines. The incorporation of [^{35}S]Met and [^{35}S]Cys is commonly employed, because these are essential amino acids that can be radiolabeled with ^{35}S. They can be easily detected using autoradiography once they are incorporated into proteins ***(1)***. It is essential that the protein to be labeled contain Met and Cys residues, and the intracellular pool of these amino acids must be depleted by incubation of cells with Met- and Cys-free medium to obtain high incorporation of labeled amino acids.

1.2.4. Generation of Engineered Cell Lines with Specific Protease Activity

A wide variety of proteins of medical and pharmaceutical importance are synthesized in vivo as precursors that are biologically inactive, e.g., proinsulin, prosomatostatin, pro-glial cell-derived neurotrophic factor, and pro-nerve growth factor ***(2)***. The biologically active peptides are produced by endopro-

Table 3
Examples of Amino Acid Sequences Recognized by Mammalian Subtilisin-Like Endopeptidases[a]

Amino Acid Sequence	Endoprotease	Prohormone
K/R-X-X-R↓	PC1	Rat proinsulin I ***(14)***; human proinsulin ***(15)***
P_4-P_3-P_2-P_1	PC2	Rat proinsulin I ***(14)***; POMC ***(16)***
	Furin/PACE	Rat proinsulin I ***(14)***
KR↓	PC1; PC2	POMC ***(17,18)***
P_2-P_1		Proneuropeptide Y and propancreatic polypeptide ***(19)***
		Proglucagon ***(8,20)***
RR↓	PC2	POMC ***(16)***
P_2-P_1	PC1?	POMC ***(16)***
	PC1	Proglucagon ***(8,20)***

[a]The amino acid sequences within the two general classes of processing sites are indicated by their single letter code. Amino acid residues within the substrate cleavage sites are denoted as P_4, P_3, P_2 and P_1 starting from the NH_2-terminal side of the cleavage site. Arrows indicate the site of cleavage.

teolytic cleavage of these proproteins by the action of endopeptidases ***(3)***. The best characterized endopeptidases cleave the proproteins at pairs of dibasic amino acids (**Table 3**). To obtain these endoproteolytic processed products within mammalian cell lines, the cell line of choice should express the appropriate endopeptidases (e.g., prohormone convertase [PC]1 and/or PC2). Most mammalian cells currently used for expression of recombinant proteins do not express PC1 or PC2, e.g., CHO, COS, HeLa, CV-1, and BHK. Since most mammalian high-level expression systems are developed for these cell lines, our laboratory has developed genetically engineered CHO cell lines that express PC1, PC2, and PC1 together with PC2 ***(4)***.

2. Materials

2.1. Radiolabeling and Immunoprecipitation

1. Minimum essential medium eagle (MEM) without L-Met (Sigma, Poole, UK).
2. Boiling buffer: 50 m*M* Na-phosphate, 1% SDS, 40 m*M* β-mercaptoethanol, 2 m*M* EDTA.
3. Specific primary antibody against the molecular species to be immunoprecipitated.
4. Protein A Sepharose CL-4B (PAS; Pharmacia Biotech, St. Albans, UK): 1 g of dried powder gives 4–5 mL of hydrated gel. Prepare a slurry with buffer (50 m*M* Tris, pH 7.0) in a ratio of 75% settled gel to 25% buffer.
5. Protease inhibitor cocktail (PIC): 2 m*M* PMSF and 0.1 μ*M* pepstatin.

6. $Trans^{35}S$-Label™ (ICN, Thame, UK) specific activity >1000 Ci/mmol.
7. Radioimmune precipitation (RIPA) buffer: 150 m*M* NaCl, 0.5% (w/v) deoxycholate, 1% (v/v) Triton X-100, 0.1% (w/v) SDS, 50 m*M* Tris, pH 7.4.

2.2. Western Blotting

1. Nitrocellulose membrane with pore size of 0.45 μm (Schleicher & Schuell, London, UK).
2. Whatman 3 MM filter paper (Maidstone, UK).
3. Trans-Blot apparatus (e.g., Hoefer).
4. Transfer buffer (TB): 25 m*M* Tris, pH 8.3, 14.4 glycine, 200 mL of methanol, 800 mL double-distilled H_2O.
5. Tween Tris-buffer saline (TTBS): 0.1% (v/v) Tween-20 in 100 m*M* Tris, pH 7.5, 0.9% NaCl.
6. Dried skimmed milk (Marvel, Premier Beverages, Stafford, UK).

2.3. Establishing Stably Transfected CHO-K1 Cells Expressing Prohormone Converting Enzymes

1. CHO-K1 (ECACC no. 85051005) cells growing in 10-cm-diameter Petri dishes (Greiner, Stonehouse, UK) at 70–80% confluence.
2. Plasmid DNA encoding PCs (provided by S. Smeekens, Howard Hughes Medical Institute Research Laboratories, University of Chicago, IL) *(3)* purified using Qiagen Maxi columns (Qiagen, Crawley, UK) was dissolved in sterile H_2O or Tris-EDTA, pH 8.0 (10 m*M* Tris-HCl; 1 m*M* ethyldiaminetetra-acetic acid [EDTA]). The eukaryotic expression vector pEE6hCMVneo (from Celltech, Slough, UK) possessing the constitutive human cytomegalovirus promoter (hCMV) and the selectable marker neomycin was employed. PC1 or PC2 were inserted in the *Eco*RI site of the polylinker downstream of the hCMV promoter (**Fig. 1**).
3. 2× HEPES-buffered saline (2× HBS): mix 8.18 g NaCl; 5.95 g HEPES-free acid, and 0.2 g Na_2HPO_4 up to 500 mL distilled water. Adjust pH to 7.1 with approximately 5 mL 1 *M* NaOH *(The pH of this solution is critical!)*. Bring the volume to 500 mL, filter-sterilize (0.2-μm filter) and store at 4°C.
4. 2 *M* $CaCl_2$; filter-sterilize and store at 4°C.
5. 15% glycerol-HBS: mix 15 mL glycerol, 50 mL 2× HBS, and 35 mL distilled water. Filter-sterilize and store at 4°C.
6. Nonselective growth medium: mix 500 mL Dulbecco's modified Eagle's medium (DMEM); GIBCO, Paiseley, UK) without sodium pyruvate, 50 mL horse serum, 25 mL newborn calf serum, 10 mL sodium pyruvate 100 m*M*, 6 mL 100× nonessential amino acids, 6 mL 5000 U penicillin, and 6 mg/mL streptomycin.
7. DMEM serum-free: DMEM without serum and supplemented with all the additives (described in **item 6**).
8. Selective medium: DMEM supplement with 50 mL newborn calf serum, 10 mL sodium pyruvate 100 m*M*, 6 mL 100× nonessential amino acids, 6 mL 5000 U penicillin, 6 mg/mL streptomycin, and 0.5–1 mg/mL of the selectable drug G418 (GIBCO).

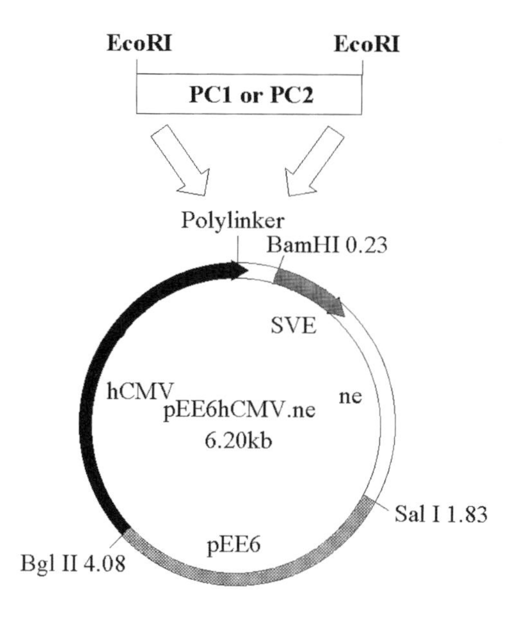

Fig. 1. Subcloning of the prohormone convertases (PC)1 or PC2 into the *Eco*RI restriction site within the polylinker region of the eukaryotic expression vector PEE6hCMVne ***(21)***. The ne expression cassette (0.23–1.83) encodes for the selectable marker neomycin. Promoters are shown in black. The hatched segment is the SV40 early polyadenylation signal. The bacterial plasmid pEE6 ***(22)*** contains an ampicillin-resistance gene. (Modified with permission from **ref. *13***).

3. Methods

3.1. Radiolabeling and Immunoprecipitation

1. Seed cells in a 25-cm^2 flask until they reach 70–80% confluency (*see* **Note 2**).
2. Wash cells once with sterile phosphate-buffered saline (PBS) and once with Met/Cys-free medium.
3. Add 3 mL of Met/Cys-free medium and incubate for 1 h.
4. Label cells for 90 min with 0.2 mCi of ^{35}S in 1 mL Met/Cys-free medium.
5. Remove medium and centrifuge at 1,000*g* for 10 min at 4°C; discard debris.
6. To the supernatant add EDTA (20 m*M* final concentration), boil for 5 min, and add PIC. Keep in an ice-water bath until immunoprecipitation.
7. Cell lysis: wash cells with PBS. Scrape in 1 mL PBS and centrifuge (200 *g* for 10 min). To the cellular pellet, add 50 μL of boiling buffer and boil cells for 5 min. Centrifuge at 10,000 *g* for 10 min at 4°C; discard pellet (debris). Dilute the supernatant two-fold with RIPA buffer containing PIC (*see* **Note 3**).
8. To preclear conditioned medium from **step 6** and cell lysate from **step 7**, add 20 μL and 4 μL of normal rabbit serum (NRS), respectively. Incubate for 1 h at 4°C on

a rotating wheel. Then add 100 μL and 30 μL of PAS to conditioned medium and cell lysate, respectively. Incubate for 1 h at 4°C on a rotating wheel. This step is optional and is advisable to reduce nonspecific immunoprecipitation.

9. Centrifuge (10,000 *g* for 1 min) and keep supernatants.
10. To the supernatants, add the appropriate dilution of the primary antibody (usually 1:10 dilution) and incubate overnight at 4°C on a rotating wheel.
11. Add 100 μL of PAS to the conditioned medium and 30 μL to the cell lysate; incubate for 2 h at 4°C on a rotating wheel.
12. Centrifuge (10,000 *g* for 2 min at 4°C) and keep supernatants (*see* **Note 4**).
13. Wash the PAS-immunocomplex twice with 1 mL RIPA buffer containing PIC and once with 1 mL PBS. Centrifuge at 10,000 *g* for 2 min and keep the supernatants and pellets.
14. Resuspend pellets in 40 μL of SDS-PAGE buffer; boil for 10 min and centrifuge at 10,000 *g* for 2 min. Samples are resolved using SDS-PAGE ***(5)***.
15. After SDS-PAGE, fix proteins by immersing the gel in isopropanol (25%)/water (65%)/acetic acid (10%) for 30 min with continuous agitation.
16. Soak the gel in the fluorographic reagent (Amplify; Amersham, Little Chalfont, UK) for 30 min in preparation for autoradiography.
17. Dry gel at 60–80°C for 3 h.
18. Expose to X-ray film (Kodak Biomax MR) for 24–72 h at –70°C . Develop X-ray film.

3.2. Western Blotting

1. Cut the nitrocellulose membrane and six pieces of Whatman 3 MM filter paper to the same size of the gel. Soak gel, nitrocellulose, and filter paper in TB.
2. Construction of the blot sandwich: place three filter papers on the positive (+) electrode of the blotting apparatus. Lay the wet nitrocellulose onto the filter papers. Place the polyacrylamide gel onto the nitrocellulose and the three filter papers on it (these papers are in contact with the negative [–] electrode). *Carefully remove bubbles from the sandwich.*
3. Transfer for 3 h at room temperature applying approximately 1 mA/cm^2 area of the gel. Usually, a current of 120 mA for a 12 × 12-cm gel for 3 h brings a good transfection for proteins between 5 and 100 kDa.
4. Disassemble the blot sandwich and immerse the nitrocellulose membrane in TTBS. Wash three times with TTBS.
5. Incubate the membrane with TTBS containing 1% (w/v) dried skimmed milk at 4°C for 5–6 h.
6. Invert membrane over 5–10 mL of the appropriate dilution of the primary antibody in TTBS containing 0.5% (w/v) dried skimmed milk and incubate for 3–4 h at 4°C.
7. Wash the membrane with three to four changes of TTBS over 15 min with gentle agitation.
8. Transfer the membrane to a dilute solution of biotinylated secondary antibody (Vectastin® ABC kit; Vector, Burlingame, CA) in TTBS. This solution can be prepared as outlined in the Vectastin ABC kit instructions (1 drop

into 10 mL of buffer) but using TTBS as diluent. Incubate for 30 min with gentle agitation.

9. Wash as in **step 7**.
10. Transfer the membrane to the Vectastin ABC-AP reagent. Incubate the membrane in this solution for 30 min with gentle agitation.
11. Wash as in **step 7**.
12. Transfer the membrane to the substrate solution. Although a variety of alkaline phosphatase substrate systems can be employed, we use the Vector® Alkaline Phosphatase Substrate Kit II, cat. no. SK-5200, which produces a black reaction product. Follow the directions included in the substrate kit to prepare the substrate solution. Incubation in substrate solution should be for 15–30 min or until suitable color develops.
13. Wash with two changes of distilled water over 10 min and allow the membrane to air-dry. Air-drying removes a slight color from the nitrocellulose but does not reduce the specific staining. Membranes should be stored in the dark.

3.3. Calcium-Phosphate Mediated Transfection

1. A day before the transfection, seed the cells to be transfected in 10-cm-diameter Petri dishes to obtain 70–80% of cell confluence the following day. Before transfection, bring all solutions to room temperature and warm growth medium to 37°C. Keep sterile conditions working in a laminar flow cabinet.
2. Mix 62 µL of 2 *M* NaCl and 10–20 µg of supercoiled DNA vector; a 15-mL sterile tube may be appropriate. Bring the volume to 0.5 mL with Tris-EDTA, pH 8.0. This must be done for each dish to be transfected. Include a mock transfection solution in which Tris-EDTA is added instead of the DNA (*see* **Note 5**).
3. Add 0.5 mL 2× HBS to a second sterile 15-mL tube.
4. Form a precipitate by adding the DNA in $CaCl_2$ to the 2× HBS solution (*see* **Note 6**).
5. Aspirate the medium from the Petri dishes and wash the cells once with 10 mL of DMEM serum-free medium.
6. Add the DNA-Ca phosphate precipitate together with 1 mL serum-free medium to the cells in the Petri dishes.
7. Incubate the dishes at 37°C for 4 h, rocking them occasionally to ensure that they do not dry out.
8. Remove the DNA-Ca phosphate precipitate from the cells by aspiration.
9. Add 3 mL 15% glycerol-HBS and incubate at 37°C for 2 min (*see* **Note 7**).
10. Add 10 mL DMEM serum-free medium and immediately remove by aspiration.
11. Add 10 mL DMEM *without* selectable markers and return dishes to the incubator.

3.4. Selection and Recovery of Resistant Cell Colonies

1. After 24 h of incubation, replace the medium with 10 mL of selection medium.
2. Feed with appropriate selective medium after 3–4 days and continue feeding until all the cells in the mock transfection are dead. Resistant colonies should appear after 2–3 weeks, depending on the selectable marker employed.
3. Mark the colonies to be picked with a marker pen on the underside of the dish and soak squares of filter papers (7 mm^2) in a sterile trypsin-EDTA solution.

4. Wash the colonies by aspirating the selectable medium and adding 10 mL (calcium/magnesium-free) sterile PBS.
5. Using sterile forceps, lay one trypsin-soaked filter paper on top of each resistant colony marked and return to the incubator at 37°C for 1 min.
6. With forceps remove each filter paper and place immediately into separate 35-mm-diameter dishes containing nonselective media. Swirl dishes gently to detach cells from the filter paper and return to the incubator (*see* **Note 8**).
7. After 24 h of incubation, feed cells with selective medium and replace it every 3–4 days until 70–80% confluence is reached. At this point cells need to be transferred to large culture flasks (e.g., 12.5-cm^2 flasks; Greiner, Stonehouse, UK) after trypsinization using 200 µL for each well, incubate for 1 min at 37°C, and immediately add growth medium.
8. After screening and selection of the best clone, expand the clone by splitting flasks one in five (*see* **Note 9**).

4. Future Trends: Generation of Mutant Cell Lines with Reduced Proteolytic Activity

Because protease inhibitors possessing specific and nontoxic characteristics are unavailable, the search for alternative methods to control protease activity in mammalian cell cultures has become increasingly necessary.

A possible strategy is to inhibit a protease by introducing the cDNA encoding for the antisense orientation, resulting in intracellular depletion of the chosen protease. The establishment of stably transfected mammalian cell lines expressing protease antisense mRNA *(7)* can be useful to block undesired proteolytic activity. However, stably transfected mutant cell lines expressing antisense mRNA for specific proteases seem to be unstable *(8)*.

An alternative elegant methodology could be to insert the cDNA encoding for peptide protease inhibitors under the control of inducible promoters. The expression of a given protease within stably transfected mammalian cell lines expressing such regulatable cassettes could be turned on or off by manipulating inducing factors, which control the expression of the antisense proteinase cDNA *(9)*.

Dexamethasone-, metallothionine-, and tetracycline-inducible expression vectors for mammalian cells are now available from commercial sources, e.g., Clontech or Invitrogen. These could also be used to express the precursor form of the neuroendocrine polypeptide 7B2, which has been shown to inhibit specifically an endopeptidase that cleaves precursor forms of hormones, neuropeptides, and growth factors (PC2; **refs.** ***3*** and ***10***). This approach could also be used to express tissue inhibitor of metalloproteinases (TIMP), an endogenous inhibitor of extracellular metalloproteinase activity. TIMP inhibits matrix metalloproteinases such as collagenase (MMP1), gelatinase (MMP2), and stromelysin (MMP3; **ref.** ***11***). Ovostatin, another proteinase inhibitor present in the chicken egg, only seems to inhibit metalloproteinases even though it is structurally related to μ-macroglobulin ***(12)***.

5. Notes

1. An important factor to take into account is that some inhibitors are not soluble in water-based solvents, and stock solutions are prepared in organic solvents. Stock solutions should be as concentrated as possible to avoid both the addition of high concentrations of organic solvent and dilution of the sample. Some inhibitors are highly toxic; therefore special precautions are required for their handling (e.g., di-isopropylfluorophosphate or PMSF). These should be weighed out in a fume cupboard.
2. Cells should be at logarithmic growth phase to increase protein synthesis and radiolabel incorporation.
3. Addition of EDTA should be avoided for proteases that do not require divalent cations for their activity. Boiling samples inactivates the proteases completely.
4. The supernatants obtained from **step 13** can be subjected to successive immunoprecipitations with different antibodies starting from **step 9**, in **Subheading 3.1.** In our hands, up to three consecutive immunoprecipitations of different molecular species can be achieved successfully.
5. In our experience, employing between 10 and 20 μg of DNA yields high transfection efficiency for CHO cells; lowering this amount results in a substantial decrease in the number of clones expressing PCs. Also, using >20 μg of DNA is detrimental for the transfection efficiency, probably because of a poor quality of DNA-Ca phosphate precipitate.
6. *The quality of the precipitate is another critical step in this method.* Although a good precipitate can be achieved by gentle agitation for 10 s using a vortex, we have dramatically improved the transfection efficiency through making the DNA-Ca-phosphate precipitate by bubbling air through the 2× HBS solution using a sterile Pasteur pipet connected to a pipette-aid. Bubbles should be made at a frequency of 1–2 bubbles/s. Add the DNA-$CaCl_2$ solution drop wise into the 2× HBS solution. Maintain the bubbling during 10–20 s after all DNA-$CaCl_2$ solution has been added; a fine-grained translucent precipitate should appear. Leave it for 10 min at room temperature.
7. An improvement in the transfection efficiency can be achieved either by incubating for longer periods of time or by increasing the glycerol concentration up to 25%. Because these drastic conditions could cause cellular death, these modifications must be adjusted for each particular cell line.
8. No more than six resistant colonies should be trypsinized at any one time; overexposure to trypsin could cause irreversible cell damage.
9. We have also generated a CHO cell line co-expressing both PC1 and PC2 after two consecutive rounds of Ca-phosphate-mediated DNA transfection. In this case, the expression vectors encoding each PC required different selectable markers for selection of stably transfected cell lines. We obtained high levels of expression of both PC1 and PC2 after transfection of a CHO cell line already expressing

PC2 (a gift of Dr. Lindberg, Louisiana State University, New Orleans, LA: **ref. *6***) with pEE6hCMV neo/PC1. Co-transfected CHO cells are grown in medium containing both methotrexate and G418.

Acknowledgments

The work described in this chapter is supported by the BBSRC (Chemicals and Pharmaceuticals Directorate). S. W. is the recipient of a BBSRC studentship.

References

1. Dorner, A. J. and Kaufman, R. J. (1990) Analysis of synthesis, processing, and secretion of protein expressed in mammalian cells. *Methods in Enzymology* **185,** 577–596.
2. Laemmli, J. H. (1970) Cleavage of structural proteins during assembly of the head of bacteriophage T4. *Nature* **227,** 680–685.
3. Castro, M. G. and Morrison, E. (1997) Posttranslational processing of proopiomelanocortin in the pituitary and in the brain. *Crit. Rev. Neurobiol.* **11(1),** 35–57.
4. Smeekens, S. P. (1993) Processing of protein precursors by a novel family of subtilisin-related mammalian endoproteases. *Biotechnology* **11,** 182–186.
5. Perone, M. J., Ahmed, I., Linton, E. A., and Castro, M. G. (1996) Procorticotrophin releasing hormone is endoproteolytically processed by the prohormone convertase PC2 but not PC1 within stably transfected CHO-K1 cells. *Biochem. Soc. Trans.* **24,** 497S.
6. Shen, F.-S., Seidah, N. G., and Lindberg, I. (1993) Biosynthesis of the prohormone convertase PC2 in Chinese hamster cells and rat insulinoma cells. *J. Biol. Chem.* **268,** 24,910–24,915.
7. Bloomquist, B. T., Eipper, B. A., and Mains, R. E. (1991) Pro-hormone converting enzymes: regulation and evaluation of function using antisense RNA. *Mol. Endocrinol.* 5, 2014–2024.
8. Rouille, Y., Westermark, G., Martin, S. K., and Steiner, D. F. (1994) Proglucagon is processed to glucagon by prohormone convertase PC2 in αTC1-6 cells. *Proc. Natl. Acad. Sci. USA* **91,** 3242–3246.
9. Gossen, M. and Bujard, H. (1992) Tight control of gene expression in mammalian cells by tetracycline-responsive promoters. *Proc. Natl. Acad. Sci. USA* **89,** 5547–5551.
10. Martens, G. J. M., Braks, J. A. M., Eib, D. W., Zhou, Y., and Lindberg, I. (1994) The neuroendocrine polypeptide 7B2 is an endogenous inhibitor of prohormone convertase PC2. *Proc. Natl. Acad. Sci. USA* **91,** 5784–5787.
11. Denhardt, D. T., Feng, B., Edwards, D. R., Cocuzzi, E. T., and Malyankar, U. M. (1993) Tissue inhibitor of metalloproteinases (TIMP, aka EPA): structure, control of expression and biological functions. *Pharmacol. Ther.* **59,** 329–341.
12. Nagase, H., Brew, K., and Harris, E. D. Jr. (1985) Ovostatin: a proteinase inhibitor in egg white that is homologous to α_2-macroglobulin. *Prog. Clin. Biol. Res.* **180,** 283–285.

13. Scott, R. E., Lam, K. S., and Gaucher, G. M. (1986) Stabilisation and purification of the secondary metabolite specific enzyme, m-hydroxybenzylalcohol dehydrogenase. *Can. J. Microbiol.* **32,** 167–175.
14. Smeekens, S. P., Albiges-Rizo, C., Carroll, R., Martin, S., Ohagi, S., Phillips, L. A., Being, M., Gardner, P., Montag, A. G., Swift, H. H., Thomas, G., and Steiner, D. F. (1992) Proinsulin processing by the subtilisin-related pro-protein convertases furin, PC2 and PC3. *Proc. Natl. Acad. Sci. USA* **89,** 8822–8827.
15. Bailyes, E. M., Shennan, K. I. J., Seal, A. J., Smeekens, S. P., Steiner, D. F., Hutton, J. C., and Docherty, K. (1992) A member of the eukaryotic subtilisin family (PC3) has the enzymic properties of the type I proinsulin-convertase endopeptidase. *Biochem. J.* **285,** 391–394.
16. Benjannet, S., Rondeau, N., Day, R., Chretien, M., and Seidah, N. G. (1991) PC1 and PC2 are proprotein convertases capable of cleaving proopiomelanocortin at distinct pairs of basic residues. *Proc. Natl. Acad. Sci. USA* **88,** 3564–35–68.
17. Jung, L. J. and Scheller, R. H. (1991) Peptide processing and targeting in neural secretory pathway. *Science* **251,** 1330–1335.
18. Rothman, J. E. and Orci, L. (1992) Molecular dissection of the secretory pathway. *Nature* **355,** 409–415.
19. Wulff, B. S., Johansen, T. E., Dalboge, H., O'Hare, M. M. T., and Schwartz, T. W. (1993) Processing of two homologous precursors, pro-neuropeptide Y and pro-pancreatic polypeptide, in transfected cell lines expressing different precursor convertases. *J. Biol. Chem.* **268,** 13,327–13,335.
20. Rothenberg, M. E., Eilertson, C. D., Klein, K., Zhou, Y., Lindberg, I., McDonald, J. K., Mackin, R. B., and Noe, B. D. J. (1995) Processing of mouse proglucagon by recombinant prohormone convertase 1 and immunopurified prohormone convertase 2 *in vitro. Biol. Chem.* **270,** 10,136–10,146.
21. Bebbington, C. R. (1991) Expression of antibody genes in monlymphoid mammalian cells. *Methods* **2,** 136–145.
22. Stephens, P. E. and Cockett, M. I. (1989) The construction of a highly efficient and verasatile set of mammalian expression vectors. *Nucleic Acids Res.* **17,** 7110.

22

Monitoring Recombinant Glycoprotein Heterogeneity

Andrew D. Hooker and David C. James

1. Introduction

Developing protein-based therapeutic agents requires both product quality and consistency to be maintained throughout the development and implementation of the production process. Differences in host cell type, the physiological status of the cell, and protein structural constraints are known to result in variations in post-translational modifications, which can affect the bioactivity, receptor binding, susceptibility to proteolysis, immunogenicity, and clearance rate of a therapeutic recombinant protein in vivo *(1)*. Glycosylation is the most extensive source of protein heterogeneity, and many recent developments in analytical biotechnology have enhanced the ability to monitor and structurally define changes in oligosaccharides associated with recombinant proteins.

Variable occupancy of potential glycosylation sites may result in extensive macroheterogeneity in addition to the considerable diversity of carbohydrate structures that can occur at individual glycosylation sites, often referred to as glycosylation microheterogeneity. Variation within a heterogeneous population of glycoforms may lead to functional consequences for the glycoprotein product. Therefore, precise determinations of product consistency throughout production processes are increasingly being carried out by biotechnology and pharmaceutical industries to meet the requirements set by regulatory authorities such as the US Food and Drug Administration and the Committee for Proprietary Medical Productions *(2)*. Indeed, detailed analyses of recombinant protein glycosylation can be seen as a competitive advantage.

The glycosylation of a recombinant protein product can be examined by

1. Analysis of oligosaccharides released from the polypeptide by chemical or enzymatic means.

From: *Methods in Biotechnology, Vol. 8: Animal Cell Biotechnology*
Edited by: N. Jenkins © Humana Press Inc., Totowa, NJ

2. Site-specific analysis of glycans associated with glycopeptide fragments following proteolysis of the intact glycoprotein.
3. Direct analysis of the whole glycoprotein.

A number of new technologies are currently being applied to provide rapid and detailed analytical analysis of glycan heterogeneity:

1. High-performance capillary electrophoresis (HPCE).
2. High-pH anion-exchange chromatography with pulsed amperometric detection (HPAE-PAD).
3. Matrix-assisted laser desorption/ionisation mass spectrometry (MALDI-MS).
4. Electrospray ionization mass spectrometry (ESI-MS).
5. Enzymatic analysis methods such as the reagent array analysis method (RAAM; 3,4).

In particular, novel mass spectrometric strategies are rapidly continuing to advance the frontiers of biomolecular analysis, with technical innovations and methodologies yielding improvements in sensitivity, mass accuracy, and resolution.

1.1. High-Performance Capillary Electrophoresis

Capillary electrophoresis has been employed in various modes in the high-resolution separation and detection of glycoprotein glycoforms, glycoconjugates, glycopeptides, and oligosaccharides, even though carbohydrate molecules do not absorb or fluoresce and are not being readily ionized ***(5–8)***. A number of approaches have been employed to render carbohydrates more amenable to analysis; these include *in situ* complex formation with ions such as borate and metal cations ***(9)*** and the addition of ultraviolet (UV)-absorbing or fluorescent tags to functional groups ***(10)***.

1.2. Matrix-Assisted Laser Desorption/Ionization Mass Spectrometry

MALDI-MS has been extensively used to determine the mass of proteins and polypeptides, confirm protein primary structure, and to characterize post-translational modifications. MALDI-MS generally employs simple time-of-flight analysis of biopolymers that are cocrystallized with a molar excess of a low molecular weight, strongly UV-absorbing matrix (e.g., 2,5-dihydroxybenzoic acid) on a metal sample disk. Both the biopolymer and matrix ions are desorbed by pulses of a UV laser. Following a linear flight path the molecular ions are detected, the time between the initial laser pulse and ion detection being directly proportional to the square root of the molecular ion mass/charge (m/z) ratio. For maximum mass accuracy, internal and external protein or peptide calibrants of known molecular mass are required. In addition to this linear mode, many instruments offer a reflectron mode which effectively lengthens the flight path by redirecting the ions toward an additional ion detector, which may enhance resolution, at the expense of decreased sensitivity. MALDI-MS is tolerant to low (mM) salt concentrations, can determine the molecular weight of biomolecules in excess of 200 kDa with a mass accuracy of ±0.1%, and is capable of analyzing heterogeneous samples with picomole to femtomole sensitivity. These properties, combined with its rapid analysis time and ease of use for the nonspecialist, have made it an attractive technique for the analysis of glycoproteins, glycopeptides, and oligosaccharides.

1.3. Electrospray Ionization Mass Spectrometry

ESI-MS is another mild ionization method, whereby the covalent bonding of the biopolymer is maintained and is typically used in combination with a single or triple quadrupole. This technique is capable of determining the molecular weight of biopolymers up to 100 kDa with a greater mass accuracy (±0.01%) and resolution (±2000) than MALDI-MS. Multiply charged molecular ions are generated by the ionization of biopolymers in volatile solvents, the resulting spectrum being convoluted to produce noncharged peaks.

ESI-MS has been extensively used for the direct mass analysis of glycopeptides and glycoproteins and is often interfaced with liquid chromatography ***(11–13)*** but has found limited application for the direct analysis of oligosaccharides ***(14)***. ESI-MS is better suited to the analysis of whole glycoprotein populations than MALDI-MS, its superior resolution permitting the identification of individual glycoforms ***(15,16)***.

2. Materials

1. P/ACE 2100 HPCE System (Beckman Instruments, High Wycombe, UK).
2. Phosphoric acid (Sigma, Poole, UK).
3. Boric acid (Sigma).
4. Trypsin, sequencing grade (Boehringer Mannheim, Lewes, UK).
5. Waters 626 Millenium high-performance liquid chromatography (HPLC) System (Millipore, Watford, UK).
6. Vydac 218TP52 reverse-phase column: C18, 2.1 × 250 mm (Hichrom, Reading UK).
7. HPLC grade water/acetonitrile (Fischer Scientific, Loughborough, UK).
8. Alpha-cyano-4-hydroxy cinnaminic acid (Aldrich, Gillingham, UK).
9. VG Tof Spec Mass Spectrometer (Fisons Instruments, Manchester, UK).
10. Vasoactive intestinal peptide, fragments 1–12 (Sigma).
11. Peptide-*N*-glycosidase F and glycosidases (Oxford Glycosystems, Abingdon, UK).
12. 2,5-Dihydroxybenzoic acid (Aldrich).
13. 2,4,6-Trihydroxyacetophenone (Aldrich).
14. Ammonium citrate (Sigma).
15. VG Quattro II triple quadrupole mass spectrometer (VG Organic, Altrincham, UK).
16. Horse heart myoglobin (Sigma).

3. Methods

This chapter describes some of the recent technological advances in the analysis of post-translational modifications made to recombinant proteins and focuses on the application of HPCE, MALDI-MS, and ESI-MS to the monitoring of glycosylation heterogeneity. These techniques, are illustrated by describing their application to the analysis of recombinant human γ-interferon (IFN-γ), a well-characterized model glycoprotein that has N-linked glycans at Asn_{25} and at the variably occupied site, Asn_{97}.

3.1. Glycosylation Analysis by HPCE

This laboratory routinely uses micellar electrokinetic capillary chromatography (MECC) to fingerprint rapidly glycoforms of recombinant human IFN-γ produced by Chinese hamster ovary (CHO) cells ***(8)*** and to quantitate variable-site occupancy (macroheterogeneity; *see* **Fig. 1**). This approach allows glycoforms to be resolved and quantified rapidly without the need for oligosaccharide release, derivatization or labeling.

1. Separations are performed with a P/ACE 2100 capillary electrophoresis system using a capillary cartridge containing a 50 μm id × 57 cm length of underivatized fused silica capillary.
2. Buffer solutions are prepared from phosphoric and boric acids using NaOH to adjust the pH.
3. Capillaries are prepared for use by rinsing with 0.1 *M* NaOH for 10 min, water for 5 min, 0.1 *M* borate, pH 8.5 for 1 h, and then 0.1 *M* NaOH and water for 10 min respectively. Prior to use, capillaries are equilibrated with electrophoresis buffer (400 m*M* borate + 100 m*M* sodium dodecyl sulfate [SDS], pH 8.5) for 1 h.
4. Voltages are applied over a 0.2-min linear ramping period at a detection wavelength of 200 nm and operating temperature of 25°C. Recombinant human IFN-γ (1 mg/mL in 50 m*M* borate, 50 m*M* SDS, pH 8.5) is injected for 5 s prior to electrophoresis at 22 kV. Between each separation, the capillary is rinsed with 0.1 *M* NaOH, water, and electrophoresis buffer for 5 min, respectively (*see* **Note 1**).

3.2. Glycosylation Analysis by MALDI-MS

Only a few reports have been published for the analysis of whole glycoproteins due to the limited resolution of this technique ***(17)***. As a result, analysis of intact glycoproteins is limited to those proteins that preferably contain one glycosylation site and are <15–20 kDa ***(15)***.

MALDI-MS has proved more useful for the identification and characterization of glycopeptides following their separation and purification by reverse-phase HPLC ***(18)***. The advantage of this approach over other methods is that site-specific glycosylation data are obtained (***19,20***; *see* **Note 2**). This approach has been successfully used to determine the differences in N-linked

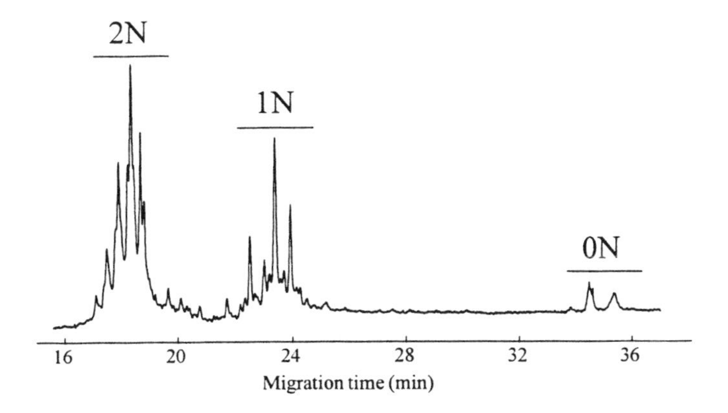

Fig. 1. Whole recombinant human IFN-γ analyzed by capillary electrophoresis. Recombinant human IFN-γ glycoforms were fingerprinted by micellar electrokinetic capillary chromatography. Peak groups represent IFN-γ variants with both Asn sites occupied (2N), one site occupied (1N), or no sites occupied (0N).

glycosylation for the Asn_{25} and Asn_{97} sites of recombinant human IFN-γ when produced in different expression systems ***(21)*** and to monitor changes at these sites during batch culture (*22*; *see* **Fig. 2**).

1. Purified IFN-γ samples are digested with trypsin (1.5 μg:50 μg) for 24 h at 30°C.
2. The glycopeptides containing the *N*-glycan populations are isolated following their separation by reverse-phase HPLC. Samples are applied in 0.06% (v/v) trifluoroacetic acid (TFA) and the peptides separated with a linear gradient (0–70%) of 80% (v/v) aqueous CH_3CN with 0.052% TFA over 100 min at a flow rate of 0.1 mL/min. Peptide peaks are detected at a wavelength of 210 nm and collected individually.
3. The glycopeptides are reduced to the aqueous phase in a Speed-Vac concentrator, lyophilized overnight and stored at –20°C.
4. 0.5 mL of the digest samples are mixed with 0.5 mL of a saturated solution of α-cyano-4-hydroxy cinnaminic acid in 60% (v/v) aqueous CH_3CN and allowed to cocrystallize on stainless steel sample disks.
5. MALDI-MS is performed with an N_2 laser at 337 nm. Desorbed positive ions are detected after a linear flight path by a microchannel plate detector and the digitalized output signal adjusted to obtain an optimum output signal-to-noise ratio from 20 averaged laser pulses. Mass spectra are calibrated with an external standard, vasoactive intestinal peptide with an average molecular mass of 1425.5.
6. Digestion of glycopeptides with peptide-*N*-glycosidase F (PNGaseF) prior to analysis by mass spectrometry confirms the mass of the core peptide. For this determination, 0.5 μL of sample and 0.5 μL of PNGaseF are incubated at 30°C for 24 h, and 0.5-μL aliquots are removed for MALDI-MS analysis.

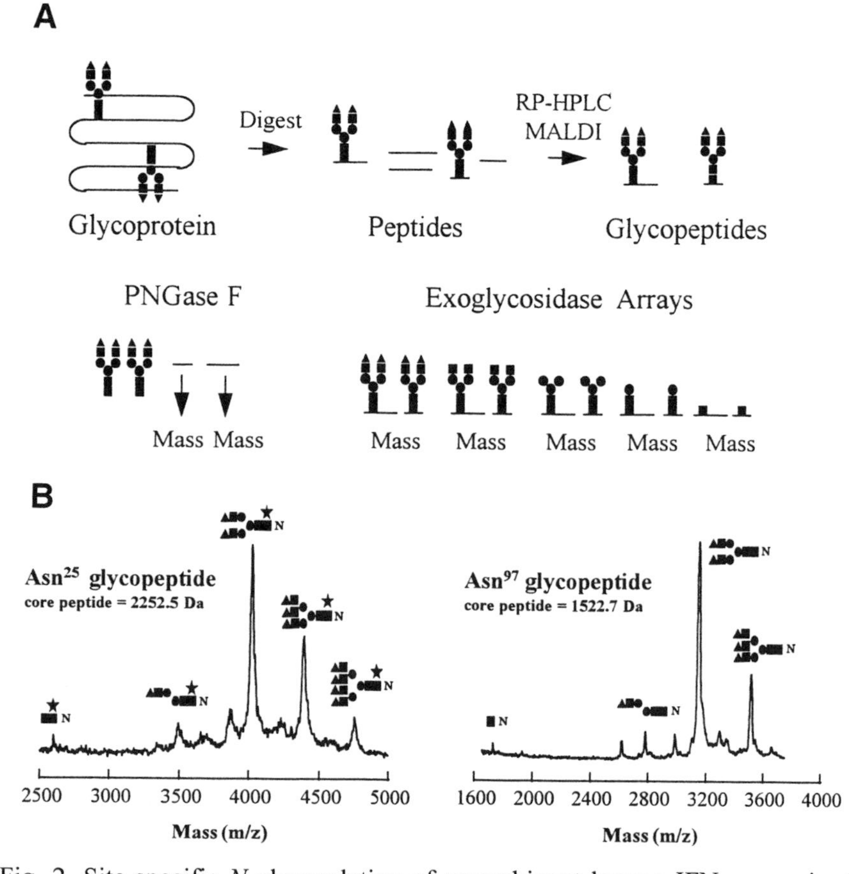

Fig. 2. Site-specific *N*-glycosylation of recombinant human IFN-γ examined by MALDI-MS analyses of glycopeptides. **(A)** The complete analytical protocol. The masses of individual *N*-glycans at a single glycosylation site were calculated by subtracting the known mass of the core peptide moiety from each component in a glycopeptide spectrum. **(B)** An *N*-glycan structure was then tentatively assigned, based on mass criteria alone. As ionization is entirely dependent on the core peptide moiety after desialylation, individual *N*-glycan structures can be quantified. Monosaccharide structures are schematically represented as: ▲, galactose (162.14); ■, N-acetylglucosamine (203.20 Daltons); ●, mannose (162.14 Daltons); and ★, fucose (146.14 Daltons).

7. Simultaneous digestion of purified glycopeptides with linkage-specific exoglycosidase arrays for the sequential removal of oligosaccharides permits the sequencing of *N*-glycans at individual glycosylation sites by MALDI-MS *(20–22)*. Sample (0.5 mL) and glycosidase (0.5 mL) (or a combination of glycosidases) are incubated at 30°C for 24 h, and 0.5-μL aliquots are removed for MALDI-MS analysis (*see* **Note 3**).

3.3. Glycosylation Analysis by ESI-MS

ESI-MS has been used to aid the analysis of glycosylation macro- and microheterogeneity and proteolytic cleavage of the C-terminal in conjunction with information obtained by HPLC and MALDI-MS of released oligosaccharides (*23*; *see* **Note 4**).

1. Spectra are obtained with a VG Quattro II triple quadrupole mass spectrometer having a mass range for singly charged ions of 4000 Daltons (**Fig. 3**).
2. Lyophilized proteins are dissolved in 50% aqueous acetonitrile, 0.2% formic acid to a concentration of 0.1 µg/µL and introduced into the electrospray source at 4 µl/min. The mass-to-charge (*m/z*) range of 600–1800 Daltons are scanned at 10 s/scan and data are summed over 3–10 min, depending on the intensity and complexity of the spectra. During each scan, the sample orifice-to-skimmer potentials (cone voltage) are scanned from 30 V at *m/z* 600 to 75 V at *m/z* 1800. The capillary voltage is set at 3.5 kV.
3. Mass scale calibrations employ the multiply charged ion series from a separate introduction of horse heart myoglobin (average molecular mass of 16,951.49). Molecular weights are based on the following atomic weights of the elements: C = 12.011, H = 1.00794, N = 14.00674, O = 15.9994, and S = 32.066 (*see* **Note 5**).
4. Background subtracted *m/z* data are processed by software employing a maximum-entropy (MaxEnt)-based analysis to produce zero-charge protein molecular weight information with optimum signal to noise ratio, resolution and mass accuracy.

4. Notes

1. Attempts to separate IFN-γ with borate alone, as used by Landers et al. (1992) for the separation of ovalbumin glycoforms ***(24)***, were unsuccessful, because SDS is required to disrupt the hydrogen-bonded dimers. Application of this technique to ribonuclease B and fetuin met with variable success. The glycoprotein microheterogeneity of a monoclonal antibody with a single glycosylation site has been mapped using a borate buffer at high pH; the glycans were enzymatically or chemically cleaved and the resulting profile used for testing batch-to-batch consistency in conjunction with MALDI-MS analysis ***(25)***.
2. MALDI-MS of free N-linked oligosaccharides, after release chemically by hydrazinolysis or enzymatically with an endoglycosidase such as PNGaseF, is also popular as it requires no prior structural knowledge of the glycoprotein of interest and is ideal for the analysis of underivatized populations of oligosaccharides. However, there is a loss of glycosylation site-specific data. Enzymatic release of oligosaccharides is preferred when an intact deglycosylated protein product is required, as *N*-glycan release by hydrazinolysis may result in peptide bond cleavage and the oligosaccharide product requires reacetylation. A drawback to the enzymatic release of oligosaccharides is that SDS is often required to denature the glycoprotein and has to be removed prior to MALDI-MS analysis. MALDI-MS has also recently been used for the analysis of IFN-γ glycoforms separated by SDS-polyacrylamide gel electrophoresis ***(26)***.

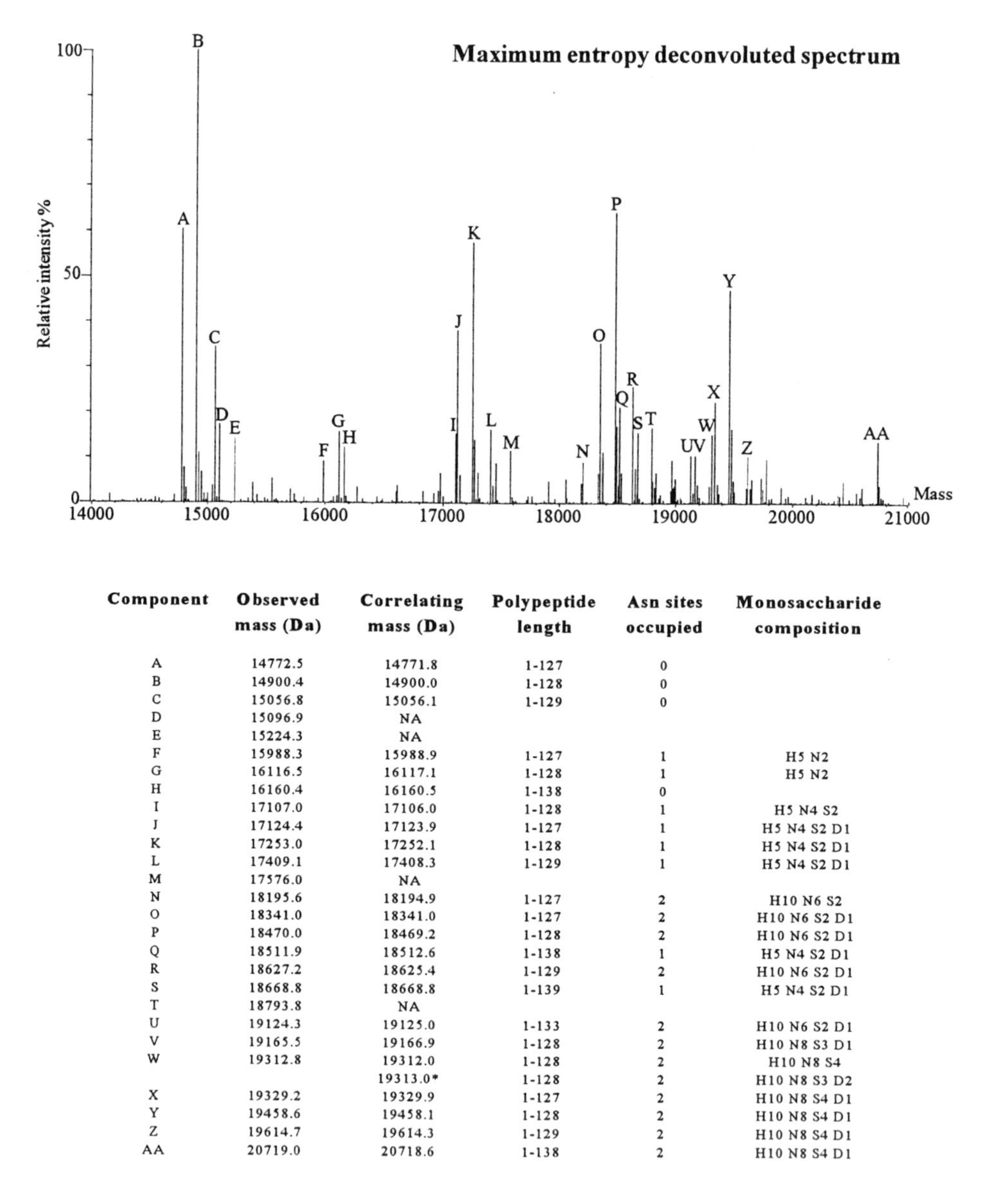

Component	Observed mass (Da)	Correlating mass (Da)	Polypeptide length	Asn sites occupied	Monosaccharide composition
A	14772.5	14771.8	1-127	0	
B	14900.4	14900.0	1-128	0	
C	15056.8	15056.1	1-129	0	
D	15096.9	NA			
E	15224.3	NA			
F	15988.3	15988.9	1-127	1	H5 N2
G	16116.5	16117.1	1-128	1	H5 N2
H	16160.4	16160.5	1-138	0	
I	17107.0	17106.0	1-128	1	H5 N4 S2
J	17124.4	17123.9	1-127	1	H5 N4 S2 D1
K	17253.0	17252.1	1-128	1	H5 N4 S2 D1
L	17409.1	17408.3	1-129	1	H5 N4 S2 D1
M	17576.0	NA			
N	18195.6	18194.9	1-127	2	H10 N6 S2
O	18341.0	18341.0	1-127	2	H10 N6 S2 D1
P	18470.0	18469.2	1-128	2	H10 N6 S2 D1
Q	18511.9	18512.6	1-138	1	H5 N4 S2 D1
R	18627.2	18625.4	1-129	2	H10 N6 S2 D1
S	18668.8	18668.8	1-139	1	H5 N4 S2 D1
T	18793.8	NA			
U	19124.3	19125.0	1-133	2	H10 N6 S2 D1
V	19165.5	19166.9	1-128	2	H10 N8 S3 D1
W	19312.8	19312.0	1-128	2	H10 N8 S4
		19313.0*	1-128	2	H10 N8 S3 D2
X	19329.2	19329.9	1-127	2	H10 N8 S4 D1
Y	19458.6	19458.1	1-128	2	H10 N8 S4 D1
Z	19614.7	19614.3	1-129	2	H10 N8 S4 D1
AA	20719.0	20718.6	1-138	2	H10 N8 S4 D1

Fig. 3. Heterogeneous glycoprotein populations directly analyzed by ESI-MS. This technique provides highly resolved mass analyses with a mass accuracy of 0.01% (1 Dalton/10 kDa). In this example, individual transgenic mouse-derived recombinant human IFN-γ components were assigned a C-terminal polypeptide cleavage site and an overall monosaccharide composition.

3. Until recently, only desialylated oligosaccharides could be analyzed successfully by MALDI-MS using 2,5, dihydroxybenzoic acid as a matrix, as negatively charged

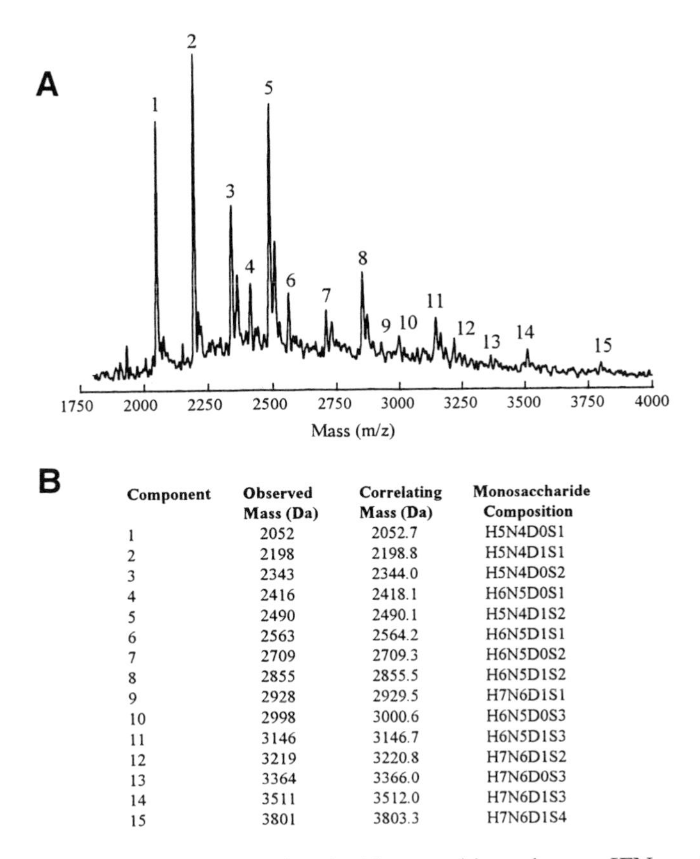

Component	Observed Mass (Da)	Correlating Mass (Da)	Monosaccharide Composition
1	2052	2052.7	H5N4D0S1
2	2198	2198.8	H5N4D1S1
3	2343	2344.0	H5N4D0S2
4	2416	2418.1	H6N5D0S1
5	2490	2490.1	H5N4D1S2
6	2563	2564.2	H6N5D1S1
7	2709	2709.3	H6N5D0S2
8	2855	2855.5	H6N5D1S2
9	2928	2929.5	H7N6D1S1
10	2998	3000.6	H6N5D0S3
11	3146	3146.7	H6N5D1S3
12	3219	3220.8	H7N6D1S2
13	3364	3366.0	H7N6D0S3
14	3511	3512.0	H7N6D1S3
15	3801	3803.3	H7N6D1S4

Fig. 4. Sialylated *N*-glycans associated with recombinant human IFN-γ released with PNGaseF, labeled with 2-aminobenzamide, and analyzed by MALDI-MS using 2,4,6,-trihydroxyacetophenone as matrix **(A)**. The masses of individual *N*-glycans are used to assign an overall monosaccharide composition, including degree of sialylation **(B)**. H, hexose; N, *N*-acetylhexosamine; D, deoxyhexose; and S, N-acetylneuraminic acid.

sialic acids interfere with the efficiency of ionization using this procedure *(27,28)*. However, advances in matrix mixtures and sample preparation schemes now promise to improve analytical protocols further. For example, sialylated oligosaccharides have recently been shown to ionize effectively, with picomole to femtomole sensitivity, as deprotonated molecular ions using 2,4,6-trihydroxyacetophenone in the presence of ammonium citrate *(29)*. Further improvements in sensitivity and resolution may be obtained by derivatization of oligosaccharides with fluorophores such as 2-aminobenzamide, as has been observed in this laboratory (**Fig. 4**). Integration of MALDI-MS peak areas obtained on analysis of sialylated glycans from IFN-γ provided quantitative information that compared favorably with analysis of the derivatized sialylated glycans by ion-exchange HPLC (unpublished data).

4. Glycopeptides may be directly analyzed by ESI-MS *(30,31)*, and the oligosaccharides sequenced following digestion with combinations of exoglycosidases *(32)* or glycoforms separated by liquid chromatography prior to analysis *(33)*. Possibly the most powerful application of this technique has resulted from its interfacing with liquid chromatography, which permits the separation and online identification of glycoproteins from protein digests *(34)*. Although ESI-MS may be interfaced with capillary electrophoresis, the separation of glycoproteins and glycopeptides under acidic conditions makes the analysis of sialylated oligosaccharides unsuitable.
5. ESI-MS is not ideally suited for the analysis of neutral and anionic oligosaccharides that have been chemically or enzymatically released from a glycoprotein of interest, as they do not readily form multiply charged ions. However, the characterization of methylated derivatives of oligosaccharides from recombinant erythropoietin by ESI-MS has been reported *(35)*.

Acknowledgments

Financial support was provided by the Chemical and Pharmaceutical Directorate of the Biotechnology and Biological Sciences Research Council (BBRSC).

References

1. Jenkins, N. and Curling, E. M. (1994) Glycosylation of recombinant proteins: problems and prospects. *Enzyme Microb. Technol.* **16**, 354–364.
2. Liu, D. T. Y (1992) Glycoprotein pharmaceuticals–scientific and regulatory considerations and the United States Orphan Drug Act. *Trends Biotechnol.* **10**, 114–120.
3. Dwek, R. A., Edge, C. J., Harvey, D. J., and Wormald, M. R. (1993) Analysis of glycoprotein associated oligosaccharides. *Annual. Rev. Biochem.* **62**, 65–100.
4. Huberty, M. C., Vath, J. E., Yu, W., and Martin, S. A. (1993) Site-specific carbohydrate identification in recombinant proteins using MALDI-TOF MS. *Anal Chem.* **65**, 2791–2800.
5. Frenz, J. and Hancock, W. S. (1991) High performance capillary electrophoresis. *Trends Biotechnol.* **9**, 243–250.
6. Novotny, M. V. and Sudor, J. (1993) High performace capillary electrophoresis of glycoconjugates. *Electrophoresis* **14**, 372–389.
7. Rush, R., Derby, P., Strickland, T., and Rohde, M. (1993) Peptide mapping and evaluation of glycopeptide microheterogeneity derived from endoproteinase digestion of erythropoietin by affinity high-performance capillary electrophoresis. *Anal. Chem.* **65**, 1834–1842.
8. James, D. C., Freedman, R. B. Hoare, M., and Jenkins, N. (1994) High resolution separation of recombinant human interferon-γ glycoforms by micellar electrokinetic capillary chromatography. *Anal. Biochem.* **222**, 315–322.
9. Rudd, P. M., Scragg, I. G., Coghill, E., and Dwek, R. A. (1992) Separation and analysis of the glycoform populations of ribonuclease β using capillary electrophoresis. *Glycoconjugate J.* **9**, 86–91.

10. El-Rassi, Z. and Mechref, Y. (1996) Recent advances in capillary electrophoresis of carbohydrates. *Electrophoresis* **17**, 275–301
11. Müller, D., Domon, B., Karas, M., van Oostrum, J., and Richter, W. J. (1994) Characterization and direct glycoform profiling of a hybrid plasminogen-activator by matrix-assisted laser-desorption and electrospray mass spectrometry–correlation with high performance liquid-chromatographic and nuclear-magnetic-resonance analyses of the released glycans. *Biol. Mass Spectrom.* **23**, 330–338.
12. Hunter, A. P. and Games, D. E. (1995) Evaluation of glycosylation site heterogeneity and selective identification of glycopeptides in proteolytic digests of bovine α1-acid glycoprotein by mass spectrometry. *Rapid Commun. Mass Spectrom.* **9**, 42–56.
13. Mann, M. and Wilm, M. (1995) Electrospray mass spectrometry for protein chacterization. *Trends Biochem. Sci.* **20**, 219–224.
14. Gu, J., Hiraga, T., and Wada,Y. (1994) Electrospray ionization mass spectrometry of pyridylaminated oligosaccharide derivatives: sensitivity and in-source fragmentation. *Biol. Mass Spectrom.* **23**, 212–217.
15. Tsarbopoulos, A., Pramanik, B. N., Nagabhushan, T. L., and Covey, T. R. (1995) Structural aalysis of the CHO-derived interleukin-4 by liquid chromatography electrospray ionization mass spectrometry. *J. Mass. Spectrom.* **30**, 1752–1763.
16. Ashton, D. S., Beddell, C. R., Cooper, D. J., Craig, S. J., Lines, A. C., Oliver, R. W. A., and Smith, M. A. (1995) Mass spectrometry of the humanized monoclonal antibody CAMPATH 1H. *Anal. Chem.* **67**, 835–842.
17. Bihoreau, N., Veillon, J. F., Ramon, C., Scohyers, J. M., and Schmitter, J. M. (1995) Characterization of a recombinant antihaemophilia-A factor (factor-VIII-delta-II) by matrix assisted laser desorption ionization mass spectrometry. *Mass. Spectrom.* **9**, 1584–1588.
18. Stone, K. L., LoPresti, M. B., Crawford, J. M., DeAngelis, R., and Williams, K. R. (1989) *A Practical Guide to Protein and Peptide Purification for Microsequencing.* Academic Press, San Diego-London, p. 31.
19. Treuheit, M. J., Costello, C. E., and Halsall, H. B. (1992) Analysis of the five glycosylation sites of human α1-acid glycoprotein. *Biochem. J.* **283**, 105–112.
20. Sutton, C. W., O'Neill, J., and Cottrell, J. S. (1994) Site specific characterization of glycoprotein carbohydrates by exoglycosidase digestion and laser desorption mass spectrometry. *Anal. Biochem.* **218**, 34–46.
21. James, D. C., Freedman, R. B., Hoare, M., Ogonah, O. W., Rooney, B. C., Larionov, O. A., Dobrolvolsky, V. N., Lagutin, O. V., and Jenkins, N. (1995) N-glycosylation of recombinant human interferon-gamma produced in different animal expression systems. *Bio/Technology* **13**, 592–596.
22. Hooker, A. D., Goldman, M. H., Markham, N. H., James, D. C., Ison, A. P., Bull, A. T., Strange, P. G., Salmon, I., Baines, A. J., and Jenkins, N. (1995) N-glycans of recombinant human interferon-γ change during batch culture of Chinese hamster ovary cells. *Biotechnol. Bioeng.* **48**, 639–648.
23. James, D. C., Goldman, M. H., Hoare,M., Jenkins, N., Oliver, R. W. A., Green, B. N., and Freedman, R. B. (1996) Post-translational processing of recombinant human interferon-gamma in animal expression systems. *Protein Sci.* **5**, 331–340.

24. Landers, J. P., Oda, R. P., Madden, B. J., and Spelsberg, T. C. (1992) High-performance capillary electrophoresis of glycoproteins: the use of modifiers of electroosmotic flow for analysis of microheterogeneity. *Anal. Biochem.* **205,** 115–124.
25. Hoffstetter-Kuhn, S. H., Alt, G., and Kuhn, R. (1996) Profiling of oligosaccharide-mediated microheterogeneity of a monoclonal antibody by capillary electrophoresis. *Electrophoresis* **17**,418–422.
26. Mortz, E., Sareneva, T., Haebel, S., Julkunen, I., and Roepstorff, P. (1996) Mass spectrometric characterization of glycosylated interferon-gamma variants by gel electrophoresis. *Electrophoresis* **17**, 925–931.
27. Stahl, B., Steup, M., Karas, M., and Hillenkamp, F. (1991) Analysis of neutral oligosaccharides by matrix-assisted laser desorption/ionization mass spectrometry. *Anal. Chem.* **63**, 1463–1466.
28. Tsarbopoulos, A., Karas, M., Strupat, K., Pramanik, B., Nagabhushan, T., and Hilenkamp, F. (1994) Comparative mapping of recombinant proteins and glycoproteins by plasma desorption and matrix-assisted laser desorption/ionization mass spectrometry. *Anal. Chem.* **66**, 2062–2070.
29. Papac, D., Wong, A., and Jones, A. (1996) Analysis of acidic oligosaccharides and glycopeptides by matrix-assisted laser desorption/ionization time-of-flight mass spectrometry. *Anal. Chem.* **68**, 3215–3223.
30. Rush, R. S., Derby, P. L., Smith, D. M., Merry, C., Rogers, G., Rohde, M. F., and Katta, V. (1995) Microheterogeneity of erythropoietin carbohydrate structure. *Anal. Chem.* **67**, 1442–1452.
31. Bloom, J. W., Madanat, M. S., and Ray, M. K. (1996) Cell-line and site-specific comparative analysis of the N-linked oligosaccharides on human ICAM/-1DES454-532 by electrospray mass spectrometry. *Biochemistry* **35**, 1856–1864.
32. Schindler, P. A., Settineri, C. A., Collet, X., Fielding, C. J., and Burlingame, A. L. (1995) Site-specific detection and structural characterization of the glycosylation of human plasma proteins lecithin:cholesterol acyltransferase and apolipoprotein D using HPLC/electrospray mass spectrometry and sequential glycosidase digestion. *Protein Sci.* **4**, 791–801.
33. Medzihradszky, K. F., Maltby, D. A., Hall, S. C., Settineri, C. A., and Burlingame, A. L. (1994) Characterization of protein N-glycosylation by reverse-phase microbore liquid chromatography electrospray mass spectrometry, complimentary mobile phases and sequential exoglycosidase digestion. *J. Am. Soc. Mass Spectrom.* **5**, 350–358.
34. Ling, V., Guzzetta, A. W., Canova-Davis, E., Stults, J. T., Hancock, W. S., Covey, T. R., and Shushan, B. I. (1991) Characterization of the tryptic map of recombinant DNA derived tissue plasminogen activator by high performance liquid chromatography-electrospray ionization mass spectrometry. *Anal. Chem.* **63**, 2909–2915.
35. Linsley, K. B., Chan, S. Y., Chan, S., Reinhold, B. B., Lisi, P. J., and Reinhold, V. N. (1994) Applications of electrospray mass-spectrometry to erythopoietin N-linked and O-linked glycans. *Anal. Biochem.* **219**, 207–217.

23

Measurement of Cell-Culture Glycosidase Activity

Michael J. Gramer

1. Introduction

Heterogeneity of cell culture-produced glycoproteins often results from the presence or absence of a few sugars found on the terminus of glycoprotein oligosaccharides. Variability in bioprocess factors can potentially lead to variability in this oligosaccharide heterogeneity *(1)*. Although stochastic events in the intracellular biosynthetic process have long been recognized as a cause of oligosaccharide heterogeneity *(2)*, more recent data has demonstrated that extracellular degradation by glycosidases can also contribute to oligosaccharide heterogeneity *(3,4)*. The purpose of this chapter is to introduce the concept and consequence of glycosidase degradation, to discuss methods for evaluating whether glycosidase degradation is significant for a particular process, and to provide some potential remedies to alleviate undesirable degradation.

2. Cause and Effect of Glycosidase Degradation

Glycosidases are enzymes that hydrolyze sugars from oligosaccharide and glycoprotein substrates. Data on mammalian glycosidases, including specificities, subcellular locations, assays, inhibitors, and purification are given in an excellent review article *(5)*. Some glycosidases of interest are described in **Fig. 1**.

Oligosaccharide structure can affect many properties of a glycoprotein, either indirectly by altering or masking the protein structure, or more directly by modifying the overall glycoprotein size, weight, and charge, and by serving as a recognition structure for receptors *(6,7)*. As a result, the purification, formulation, and therapeutic efficacy of a glycoprotein can be strongly influenced by oligosaccharide structure, which, in turn, can be affected by glycosidase degradation. For example, treatment of glycoproteins with sialidase, β-galactosidase, β-hexosaminidase, and fucosidase can incrementally alter the rate and

From: *Methods in Biotechnology, Vol. 8: Animal Cell Biotechnology*
Edited by: N. Jenkins ©Humana Press Inc., Totowa, NJ

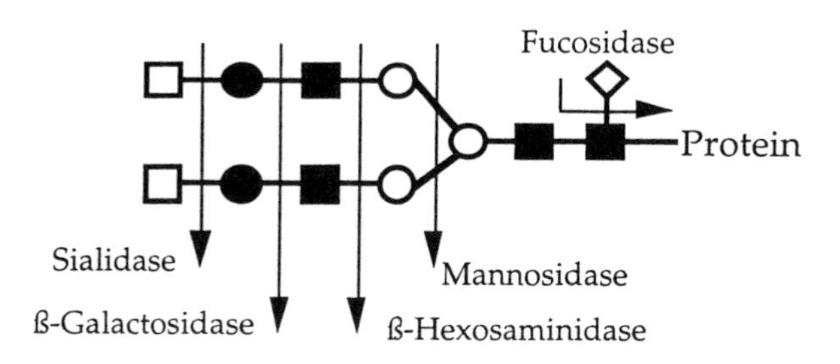

Fig. 1. Some glycosidases of interest and a sample substrate. Shown is a schematic of a typical complex-type N-linked oligosaccharide attached to a glycoprotein. The arrows show cleavage points for some of the endogenous mammalian glycosidases including sialidase, β-galactosidase, β-hexosaminidase, mannosidase, and fucosidase, which hydrolyze sialic acid (open squares), galactose (closed circles), *N*-acetylglucosamine (closed squares), mannose (open circles), and fucose (open diamond), respectively. The relevant enzymes are primarily exoglycosidases, which hydrolyze only exposed, terminal sugars. For example, with the substrate shown above, sialic acid must be removed before the galactose can be hydrolyzed by galactosidase. The specific linkages between the sugars are not shown here; however, a glycosidase may hydrolyze different linkages at different rates.

Table 1
Factors Important in Understanding the Potential for Degradation of Glycoproteins by Glycosidases in the Cell Culture Supernatant

1. Level of glycosidase activity in the host cell
2. Rate and mechanism of glycosidase release
3. Stability of the glycosidase under bioprocess conditions
4. Activity of the glycosidase under bioprocess conditions
5. Kinetics of sugar hydrolysis from the glycoprotein substrate of interest

mechanism of glycoprotein clearance from the circulatory system ***(8–11)***, and the therapeutic efficacy of glycoproteins such as granulocyte/macrophage colony-stimulating factor ***(12)***, glucocerebrosidase ***(13)***, and erythropoietin ***(14)*** is affected by glycosidase treatment.

Table 1 lists the factors important in evaluating whether degradation of glycoproteins by glycosidases in the cell culture supernatant is an issue for a particular process. For glycosidase degradation to be significant, a glycosidase must be synthesized by the host cell and released into the culture supernatant, where the enzyme must be stable and active toward the glycoprotein product. Some data addressing these factors is available. The potential for degradation of cell culture-produced glycoproteins by glycosidases was first recognized in 1991 ***(15)***. Since then, glycosidase activity measured with artificial substrates has been demonstrated in the cell lysates and/or cell culture supernatants in a number of mammalian ***(3,4,16–19)*** and insect ***(20)*** cell lines that are often used for glycoprotein production. Further work has demonstrated that Chinese hamster ovary (CHO) cell sialidase can accumulate to high activities and remove sialic acid from glycoproteins in cell culture supernatants ***(3,4,19,21)***. CHO cell fucosidase was also present extracellularly at high activities, but this enzyme is not expected to contribute to glycoprotein heterogeneity since fucose was only hydrolyzed from the oligosaccharide after the protein portion of the glycoprotein had been removed (*see* **Note 1; ref. *17***).

3. Measurement of Glycosidase Activities Using Fluorescent Substrates

Glycosidase activity is most conveniently measured using artificial fluorescent substrates consisting of 4-methylumbelliferone (4MU) attached to the sugar of interest ***(5,16,22)***. For example, sialidase (neuraminidase) activity is measured with 2'-(4-methylumbelliferyl)-α-D-*N*-acetylneuraminic acid (4MU-NeuAc), while β-galactosidase activity is measured with 4-methylumbelliferyl-β-D-galactoside (Sigma, St. Louis, MO). The fluorescence of the 4MU moiety is shifted and considerably enhanced after the sugar is hydrolyzed from it (**Fig. 2**), allowing very sensitive (as low as 1 nmol/h/mL) determination of enzyme activity in crude biological samples.

3.1. Materials/Stock Solutions

1. Standard: 100 μM 4MU in water.
2. Assay buffer: 1 *M* buffer (acetate, phosphate, and so forth) in water, pH adjusted with NaOH or HCl.
3. Substrate: 4 m*M* 4MU-sugar substrate in water.
4. Enzyme 1–1000 nmol/h/mL activity in saline or weakly buffered solution.:
5. Stopping buffer: 0.2 *M* glycine buffer adjusted to pH 10.4 with NaOH (*see* **Note 2**).

3.2. 4MU Glycosidase Assay Method

1. Buffer, water, and enzyme are added to 1.5-mL plastic microcentrifuge tubes in the proportions shown in **Table 2**, which will give a 1-m*M* substrate in a 0.1-*M* buffer solution.
2. The assay is initiated by adding the substrate, and the contents are mixed with a vortexer and placed at 37°C.
3. After 30 min, 900 μL of the stopping buffer is added to each tube to stop the reaction. The samples can be centrifuged to remove any debris, and then further diluted 10× in the glycine buffer (to get in the linear range of detection, which is from about 0 to 1 μ*M*).
4. The fluorescence of each sample is then determined at an excitation of 362 nm with emission at 450 nm.

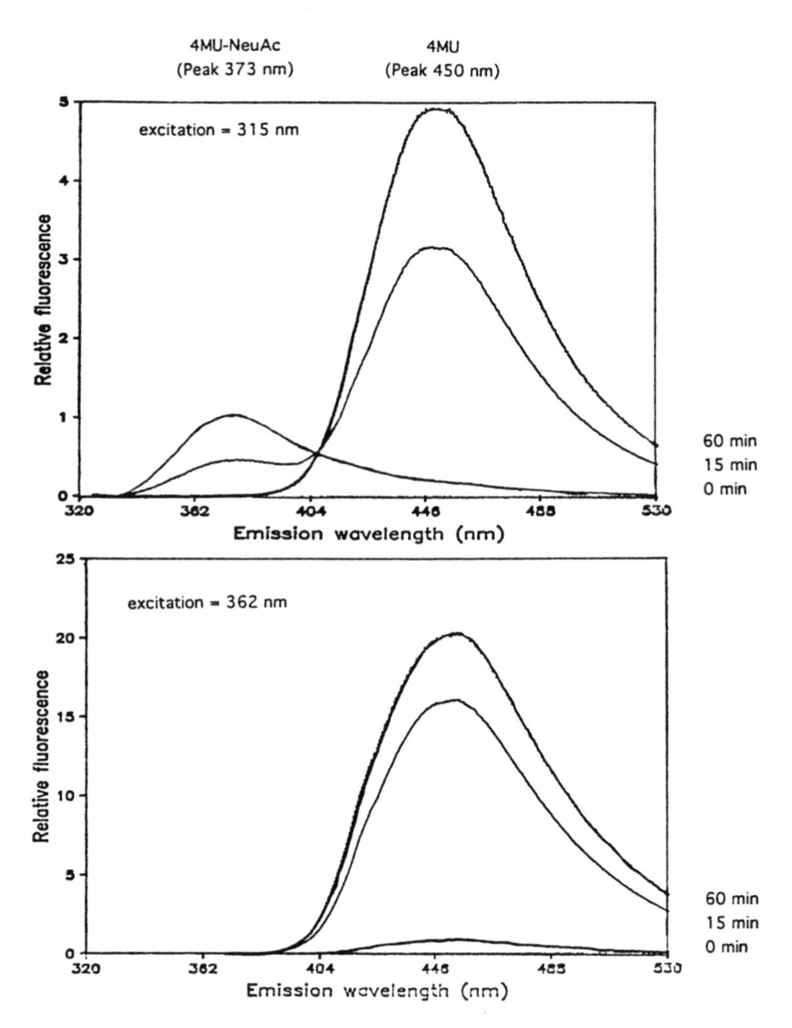

Fig. 2. Fluorescence spectrums demonstrating hydrolysis of 4MU-NeuAc by sialidase. Three times are shown including before addition of sialidase (0 min), shortly after addition of sialidase (15 min), and after complete hydrolysis of 4MU-NeuAc by sialidase (60 min). Intact 4MU-NeuAc has an excitation maximum of 315 nm (top graph), and free 4MU has an excitation maximum 362 nm (bottom graph). The fluorescent signal from 4MU-NeuAc can be seen in the top graph at an emission wavelength of 373 nm, and this peak disappears with time. The 4MU peak can be seen in both graphs at an emission wavelength of 450 nm, and this peak increases with time. The 4MU released is measured at an excitation of 362 nm with emission at 450 nm since the fluorescence of intact 4MU-NueAc does not interfere at these wavelengths, as seen in the bottom graph. The small peak of 4MU at 0 min in the bottom graph is due to the presence of about 1–2% free 4MU in the commercially available 4MU-NeuAc substrate.

Table 2
Typical Glycosidase Assay Contents Using 4MU Substrates

	Volume of stock solution added (μL) in a 100-μL glycosidase assay			
Tube	Assay buffer	Substrate	Enzyme	Water
Sample	10	25	10	55
Control 1	10	25	0	65
Control 2	10	0	10	80

5. The concentration of 4MU released by the enzyme is determined by subtracting the fluorescence of control 1 and control 2 from the sample fluorescence, and by subsequently comparing that number with the fluorescence of 4MU standards diluted in glycine.
6. The activity is calculated and is often reported in units of nmol 4MU hydrolyzed per hour per mL of enzyme stock solution (or /mg protein) added to the assay (*see* **Note 3**).

4. Measurement of Glycosidase Activity Toward a Glycoprotein

Measuring the activity of a glycosidase towards a natural substrate is very similar in principle to measuring the activity toward a 4MU substrate. However, it is much more difficult to quantify the release of sugar. Substances in crude biological samples often interfere with the assays, or the assays are not sensitive enough. Here is a general method to ensure success for this assay:

1. Purify the enzyme and substrate.
2. Incubate the enzyme and substrate in a pH-buffered solution at 37°C.
3. Stop the reaction.
4. Remove the protein.
5. Assay for released sugar.
6. Calculate the activity.

Purification of many of the glycosidases is straightforward using a one- or two-step process with commercially available affinity columns (Sigma) specific for each type of glycosidase ***(5)***. For example, α-L-fucosidase has been purified 11,200-fold to apparent homogeneity from CHO cell culture supernatant using a DEAE-Sepharose ion-exchange column (5.4-fold purification) followed by *N*-(ε-aminocaproyl)-β-L-fucopyanosylamine affinity chromatography (2100-fold purification) ***(17)***. Unfortunately, no affinity column currently exists for sialidase, and a more extensive purification procedure may be necessary for this enzyme ***(19)***.

The assay contents for measurement of sugar hydrolysis from a glycoprotein are identical to those shown in **Table 2**, with the glycoprotein substrate replacing the 4MU-sugar substrate *(3,17)*. At the end of the reaction time, the reaction is stopped by adjusting the pH to a value at which the enzyme is not active, by boiling, or by the addition of a potent glycosidase inhibitor. The samples can then be passed through centrifugal ultrafiltration units (such as the 30-kDa Millipore [Bedford, MA] Ultrafree-MC) to remove protein that may interfere in the subsequent assay. The initial rates of hydrolysis typically follow Michaelis-Menten kinetics with a K_m on the order of 0.1–10 m*M*.

The most general method of determining the amount of sugar released is by using a system designed for carbohydrate analysis. Examples are the HPAE-PAD method on a Dionex (Sunnyvale, CA) chromatography system *(3,17,23)* and the FACE method on a Glyco (Novato, CA) electrophoresis system *(24)*. These methods can be used to measure any of the sugars released; the sugar is identified and quantified by comparing the retention time/mobility and integrated area of each peak/band seen with that of known sugar standards. The initial amount of sugar on the protein can be quantified by chemical hydrolysis or enzymatic release with a commercially available glycosidase.

Enzymatic or chemical assays specific for some of the sugars are also available such as the thiobarbituric acid assay for sialic acid *(25)*, the galactose dehydrogenase method for galactose *(26)*, and the fucose dehydrogenase method for fucose *(27)*. These types of assays range in sensitivity and susceptibility to interferences; additional controls consisting of sugar standards in the assays with substrate and inactivated enzyme should be used to ensure that no interfering substances are present. Other methods involving high-performance liquid chromatography or radioactivity have also been used for glycosidase assay analysis *(5)*.

5. Determining Whether Glycosidase Degradation Is Significant

Determining the extent of glycosylation heterogeneity associated with biosynthesis vs degradation can be a very difficult study. Simply incubating a glycoprotein in conditioned cell culture medium is not adequate to assess degradation potential since the enzyme activity may not be stable. Glycosidase activity should be assessed with the 4MU substrates in cell lysates, cell culture supernatants, and downstream processing steps (especially those involving concentration and acidic pH). The finding of no activity toward the 4MU substrates (<~1 nmol/h/mL) strongly suggests that degradation by glycosidases will be insignificant. The finding of significant activity toward the 4MU substrates provides strong motivation to assess the glycoprotein specificity of the glycosidase activity further.

If the kinetics of sugar hydrolysis by a particular glycosidase activity are known for both the 4MU substrate and the glycoprotein product, then the level of glycoprotein degradation can be estimated by measuring the 4MU glycosidase activity throughout all bioprocessing conditions. For example, the glycoprotein gp120 was added to the supernatant of a CHO cell line that did not produce gp120 ***(3)***. The sialidase activity was measured with 4MU-NeuAc over a 6-d culture period, and the gp120 was purified and analyzed for sialic acid content. Using kinetic parameters previously determined for purified CHO cell sialidase toward gp120 and 4MU-NeuAc, the actual level of sialic acid removal from gp120 ***(14%)*** was accurately predicted from the activity measurements with 4MU-NeuAc. However, determining the actual amount of degradation of a glycoprotein that is being secreted by the cell is more difficult since heterogeneity associated with biosynthesis can potentially change throughout the course of cell culture.

A more direct method of distinguishing the mechanisms of heterogeneity is to compare the glycosylation pattern of a glycoprotein produced in the presence and absence of a glycosidase inhibitor. In the example above with gp120, growth of CHO cells was not affected by the presence of a sialidase inhibitor (2,3-dehydro-2-deoxy-*N*-acetylneuraminic acid), and degradation of gp120 was completely suppressed ***(3)***. However, this method is not absolutely definitive in situations of glycoprotein secretion since the inhibitor could conceivably affect glycoprotein biosynthesis.

Another method of quantifying degradation is to label a glycoprotein biosynthetically (for example, with a pulse of [^{35}S]methionine) and follow the heterogeneity of that glycoprotein throughout the course of cell culture. This method was used along with isoelectric focusing to demonstrate the loss of both sialic acid and phosphate on recombinant human deoxyribonuclease produced in batch CHO cell culture ***(21)***. However, this method requires previous knowledge of the oligosaccharide structural detail for each band in the gel.

Lectin blotting has been used to demonstrate the loss of sialic acid qualitatively from recombinant antithrombin III due to degradation during batch CHO cell culture ***(4)***. The actual amount of sialic acid removed cannot be determined quantitatively with this method because of the potential for changes in intracellular biosynthesis throughout the course of batch culture. However, this method could be coupled with addition of a glycosidase inhibitor for more quantitative results.

6. Minimizing Degradation

A number of methods could be used to minimize undesirable degradation by a glycosidase. Bioprocess conditions can be adjusted to minimize degradation based on a fundamental understanding of the factors in **Table 1**. For example, if the mechanism of glycosidase release is cell lysis, then avoiding high-density and low-viability cultures will minimize degradation. However, adjusting bioprocess conditions may be contrary to other optimization goals. Other more specific methods of minimizing degradation are available, including addition of a glycosidase inhibitor *(3)*, isolating a cell line with reduced glycosidase activity *(3)*, engineering a cell line to produce sugar linkages resistant to hydrolysis *(3,28)*, and genetically engineering a cell line with antisense antiglycosidase DNA *(29)*.

7. Notes

1. All mammalian cells synthesize glycosidases, but the level of endogenous glycosidase activity is cell line dependent. Glycosidases are found in the lysosomes, cytosol, plasma membrane, and Golgi apparatus. The mechanism of release depends on the intracellular localization; plasma membrane glycosidases are naturally secreted, lysosomal glycosidases are secreted due to a missorting effect, which is enhanced by the accumulation of ammonium ion, and cytosolic glycosidases escape as a result of damage to the cellular membrane. The stability and activity of each glycosidase in the culture supernatant will vary, however, most glycosidases have optimal activity from pH 4.0 to 6.0. Generally, cell culture medium components do not significantly inhibit glycosidase activity. The rate of sugar hydrolysis will be substrate dependent; protein structure and oligosaccharide heterogeneity can affect the rate of hydrolysis by a glycosidase.
2. Stock solutions of assay buffer, standard (kept in the dark), and the stopping buffer can be stored at room temperature for at least 6 months. The enzyme stock solution can be prepared from cell lysates *(16)*, cell culture supernatants (which can be concentrated by ultrafiltration if necessary), or a purified enzyme solution. The enzyme solutions are most conveniently stored in frozen aliquots; however, the stability of glycosidase activity under storage conditions should be assessed. The 4MU substrates can be placed in a refrigerator for short-term storage (a few weeks) or in a freezer for longer term storage. The 4MU moiety is somewhat hydrophobic, so the 4MU-sugar substrates that do not contain a charged sugar may require heating (60°C for 1 h) for solubilization. The sialidase substrate spontaneously hydrolyzes (particularly at acidic pH values) and typically comes with about 1–2% free 4MU, which can be removed by extraction with 1–3 equal volumes of ethyl acetate for measurement of low levels of sialidase activity *(22)*.
3. The assay samples are stable for a few days at room temperature in the stopping buffer, but the fluorescence of 4MU in glycine will diminish over a few weeks. It is essential to have a different control 1 for each assay pH when measuring sialidase activity since pH affects the spontaneous rate of 4MU-NeuAc hydrolysis. The fluorescence of the control 2 sample is often negligible. The assay is highly reproducible, and triplicate samples are adequate for statistical analysis. The assay is also quite flexible; the parameters given here are adjustable to optimize activity measurement under the conditions of interest. The hydrolysis typically follows Michaelis-Menten kinetics with K_m values in the range of 0.01–1 m*M*. The release of 4MU should be linear with time. This can be

checked by removing 5 μL from the assay contents and diluting it directly into 495 μL glycine every 15 min for 1 h. If the activity falls off with time, it is possible that the enzyme is not stable, that the released sugar is acting as an inhibitor, or that a significant fraction (>5%) of the substrate has been cleaved.

References

1. Andersen, D. C. and Goochee, C. F. (1994) The effect of cell-culture conditions on the oligosaccharide structures of secreted glycoproteins. *Curr. Opin. Biotechnol.* **5,** 546–549.
2. Kornfeld, R. and Kornfeld, S. (1985) Assembly of asparagine-linked oligosaccharides. *Annu. Rev. Biochem.* **54,** 631–664.
3. Gramer, M. J., Goochee, C. F., Chock, V. Y., Brousseau, D. T., and Sliwkowski, M. B. (1995) Removal of sialic acid from a glycoprotein in CHO cell culture supernatant by action of an extracellular CHO cell sialidase. *Bio/Technology* **13,** 692–698.
4. Munzert, E., Muthing, J., Buntemeyer, H., and Lehmann, J. (1996) Sialidase activity in culture fluid of Chinese hamster ovary cells during batch culture and its effect on recombinant human antithrombin III integrity. *Biotechnol. Prog.* **12,** 559–563.
5. Conzelmann, E. and Sandhoff, K. (1987) Glycolipid and glycoprotein degradation. *Adv. Enzymol.* **60,** 89–216.
6. Goochee, C. F., Gramer, M. J., Andersen, D. C., Bahr, J. B., and Rasmussen, J. R. (1992) The oligosaccharides of glycoproteins: factors affecting their synthesis and their effect on glycoprotein properties, in *Frontiers in Bioprocessing II* (Todd, P., Sikdar, S. L., and Bier, M., eds.), American Chemical Society, Washington, DC, pp. 199–240.
7. Varki, A. (1993) Biological roles of oligosaccharides: all of the theories are correct. *Glycobiology* **3,** 97–130.
8. Furbish, F. S., Steer, C. J., Krett, N. L., and Barranger, J. A. (1981) Uptake and distribution of placental glucocerebrosidase in rat hepatic cells and effects of sequential deglycosylation. *Biochim. Biophys. Acta* **673,** 425–434.
9. Steer, C. J. and Clarenburg, R. (1979) Unique distribution of glycoprotein receptors on parenchymal and sinusoidal cells of rat liver. *J. Biol. Chem.* **265,** 874–881.
10. Stockert, R. J., Morell, A. G., and Scheinberg, I. H. (1976) The existence of a second route for the transfer of certain glycoproteins from the circulation into the liver. *Biochem. Biophys. Res. Commun.* **265, 68,** 988–993.
11. Winkelhake, J. L. and Nicolson, G. L. (1976) Aglycosylantibody: effects of exoglycosidase treatments on autochthonous antibody survival time in the circulation. *J. Biol. Chem.* **251,** 1074–1080.
12. Donahue, R. E., Wang, E. A., Kaufman, R. J., Foutch, L., Leary, A. C., Witek-Giannetti, J. S., Metzger, M., Hewick, R. M., Steinbrink, D. R., Shaw, G., Kamen, R., and Clark, S. C. (1986) Effects of N-linked carbohydrate on the in vivo properties of human GM-CSF, *Cold Spring Harbor Symp. Quant. Biol.* **51,** 685–692.
13. Murray, G. J. (1987) Lectin-specific targeting of lysosomal enzymes to reticuloendothelial cells. *Methods Enzymol.* **149,** 25–42.

14. Fukuda, M. N., Sasaki, H., Lopez, L., and Fukuda, M. (1989) Survival of recombinant erythropoietin in the circulation: the role of carbohydrates. *Blood* **73,** 84–89.
15. Gramer, M. J. and Goochee, C. F. (1991) Potential for degradation of glycoprotein oligosaccharides by extracellular glycosidases. AICHE National Meeting, Los Angeles, CA.
16. Gramer, M. J. and Goochee, C. F. (1993) Glycosidase activities in Chinese hamster ovary cell lysate and cell culture supernatant. *Biotechnol. Prog.* **9,** 366–373.
17. Gramer, M. J., Schaffer, D. V., and Sliwkowski, M. B. (1994) Purification and characterization of α-L-fucosidase from Chinese hamster ovary cell culture supernatant. *Glycobiology* **4,** 611–616.
18. Gramer, M. J. and Goochee, C. F. (1994) Glycosidase activities of the 293 and NS0 cell lines, and of an antibody-producing hybridoma cell line. *Biotechnol. Bioeng.* **43,** 423–428.
19. Warner, T. G., Chang, J., Ferrari, J., Harris, R., McNerney, T., Bennett, G., Burnier, J., and Sliwkowski, M. B. (1993) Isolation and properties of a soluble sialidase from the culture fluid of Chinese hamster ovary cells. *Glycobiology* **3,** 455–463.
20. Licari, P. J., Jarvis, D. L., and Bailey, J. E. (1993) Insect cell hosts for baculovirus expression vectors contain endogenous exoglycosidase activity. *Biotechnol. Prog.* **9,** 146–152.
21. Sliwkowski, M. B., Gunson, J. V., and Warner, T. G. (1992) Sialylation and phosphorylation as a function of culture conditions for recombinant human deoxyribonuclease produced by CHO cells. *J. Cell. Biochem.* **16(Suppl.),** 150.
22. Warner, T. G. and O'Brien, J. S. (1979) Synthesis of 2'-(4-methylumbelliferyl-α-D-N-acetylneuraminic acid and detection of skin fibroblast neuraminidase in normal humans and in sialidosis. *Biochemistry* **18,** 2783–2787.
23. Hardy, M. R., Townsend, R. R., and Lee, Y. C. (1988) Monosaccharide analysis of glycoconjugates by anion-exchange chromatography with pulsed amperometric detection. *Anal. Biochem.* **170,** 54–62.
24. Starr, C., Masada, R. I., Hauge, C., Skop, E., and Kock, J. (1996) Fluorophore-assisted-carbohydrate-electrophoresis, FACE, in the separation, analysis, and sequencing of carbohydrates. *J. Chromatogr.* **720,** 295–321.
25. Skoza, L. and Mohos, S. (1976) Stable thiobarbituric acid chromaphore with dimethyl sulphoxide. *Biochem J.* **158,** 457–321.
26. Urbanowski, J. C., Wunz, T. M., and Dain, J. A. (1980) A colorimetric procedure for measuring the enzymatic hydrolysis of terminal galactose from GM_1 ganglioside. *Anal. Biochem.* **105,** 461–467.
27. Cohenford, M. A., Abraham, A., Abraham, J., and Dain, J. A. (1989) Colorimetric assay for free and bound L-fucose. *Anal. Biochem.* **177,** 172–177.
28. Minch, S. L., Kallio, P. T., and Bailey, J. E (1995) Tissue plasminogen activator expressed in Chinese hamster ovary cells with α(2,6)Galβ(1,4)GlcNAc-R linkages. *Biotechnol. Prog.* **11,** 348–351.
29. Sato, K. and Miyagi, T. (1996) Involvement of an endogenous sialidase in skeletal muscle cell differentiation. *Biochem. Biophys. Res. Commun.* **221,** 826–830.

Index